Vittorio Ingegnoli

Landscape Ecology: A Widening Foundation

Springer

Berlin
Heidelberg
New York
Barcelona
Hong Kong
London
Milan
Paris
Tokyo

Vittorio Ingegnoli

Landscape Ecology:
A Widening Foundation

With 127 Figures

 Springer

DR. VITTORIO INGEGNOLI

Professor by Contract of Landscape Ecology
University of Milan
Department of Biology
Via Celoria 26
20133 Milan
Italy

e-mail: vingegnoli@iol.it

ISBN 3-540-42743-0   Springer-Verlag Berlin Heidelberg New York

Die Deutsche Bibliothek - CIP-Einheitsaufnahme

Ingegnoli, Vittorio:
Landscape ecology : a widening foundation / Vittorio Ingegnoli. - Berlin ;
Heidelberg ; New York ; Barcelona ; Hong Kong ; London ; Milan ; Paris ;
Tokyo : Springer, 2002
  ISBN 3-540-42743-0

Springer-Verlag Berlin Heidelberg New York
a member of BertelsmannSpringer Science + Business Media GmbH
http://www.springer.de
© Springer-Verlag Berlin Heidelberg 2002
Printed in Germany

Production: Goldener Schnitt, Sinzheim
Cover design: design & production, Heidelberg
Typesetting: Camera-ready by Vittorio and Elena Ingegnoli
SPIN 10792984 31/3130 5 4 3 2 1 0 - Printed on acid free paper

Dedicated to Elena, my Wife

Foreword

Exactly 25 years ago on a warm autumn afternoon a young ecologist walked slowly through a tiny oak wood, and perched on a log to reflect. He had measured and seemingly knew "all" the species present - trees, mosses, mushrooms, birds and more. The research, based on this and other woods in the landscape, was the first rigorous test to see if island biogeographic theory was of use in heterogeneous land. Unexpectedly, an interior-to-edge model was found to be more useful. But on this beautiful sunny day he gazed out through the trees at the surrounding bean and maize fields.

Suddenly a terrible thought hit him. The land surrounding the other woods differed slightly from this scene. Here there were two bean fields plus a maize field, meadow, hedgerow and farm road, but the other comparably sized woods studied had different mixes of these land uses. Wouldn't the surroundings seriously affect the species in the woods? Had he done "bad science" (an awful feeling for a scientist)? Immediately he went to all his ecology books, searching for discussions of patchiness, mosaic pattern, interactions between ecosystems, and the like. Nothing. Surprise was a new ingredient to ponder. Then for 3 months every spare moment found him in the university library digging deeper, collecting tidbits and clues. A new feeling took over, challenge and excitement. The spatial arrangement of ecosystems and land uses is important ecologically! A giant but approachable scholarly frontier. Furthermore, a sense emerged that solutions to most environmental issues lay within the frontier.

Events then cascaded. Research led to publication of a patch-corridor-matrix model in 1979. He met like-minded researchers and spoke at a 1981 Dutch conference on landscape ecology, a term he had never heard, yet one with a nice ring. In 1983 landscape ecology was introduced to North America at an Illinois conference he and colleagues organized. As the field accelerated thereafter its diverse roots and foundations became clearer. Especially important were the widely known geography-vegetation traditions in Germany, spatial animal-movement-habitat studies in The Netherlands and in Australia, diverse land planning and evaluation approaches in Europe and the Mediterranean Region, hedgerow network studies in Canada, and patch-corridor-matrix, pattern-process-change, island biogeography and spatial modeling of heterogeneous landscapes across the USA.

These roots led to extensive research and abundant literature centered around the ecology of landscapes. Modern landscape ecology, with a highly useful body of theory and principles, is the result. With leading figures worldwide, the field is growing rapidly and today embraces a richness of perspectives, which provide hybrid vigor.

Into this dynamic field arrives the wonderful book in your hand written by Vittorio Ingegnoli. This is the seventh landscape-ecology book on my shelf authored or edited by Sandro Pignatti, Almo Farina, and Dr. Ingegnoli, all leaders in Italian landscape ecology. These books reveal a richness of theory, perceptive syntheses, many useful Italian and worldwide applications, and distinct areas where Italian research leads the way. Perhaps only Australia and the USA rival Italy in the number of valuable landscape-ecology books available.

I am the ecologist who perched on a log, and am not only an admirer of this Italian work, but also have learned much from Italian landscapes. As a child I ran around the Ponte Vecchio, liked Giotto, wondered where Dante got his information, and explored from the Alps (where my uncle served just before World War I) to Taormina (where my aunt is buried). Much later, my workshop-seminars in Florence for students and professionals were followed, in turn, by publishing a landscape-ecology article in Italian, doing transects to the most remote spots in Venice, and receiving a Universita degli Studi di Firenze medal and honorary membership in the Societa Italiana di Ecologia del Paesaggio. Although I have published on the spatial patterns produced by different processes, from atop a tower in San Gimignano, Dr. Ingegnoli taught me a new pattern, the distinctive stable fine-scale result of centuries of trial-and-error by rural landowners in Europe and elsewhere.

Every landscape ecologist will appreciate what Vittorio Ingegnoli has done for us in this book. The diverse European and North American approaches and perspectives have finally coalesced. An amazingly rich palette of methods, both for analytic understanding and to evaluate landscapes for appropriate land uses, is lucidly revealed. An impressively wide range of theories and simple useful equations for landscape ecology is laid out. A promising approach to understand landscape "pathology" or degradation is introduced. Nice new case-study applications in Italy, Africa, and across Europe are presented. The "widening foundation", whereby landscape-ecology principles are increasingly used in fields related to land use, is illustrated from urban planning and conservation biology to agricultural land and economic/ecological sustainability. Modern landscape ecology has come a long way in a short time.

Yet opportunities abound for even more rapid progress in this "Decade of Landscape Ecology". I see three groups of opportunities.

First, the *significant gaps* in today's landscape ecology include: (a) ecological flows across the land; (b) adjacencies and neighborhood configurations; (c) stream/river corridor width and design; (d) cluster of small patches versus a corridor for species movement; (e) importance of plants and vegetation in landscape ecology; (f) ecologically optimum network forms; and (g) the roles of spatial patterns produced by nature, planning/design, and lack thereof.

Second, *new frontiers* evolve naturally from landscape ecology. (a) "Spatially meshing nature and culture" is of unparalleled importance, but remains shrouded. (b) In contrast, "road ecology" is emerging with clear focal areas, such as traffic disturbance/noise effects on natural communities, the road-effect zone, development of theory, and meshing ecological flows and biological diversity with safe and efficient transport.

Third, we *the people* highlights how landscape ecologists might work. Begin by considering the change from how we spend our time to how we get (and got) our inspiration. Also, look beyond our colorful land-use-and-ownership models to see the key fine-scale attributes of a landscape, such as little-road systems, hedgerow habitats, groundwater flows, and human paths and animal routes across heterogeneous land. Make sure that both solid long-half-life empirical insight and promising short-half-life quantitative models move forward arm in arm. Spend more time with colleagues in landscapes outside one's own continent. Write rather than edit concise books on landscape ecology, its major areas and applications. Outline visions for the future, in addition to improvements for the present. Indeed, imagine landscape ecology at the end of the decade.

In conclusion, one does not have to be a "rocket ecologist" to see that society is wasting land. Unplanned suburbanization rampages across the land. Houses "grow" on good agricultural soil in a world with hungry bellies. Increasingly, busy highways degrade nature in wide swaths across the land.

Yet, in spots, we actually see landscape ecology principles being used for solutions, reversing the trend of wasting land. Will these spots grow in number and size? Certainly. And coalesce? Hopefully. The science and applications of landscape ecology will be at the heart of success. Indeed, this book, *Landscape Ecology: A Widening Foundation*, is well named. All students and problem-solvers concerned with changing land should carefully absorb its pages.

Harvard University,
November 2001   Richard T. T. Forman

Preface

Some twenty years after the foundation of the IALE (International Association for Landscape Ecology), during the Fifth World Congress (1999 - Snowmass Village, Colorado), President John Wiens noted that the variety of topics and approaches represented in the literature testifies to the diversity of landscape ecology as a discipline. This diversity is at once the great strength and the potential weakness of landscape ecology. In fact, different research traditions and cultures have been brought to landscape ecology through different ways.

At present, if we compare the main topics of landscape ecology, we can recognise four principal disciplinary models: (1) geographical, (2) chorological, (3) matrix configured, (4) holistic. The first model is closely associated with the interrelation between natural and human components, from the point of view of geomorphology, botany, architecture, etc., and has been led by geography toward an interdisciplinary science. The second model is rooted in population ecology and zoology, driven by the need to develop spatial characters and scale processes in general ecology. The third disciplinary model is related to the attempt to study the ecology of land ecosystems. The last model derives from an holistic view, dealing with landscapes as open, adaptive, dissipative, self-transcendent systems of natural and human elements. Therefore, each point of view defines the landscape in a different manner.

At the same time, the urgent need for a sustainable environment has made landscape ecology increasingly accepted by the policy makers of nature conservation and restoration and of territorial planning. But what should landscape ecology be? Is it an interdisciplinary method necessary to study the environment on a landscape scale? Or, is it a level of biological organisation? And, how does the inseparability of landscape and culture affect the contents of this discipline?

There is no doubt that the above mentioned problems require a widening foundation of landscape ecology in order to reach a manifold but unique definition of landscape and to recognise what is important about landscapes.

I agree with Richard Forman: it is important to start from the solid scientific and spatial foundations of landscape ecology. If that is done, then applications related to the land can go in any direction: sustainability, biological conservation, territorial planning, etc. If landscape ecology principles can be unified to create a cultural ecological utopia, that is fine, but it can not be a confused one, since that suggests that the scientific and spatial principles have been inadequately used or understood. In this framework, trying to converge towards such a unification seems to be very difficult. In my opinion it is necessary to arrive at a new disciplinary model capable of utilising and integrating the best of all the others. The basis of this integration is not impossible, if we give a widening foundation to our discipline, in which correct epistemological references, linked to the new scientific paradigms and a more adequate ecological theory, permit the complex self-organising system that we call landscape to be studied in a way that could be available to respond to the challenge of sustainable development and biological conservation.

The following main steps seem to be necessary to reach the integration: (1) to specify that the landscape, as a system of biogeocoenosis (Forman and Godron 1986), *is a proper biological system*, enhancing that even culture does not implicate the subjection of nature to the dominance of man. As a matter of fact the distinction between a living being and his environment has no substantial significance. (2) Given that a landscape is much more than a set of spatial heterogeneous characters, to specify the *intrinsic biological* characters of the landscape (structural and dynamic) as being different from the *exportable* characters (mainly chorological). (3) Remembering the expressions of Forman and Naveh, who mentioned the "interweaving of the ecosystems in a landscape", to refer to a new, more complex structural model, based on the concept of tissue, which could be named *ecotissue* (Ingegnoli 1999). Actually, all the other disciplinary models present their spatial schemes based on the concept of mosaic. But in this way it is impossible to integrate completely their differences. Finally, (4) to consider landscape ecology as a discipline like medicine, *biologically based* but *transdisciplinary*. In fact, since the landscape is a biological system, it is the physiology(ecology)/pathology ratio which permits a clinical diagnosis of a landscape.

Moreover, we have to note that a discipline of landscape ecology with a wider basis may arrive to change many principles of traditional ecology, especially vegetation science, leading to a unified ecology. If vegetation science remains principally based on phytosociology, neglecting the underlying principle of order and allowing an endless analysis (Pignatti 1998), it makes every attempt at synthesis difficult. The main objective of vegetation science must concern description, synthesis and *diagnosis* of the complex adaptive system of vegetation. For this purpose, it is necessary to integrate the floristic description, biodiversity and synecology analyses, following the holistic, unifying landscape ecological approach.

Section I of this book develops the *theoretical principles* concerning all that has been mentioned above, trying to demonstrate the possibility of widening our discipline. In doing this, it will be possible to clarify the relationships among landscape ecology, landscape sciences and their applications.

Section II of this book is related to the question of a proper *methodology* of landscape study. In fact, any application needs a correct analysis and evaluation of the landscape (or of some parts of it) together with a good diagnostic phase. In coherence with the first section's principles, new ecological indexes are presented, like the BTC (Biological Territorial Capacity), which is a quantity able to express the level of complexity and the capacity of self re-equilibrium of a vegetation ecotope. Moreover, it is generally necessary to control the results using these landscape ecological indexes and comparing this information with the characters derived from other scales. As in medicine, the environmental evaluation needs comparisons with "normal" patterns of behaviour of a system of ecosystems. Therefore, the main problem becomes how to recognise the normal state of an ecological system, and/or how to identify the levels of alteration of that system, which may threaten human health, even in the absence of pollution.

Strictly linked to methodology are the applications. The main three chapters of landscape ecological applications are synthesised in: (1) environmental sustainability, (2) biological conservation, (3) environmental design and territorial planning. In *environmental sustainability* - at present - every ecological system is defined as an ecosystem, generating an ambiguous and limiting reference; that is why, at a recent scientific meeting of German and Italian ecologists at the German Congress Center Villa Vigoni (Lake Como, 1999 - chairman Wolfgang Haber and Sandro Pignatti), my working group underlined the importance of healthy natural and cultural landscapes among the major goals for sustainable development. In *biological conservation* the need for landscape ecology was so strong as to be one of its founding disciplines; for example, biodiversity has to be intended not only as α, β, γ, (*sensu* Whittaker) but also as landscape biodiversity. In *environmental design and territorial planning* the main question remains how to design with nature: landscape ecological principles and methods must be added to the traditional ones. In the interaction between urbanisation and surrounding landscapes the need for new methodological criteria becomes crucial, especially because the most of the human population in the biosphere is going to live in urban and suburban landscapes.

In conclusion, I hope that this book will be of interest to a broad audience in suggesting a new disciplinary perspective in landscape ecology and to create new professionals, such as environmental physicians, because the challenge of sustainability is a continuous one.

Milan, October 2001   Vittorio Ingegnoli

Acknowledgements

The recent evolution of my thinking has been influenced by discussions with Richard T.T. Forman, Zev Naveh, Sandro Pignatti, Jürgen Ott, Frank B. Golley, Almo Farina, Wolfgang Haber, Janusz B. Falinski, Franco Pedrotti and Renato Massa. Good discussions also took place with many students of my graduate classes at the Universities of Milan and Camerino. I thank each of them warmly.

Encouragements to persist in my studies came from many colleagues and friends especially Marc Antrop, Jacques Boudrie, Robert G.H. Bunce, Françoise Burel, Roberto Canullo, Gianumberto Caravello, Giuseppe Chiaudani, Fiorenza De Bernardi, Marco Ferraguti, Carlo Ferrari, Gioia Gibelli, Annalisa Calcagno Maniglio, Felix Müller, Michel J. Samways and Jim Sanderson. I am grateful to you.

Many sincere thanks for reviewing portions of this book to Rita Colantonio Venturelli, Giuseppe Gibelli, Alessandro Ingegnoli, Jürgen Ott, Emilio Padoa Schioppa, Franco Pedrotti, Sandro Pignatti. A special appreciation to Elena Giglio Ingegnoli, who reviewed the entire manuscript and wrote the glossary with an exceptional competence. The figures, except a few, were drawn by Roberta Donati, to whom goes my thankfulness.

Vittorio Ingegnoli

Contents

6 Theoretical Influence of Landscape Ecology

Section II Landscape Methodology and Applications

7 Landscape Analysis

8 Landscape Components Evaluation

9 Landscape Criteria of Evaluation and Diagnosis

10 Landscape Ecology and Sustainability

11 Landscape Ecology and Conservation Biology

12 Environmental Design and Territorial Planning

13 Examples of Application

Glossary   323

ELENA GIGLIO INGEGNOLI

References   335

Index   347

Section I
Principles of Landscape Ecology

1 The Landscape as a Specific Living Entity

1.1
The Emergence of the Concept of Landscape

1.1.1
The Concept of Landscape in the Ancient World

The earliest written word meaning landscape was cited in the Book of Psalms about three thousand years ago (Naveh and Lieberman 1984). This word - *noff* in Hebrew - concerned mainly the perception of a landscape, giving importance to the visual aspect. But we can recognise a visual character only if we are conscious of the real organised consistency of the object, that is, after a first conceptualisation of it. This fact has to be well understood, because the apparent ambiguity of the term landscape, which means both "perception of aesthetic view" and "mosaic of interacting natural elements", seems to be a source of discussion even in landscape ecology today.

At the beginning of Western civilisation (near the end of the Palaeolithic age), the concept of life was limited to the individual organism and the population (if small), the other elements of life organisation were seen as an indistinct whole. And in this whole, biotic and abiotic components probably had a similar importance.

The first emergence of the concept of landscape as "mosaic of interacting natural elements" was indeed very ancient, dating from the period in which humans learnt how to combine diverse landscape elements to choose an optimal living site. This necessity became prominent when human populations evolved into proto-agricultural societies. In that period, man was obliged to gather information on the entire ecological mosaic which formed his territory in order to be able to plan successfully his fields, orchards and settlements. Even defensive aspects, regarding both natural and human disturbances, played an important role in this sense. During the agricultural revolution of the early Holocene, people knew how to modify ecological mosaics to raise production and live on resources more predictable than those coming from gathering and hunting.

It becomes clear that the emergence of the concept of landscape had the significance of an event of consciousness in the human population regarding new linkages between man and nature. It was the beginning of a true mutualism with natural elements, as components of the landscape, that led to great transformations in cultural and natural fields. Religion was focused on by the worship of the Sun, the culture was dominated by regional characters, and the environment was changed by men on a wide spatial scale. New landscapes were created, with a huge amount of heterogeneity, following actively the great space- climatic-environmental changes that were also occurring.

Thus, in Europe, we can find the presence of many agricultural settlements referring to the Boreal period (6800-5500 B.C., beginning of the Neolithic Age) - characterised by near-arid climate and the presence of pine trees (*Pinus nigra, P. sylvestris, P. pinaster, etc.*) and hazels (*Corylus avellana*) - especially near the Mediterranean sea (e.g. La Font des Pigeons, at Chateauneuf les Martigues, near the mouth of the Rhone, in France). The Neolithic settlement of Martigues was localised at the confluence of many diverse types of ecological systems (Renault-Miskovsky 1986), showing an accurate choice of the area, in accordance with the heterogeneity of the local landscape.

In Europe, the first representations of topographic maps of a territory date back to the Chalcolithic Age, at the beginning of the sub-Boreal period - characterised by a cooler climate and dominated by oak and alder trees (*Quercus robur, Alnus glutinosa,* etc.). This fact is very important, because it demonstrates a high level of conceptualisation of the elements of the landscape. The map of Bedolina (Val Camonica, Lombardy; Anati 1985) shows the representation of cultivated fields, paths, rivers, canals and houses as topographic drawings on the smooth local rocks (Fig.1.1).

Fig. 1.1. One of the oldest maps, representing many elements of an agricultural landscape in the alpine valley of the Camuni people (ca. 4000 years before present)

So, it is not insignificant that a map of a landscape appeared at least one millennium before its denomination with a written word. As pointed out by Ingegnoli (1981), the parallel evolution of the sub-alpine peoples with the local landscapes after the last glaciation (Dryas period, 9000-8000 B.C.) led to sub-alpine or mountain agricultural landscapes, in which the natural and the human patches were so well integrated and defined as to remain extremely stable during four or five millennia.

1.1.2
The Concept of Landscape in the Last Twenty Centuries

In the Roman world, the term landscape was *regio-regionis,* which emphasised a geographic aspect, while the term for scenery was *prospectus* (Mariano 1958). A distinction between the visual-artistic and the geographic-ecological meaning appeared in the definition of a landscape painter: *"pictor topiarius, qui regiones formas pingit"* that is "country painter, who paints the shapes of a landscape".

Moreover, at the beginning of the Christian Age some prodromes of ecology appeared in the scientific field of agronomy. Well known treatises, like *De re rustica* (Columella, first century of the Christian Era), were not only practical but also theoretical, as shown by the description of the process of soil fertility, etc.

They presented some ecological notes concerning the man-landscape relationships, too, such as the assertion that "no field is tilled without profit if the owner, through much experimentation, causes it to be fitted for the use which it can best serve" and that it "is important to plan the right proportion between the farmstead and the territory":

> Ne villa fundum quaerat neve fundus villam; *and*: Quod incohatur sicut salubri regione ita saluberrima parte regionis debet constitui. (Columella, Book I).

It means: "the buildings may not seek for land, nor the land seek for buildings" and "as a building which is begun should be situated in a healthful landscape, so too in the most healthful part of that landscape".

The emergence of the Renaissance in the XIV century in Italy reintroduced the concept of landscape with new interest. An extraordinary cultural enterprise was the famous climb of Mont Vantoux in Provence of Francesco Petrarca (1336), from the top of which the landscape was described not only as a visual aspect but even as an environmental and cultural structure (Pignatti 1996), and with the consciousness of the sense of perspective. Perspective rules are important, because they cause a deep consciousness of the point of view and of the scale, and this is crucial, even for science. During Roman times, perspective was known, but without all its complex geometrical rules, while in the Medieval age the sense, even intuitive, of perspective seems to have been lost.

The artistic world of the Renaissance was the source of many fields of science, and many artists, especially painters, studied and drew the landscape in a perspective representation (Piero della Francesca). While drawing, Leonardo da Vinci began asking why the elements of the landscape had been shaped in that particular way and what had been the reasons for their development (Clark 1976). Leonardo compared the elements of a landscape with the components of an animal body, recognising not only a structural sense, but also a functional one. Through those studies (Fig. 1.2), Leonardo founded the discipline of geomorphology and was conscious of the concept of vegetation.

Fig. 1.2. A landscape drawn by Leonardo da Vinci (about 1500) representing a storm in a pre-alpine landscape. While drawing, Leonardo began asking why the elements of the landscape had been shaped in that particular way and what had been the reasons for their development. The interest of painters was therefore not only aesthetic

As shown by the preceding observations, many aspects of the artistic notion of landscape are useful in understanding landscape ecology. Forman and Godron (1986) underlined in this sense other aspects: (a) the presence of a spatial scale, (b) the presence of two or more visual objects, (c) the subject matter, such as vegetation and animals, people, settlements, etc., (d) the connections between human and natural patches.

The "Grand Tour", which all cultured Englishmen and northern Europeans made through the Alps to Italy in the seventeenth and eighteenth centuries, brought them in contact with rugged, picturesque scenery. Many painters portrayed these landscapes (e.g. Salvator Rosa, Claude Lorrain), and all those influences led to English landscape architecture. The landscape garden was a product of the Romantic movement. Its forms were based on direct observation of nature and the principles of painting.

In the late eighteenth century, Humphry Repton (1803) published a theory of landscape gardening, but the great success of this school of landscape design depended principally on ecological reasons. An explanation can be found outside the cultural aims of the movement, in a sort of unconscious reaction to the extreme transformations of the territory due to industrialisation and urbanisation. These large changes in Western landscapes signaled the end of French formal gardening, in favour of a natural-like environment.

The non-classical principle of wilderness, particularly strong in the United States, was opposed against the French idea of a deterministic (Cartesian) nature. Thoreau wrote (1854): "In wilderness lies the preservation of the world". The concept of landscape was strictly linked to the concept of wilderness. Like the poet Wordsworth and the painter Constable in England, and like Emerson and Thoreau in the United States, Frederick Law Olmsted (1865, 1870) felt the great moral appeal in natural landscape beauty and wilderness and the necessity to preserve natural areas. The first preservation area of natural wilderness and scenic landscapes was created in Yosemite Valley, in California (1864), a few years later designated the first National Park in the world by the government of the United States (1872).

1.1.3
The Concept of Landscape as a Scientific Object

As we see from these notes, the concept of landscape emerged in Western civilisation step by step over five-six millennia (from the Neolithic to the Iron Age) and was gradually brought into scientific disciplines beginning in the first century (*agronomi, agrimensores, etc.*) until the nineteenth century (landscape planning and design, nature preservation, etc.), with an enrichment of its complex meaning during the Renaissance. But it has been only during the last 160 years that the concept of landscape has become by itself a scientific object and has entered directly into science.

The biogeographer Alexander von Humboldt, in Germany, was the first to give a scientific definition (1840) to the concept of landscape (*landschaft*) as

Der Totalcharakter einer Erdgegend, that is, the total character of a given territory, meaning both the perceptive and the natural aspects. A geographer, Von Richtofen (1890), following this school, linked the concept of landscape to the possibility to connect different aspects of nature. Dokuchaev (1898), the founder of pedology (soil science), adopted in the Russian language the term *Landschaft*, with the same significance as Von Humboldt.

After the beginning of Ecology as a science, due to the German naturalist (and painter) Haeckel (1869), and after the concept of biocoenosis (Moebius, 1877), another German, the geomorphologist (and physician) Passarge, wrote about the landscape physiology (1912). Passarge described many modern observations based on a climatic geomorphology which could be integrated with natural sciences, allowing the foundation of landscape ecology.

1.2
The Development of Landscape Ecology

1.2.1
The Emergence of Landscape Ecology as a Discipline

A German bio-geographer, Carl Troll, wrote (1939) that all the methods of natural science - from the science of forest vegetation and biological aerial photo interpretation to geography as "landscape science" and "ecology" - meet here, in a new discipline named "landscape ecology" (*Landschaft Oekologie*). Notwithstanding the effort of Troll, trying to link geological and biological fields, the study of landscape continued to proceed along two branches for about 30-40 years. From the point of view of geomorphology, a basic reference for landscape analysis was the biostatic-rexistatic theory by Erhart (1956), which considered pedogenesis as a cyclic process. Bertrand (1968) defined the landscape as an area characterised by dynamic geographical elements (geophysical, biological, anthropic) which makes a geographical system able to evolve as a whole. He followed the cyclic process of Erhart and defined three landscape units:

1. Geotope, a minimum unit of a few metres, e.g. a muddy patch.
2. Geofacies, a unit of a few thousand metres characterised by a toposequence, e.g. an oriented slope.
3. Geosystem, a unit of some square kilometres, integrating diverse geofacies, e.g. the two slopes of a valley.

Soil scientists have underlined the concept of "soil chain" or "topo-sequence", which is manifested in geofacies with climatic uniformity, so that pedogenetic processes are conditioned only by the morphology, e.g.: summit, shoulder, backslope, footslope, toeslope, etc. Geomorphology has been seen as a factor regulating many others landscape functions (Leser 1978), thus assuming a strong ecological significance. And Tricart and Kilian (1979) noted that pedogenesis and morphogenesis are not cyclic, but they are rather contemporaries and interfere with each other. The analysis of the soil chains and the study of the many forms of humus (Finke 1972; Vos and Stortelder 1992) led to a convergence with biological studies of the landscape. From the end of the 1960s, we assisted in a revaluation of the theories of Troll (1939, 1950) on the landscape, and of Watt (1947) on the pattern process of environmental mosaics. Mac Arthur and Wilson (1967, 1972) presented their theory on island biogeography and geographical ecology, and phytosociologists such as Giacomini (1958), Tüxen (1968), and Gehu (1979), whose analysis derives from the concept of vegetational association of associations, underlined the importance of syn-associations: Phytosociology changes from the identification of the relationships among the plant communities which make up the plant landscape, to the serial and chain types, according to which they appear inside the same territory.

During the same time, European and American landscape architects developed their studies on the landscape in new ways, less visual and more ecological, as shown in the book of Buchwald and Engelhart (1968), or in the book *Design with Nature*, by Ian Mac Harg, published in 1969 by the Natural History Press of New York and mentioned by E.P.Odum.

1.2.2
The Foundation of the IALE

In summary, the ecological studies of the landscape derived from polygenic lines and landscape ecology remained mainly an interdisciplinary science, at least until the international Congress of Veldhoven (The Nederlands) in 1981. After this congress, in 1982, the IALE (International Association for Landscape Ecology) was founded and a series of new, specific books were published.

As Isaak Zonneveld pointed out (1989), this very complex field of landscape studies began to need theories, in the sense of verified hypotheses or - at least - firm sources of generalised empirical knowledge. The difficulty of doing inductive studies (from details to the whole) is often the demand for deductive reasoning, based on diagnostic characteristics of the situation and on the use of basic theories.

In the 1980s, some of the first basic books on landscape ecology were published. In his theory, Van Leeuven (Tjallingi and de Veer 1982) proposed a tool, a sort of language for studying the landscape, which considers time and space relationships and the differences between homeostasis and homeorhesis. Naveh and Lieberman (1984) recalled the principal aspects of general systems theory (holism, cybernetics and information). Naveh underlined the importance of hierarchy and defined the landscape as the level of life organisation forming a total human ecosystem. Merriam (Brandt and Agger 1984) enhanced the concept of connectivity, essential in the study of the chorological relationships between landscape patches.

The very first complete book of landscape ecology as a branch of general ecology was published in 1986, by R.T. Forman (Harvard, USA) and M. Godron (Montpellier, France). The ecological background was very practical, but they also articulated landscape structure and landscape dynamics in a sequence that allowed a true disciplinary field to be developed. *Quantitative Methods in Landscape Ecology*, edited by Monica Turner and Robert Gardner in 1991, completed this period of the emerging new chapter of ecology, by providing more tools for applications. Several studies followed the development of the theoretical advance of landscape ecology. For instance, many of them suggested that the landscape has critical thresholds at which ecological processes will show dramatic qualitative changes. These changes have important implications for nature conservation and territorial planning. This could be the case of habitat fragmentation, which may progress with little effect on a population until the critical pathways of connectivity are disrupted.

Conversely, several studies remained strictly linked to the definition of Troll, affirming landscape ecology not as a true science, but as a specific outlook of integrated research, since most work is based on borrowing from other disciplines, such as general ecology, physical geography, biogeography, community ecology, phytosociology, or even applied sciences such as landscape planning or forestry.

The growth of the IALE, now in about 40 countries, and the questions posed by many fields of application, now requires more efforts in order to clarify the disciplinary fundamentals of landscape ecology. After spreading to several German universities including Munich and Hannover (Haber, Buchwald, Finke), and Dutch universities such as Wageningen (Zonneveld, Vink, Vos), landscape ecology manifested a wide growth in the USA in the 1980s, due especially to Forman, Golley, Harris, Nassauer, Turner and Wiens (US-IALE, open: 1987), and at the same time in other European countries, e.g. in Italy, with Caravello, Farina, Ingegnoli and Pignatti (SIEP-IALE, open: 1989), and Britain (Bunce, Dover, Dennis Green, Malanson, Samways; UK-IALE). This has enhanced the discussion among scientists, producing a very large literature on our discipline.

As we tried to expose briefly, the development of landscape ecology has advanced, but it still remains problematic. Researchers are divided into considering it a new discipline covering all the aspects of environmental heterogeneity (Wiens et al. 1993) or a new specific branch of ecology, or as a synthesis of integrated interdisciplinary studies. The use of the concept of landscape seems to remain ambiguous, with many meanings. However, nature is heterogeneous at all spatial scales, even within patches that we recognise as "homogeneous" habitat for a certain species. Therefore we are convinced, with Sanderson and Harris (2000), that landscape ecology is not a new discipline, rather a chapter of general ecology, effective at a typical range of scales, but also very important because of its capacity to renew the entire discipline of ecology toward a true unified discipline with a large range of applications. But further clarification is   needed.

1.3
Main Tendencies in Landscape Ecology

1.3.1
Discerning for Unifying: The Four Main Disciplinary Models

The IALE was founded about twenty years ago, and during this period the variety of approaches and topics in landscape ecology has testified to the substantial diversity of our field. This fact gives landscape ecology a great strength, but also a potential weakness. The main reasons for this situation can be readily understood: researchers in landscape ecology derive from different cultural backgrounds as well as academic disciplines, such as geography and architecture, botany and zoology. Thus different approaches are unavoidable.

Today, comparing the main topics of landscape ecology, four main disciplinary models can be recognised: (a) geographical, (b) chorological, (c) matrix configured, (d) holistic. Note that even if they have not been expressly recognised as models by the majority of scientists, they have nonetheless played a basic role in both the foundation and the advances in the field of landscape ecology. We present these models in a synthetic table (Table 1.1), after ranking their main characters as follows:

1. The first one, *geographical,* is closely associated with the interrelation between natural and human components, from the point of view of geomorphology, botany, architecture, social sciences, etc., and has been led by geography toward an interdisciplinary science. The first to emerge (at about the middle of the century), but still very rich, it defines the landscape as a geographical context for ecological communities and human populations. "Landscape ecology is the study of spatial variation in landscapes at a variety of scales. It includes the biophysical and societal causes and consequences of landscape heterogeneity. Above all, it is broadly interdisciplinary" (Wiens and Moss 1999). Its main structural reference is a geographical mosaic. We may derive this vision, even partially, from many authors, such as Troll (1950), Buchwald and Engelhart (1968), Finke (1972), Leser (1978), Jongman (1999), etc.
2. The second one, *chorological,* is rooted in population ecology and zoology, driven by the important need to consider spatial characters and scale processes in general ecology. Emerging in the 1960s, it is based on spatial configuration and processes applicable at any level of the organisational hierarchy. The main structural model is a species specific mosaic, which is linked with a generic definition of landscape as an environmental system. "Landscapes are characterised by their spatial configuration. It is this locational pattern, and the way it affects and is affected by spatially dependent processes, that is the subject of study of landscape ecology" (Wiens and Moss 1999). We may derive this vision, even partially, from many authors, such as MacArthur and Wilson (1967, 1972), Hanski (1983), Wiens (1995, 1999) and Farina (1998, 2000).

Table 1.1. Main disciplinary models in landscape ecology

Main topics	Geographical	Chorologic	Matrix configurated	Holistic
Definition of the landscape	Geographical context for ecological communities and human populations.	Environmental system of ecological fields.	System of interacting ecosystems that is repeated in similar form throughout.	Interwoven of natural and cultural entities of the total living space.
Structural scheme of the landscape	Geographical area with a characteristic arrangement of ecosystems.	Fuzzy-edged mosaic, species specific	Mosaic of patches and corridors on landscape matrix.	Ecological holarchy of natural and technological ecotopes.
Essential characters of the landscape	Physiographic factors, plant factors, faunal factors, human factors and their interdependence.	Spatial configuration and spatially dependent processes	Spatial relationships among landscape elements, interactions among them, changes in the structure and functions of the ecological mosaic.	Single whole in nature as an open, adaptive, non-equilibrium, self-transcendent system.
Scale	Geographical scale.	Any level of organisational hierarchy	From element to region	From ecotope to ecosphere
Main contributions in landscape models and indices	• Zoning criteria • Land-use mosaic • Visual analysis • Human suitability • Biotopes distribution • Plant association systems • Soil mosaic • Land features	• Metapopulation • Island biogeography • Source-sink • Faunal perception • Ecotonal pattern • Supplementation • Spatial statistic	• Heterogeneity • Patch fragmentation • Patch orientation • Grain size • Boundary form • Perimeter/area ratio • Contrast, connectivity • Ecological web • Landscape principles for application	• Ecotope definition • Ordination of landscapes • Homeorhetic process • Land protection conflict • Differentiated land use concept
Disciplinary field	Interdisciplinary	Ecological, strongly zoological	Ecological, general	Ecological, includes human ecology; transdisciplinary
Epistemology	Traditional scientific paradigm still strong	Refusal of the abstracting capacity of human perception	Land ethics and pragmatism	Holon concept, "Gestalt" principles, autopoiesis

3. The third disciplinary model, *matrix configured*, is related with the attempt to study the ecology of land ecosystems. Emerging in the 1970s, it is very useful and has a strong practical background in a correct scientific framework. Landscape structure and landscape dynamics are articulated in a sequence that gives the possibility to develop a true disciplinary ecological field. It considers the landscape as a level of biological organisation: it defines a landscape as "a system of interacting ecosystems that is repeated in similar form throughout" (Forman and Godron 1986). The main structural model is a mosaic of patches and corridors in a landscape matrix. We may derive this vision, even partially, from many authors, including Forman and Godron (1981, 1986), Odum (1989), Ingegnoli (1993), Forman (1995), Burel and Baudry (1999), Sanderson and Harris (2000).

4. The last one, *holistic,* derives from an holistic view, dealing with landscapes as open, adaptive, dissipative, self-transcendent systems of natural and human elements. Emerging at the end of the 1970s, it is based on the most advanced scientific paradigms of the last century, which led to a landscape structured in an ecological holarchy of natural and technological ecotopes. The concept of trans-disciplinarity substitutes the one of inter-disciplinarity, giving prominence to ecology and including human ecology. We may derive this vision, even partially, from many authors, e.g. Haber (1980), Naveh and Lieberman (1984, 1994), Ingegnoli (1993), Jongman and Mander (2000), Brandt et al. (2000).

1.3.2
The Need To Clarify

Even if the majority of authors do not point out the confusion regarding landscape ecological definitions, there is no doubt that each point of view defines the landscape in remarkably different ways. This is presently the greatest limit in landscape ecology. Today, the urgent need for a sustainable environment has led to the increasing acceptance of landscape ecology by nature conservation and restoration policymakers and territorial planners. Therefore, as we pointed out in the Preface, we need to clarify exactly what is landscape ecology. Many authors define this field of ecology as a multidisciplinary method needed to study the environment at a landscape scale. Others define landscape ecology as a discipline necessary to study the landscape as a level of biological organisation. Moreover it is not clear if the inseparability of landscape and culture affects the contents of this discipline.

There is no doubt that if landscape ecology is to become a unified discipline then we need to recognise what is important about landscapes and what it is wrong about general ecology. Trying to converge toward such a unification seems to be very complex, especially for historical reasons. At the same time, the wealth of theoretical and methodological information contributed by these four approaches to the development of landscape ecology represents an indisputable advantage: it is possible to propose a new disciplinary model able to accept, utilise and integrate the best of the results of each one. We have to consider the basic importance of the consequences of the new scientific paradigms and the review of the hierarchy of levels of life organisation. Some considerations will follow in the next three paragraphs.

1.4
The Principal Limits of General Ecology

1.4.1
Contrasts and Scalar Sequence in Developing Ecological Chapters

Even if the bio-geologist C. Troll had been using the term "landscape" since 1939, and although landscapes are accessible from commonplace experience, they had been neglected for most of the past century, ignored by almost all ecologists. Only recently has general ecology began to consider the landscape (Odum 1989; Allen and Hoekstra 1992) as a level of the biological hierarchy. What happened? It is important to try to answer this question, as it is linked with the remarkable limitations of general ecology. In reality, ecology was born as "autoecology", that is, as the study of the relationships between an organism and its environment, and evolved, step by step, from the individual organism level of biological organisation towards the higher levels, developing by separate and sometime even contrasting chapters. It is possible to recognise at least five periods.

1. *Auto-Ecology*. From 1840, e.g. Liebig (1840), Darwin (1859), Haeckel (1869), Shelford (1913).
2. *Population and Community Ecology*. From 1877, e.g. Moebius (1877), Clements (1916), Gleason (1922), Hutchinson (1978).
3. *Ecosystem Ecology*. From 1935, e.g. Tansley (1935), E.P. Odum (1953, 1971, 1983), Whittaker (1975), Ricklefs (1973), Brewer (1988), Begon et al. (1990)
4. *Landscape Ecology*. From 1950, e.g. Troll (1950), Tüxen (1968), Buchwald and Engelhart (1968), Forman and Godron (1986), Naveh and Lieberman (1984, 1994), Forman (1995).
5. *Global Ecology*. From 1965, e.g. Lovelock (1979), Margulis and Lovelock (1974), Rambler, Margulis and Fester (1989).

We may observe a clear proportionality between dates and hierarchic levels of biological organisation, from small to large scales, that is from the organism to the biosphere (ecosphere) levels.

But the new chapters are difficult to integrate into an unified ecology: too much reductionism in an holistic discipline! The entire development of ecological theories has shown separations and contrasts. Scientists know the famous dispute between population ecologists and ecosystem ecologists: the concept of community (a complex of populations) and the concept of ecosystem (a system of ecological processes) have been reassembled by Odum (1971) through a compromise: the community as the biotic part of the ecosystem.

Besides, we have to underline that many people mistake very often the concept "ecological system" with the concept of "ecosystem". Many of the various levels of biological organisation are "ecological systems", but they generally can not be reduced to the ecosystem level. In proposing the term "ecosystem", Tansley (1935) did not want to refer exclusively to biotic entities (the previous population-community school): to him, the fundamental ecological unit included both organisms and physical components. The problem is that this fundamental unit represents an ambiguous concept (Fig.1.3).

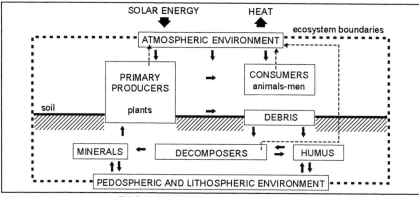

ECOSYSTEM FUNCTIONAL MODEL

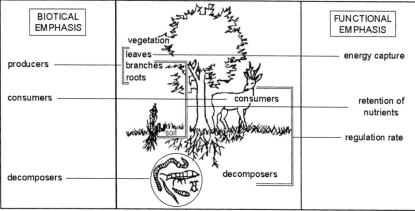

THE AMBIGUITY OF THE CONCEPT OF ECOSYSTEM

Fig. 1.3. Ambiguity of the ecosystem. A simple functional model is compared with the dualism of its components. To the *left* the biotical interpretation, to the *right* the functional one. In the *centre* the reality of an ecological system

Biotic and functional points of view, in fact, groups the elements composing an ecological unit in two different ways. The population-community approach tends to consider ecological systems as networks of interacting populations. So, the biota are the eco-systems and abiotic components are external influences. A typical subdivision consequently results in primary producers, consumers, and decomposers. The process-functional approach tends to view ecosystems in the sense of Tansley, emphasising energetics as the central focus. Thus, a second subdivision shows energy capture, nutrient retention, and rate regulations.

On the other end, O'Neill and his group (1986) underlined that these two different points of view have never been completely compatible. As a matter of fact, as Allen and Hoekstra wrote (1992), the conservation of matter and energy (which is a fundamental basis for the concept of ecosystem) has no relevance for the concept of community; furthermore, a portion of an ecosystem, for example a species pathway, may correspond to different portions of a community, because each organism may be an autonomous member of a community.

1.4.2
Too Much Reductionism

The emergence of landscape ecology was due, among other things, to the limitations of general ecology concerning heterogeneity and the multiple dimensions of the environment: it does not present any effective possibility to study it. Many practical scientists and technicians, such as conservation biologists, vegetationists, foresters, biogeographers, agronomists and landscape architects, have underlined those limitations, ranking some basic questions. It could be useful to expose some of these questions:

- Which is the role of spatial-temporal scales on ecological processes, especially on disturbances ? Are ecological processes dependent on scale ?
- What is the deep nature of phenomena producing heterogeneity ?
- How is a sustainable land use compatible with ecological principles ?
- What is the variation of exchanges of organisms, matter and energy among the elements of an ecological mosaic in relationship to its structural configuration ?
- How can we evaluate and manage natural and semi-natural patches, preserving them in a mosaic of human ecosystems ?
- How can we transform semantic information into pragmatic information, integrating natural and human parts of ecology ?

In fact, general ecology, at least the traditional form, is in trouble when dealing with problems of applications on a territorial (landscape) scale, because the concept of ecosystem can not be effective for all the scales, especially when various spatial schemes interact. Remember that the ecology of ecosystems considers the environmental heterogeneity as background noise (Blondel 1986) with a secondary importance.

But the reality is very different: the heterogeneity and variability of the environmental mosaic represent a very important component in establishing the distribution of organisms and their interactions, in an evolutionary sense too.

General ecology seems to accept the concept of adaptation only in a traditional way: responses of biological systems (e.g. organism, population, ecosystem) to environmental requirements. The consequent concept of evolution is seen as a progressive optimisation of adaptation, in a strictly Darwinian definition. In this case, the structuring of biological systems is based on the concept of cause and the leading forces on the concept of fighting and impact. For the same reasons, the great importance of disturbances as structuring forces and the active responses of biological systems to disturbances were not considered at all. Integration and co-operation, too, were not recognised as strictly linked to the concept of adaptation.

The limitations of this approach are mainly concerned with the ignorance of scale, therefore with reductionism. If we want to reach a new vision of evolution we have to avoid the famous dilemma by Monod (1970) between chance and necessity, because evolution is not due to purely random events nor to a mechanical ontogenesis but "to the demand of the accumulation of order in a self-organising system". This is linked to the principle of the irreversibility of time and of indeterminacy combined with realism (see Sect. 2.2).

1.4.3
Problems in Defining the Landscape and Studying Complex Systems

Very few text books of general ecology mention the word landscape, not in the well known books of E.P.Odum, at least before 1989, nor in other texts such as Ricklefs (1973), Colinvaux (1993), etc. The best interpreter of the concept of landscape in general ecology was Whittaker (1975), even if he limited it to the definition in the sense of a geographical context for communities:

> Communities are the contexts in which species survive and evolve. The statement of context is incomplete, however, if we do not consider the full range of environments and communities in which a species occurs. The communities of a landscape form a pattern that corresponds to the landscape's pattern of environments. (Whittaker 1975)

On the other hand, Brewer (1988), at the beginning of the chapter on community and ecosystem ecology, still defines a landscape as a perceptive context of ecosystems. Moreover, many ecologists talk about a landscape simply as a larger-version ecosystem! The question depends again on the ambiguity of the concept of ecosystem, when thought applicable at any scale.

Ecosystem functions, e.g. trophic pyramid, are not able to describe the true dynamic of the landscape. That is why Sanderson and Harris (2000) asserted that the word and concept of landscape must be explicitly defined as being distinct from ecosystem. In fact, for pure and simple logic and lucidity, they define the landscape as consisting of two or more ecosystems in close proximity.

No doubt that the pretension to analyse the landscape with the limitations of general ecology, which we tried to discuss briefly, leads to poor methodologies, and hardly reaches the goal of understanding landscape structure and processes.

The integration of traditional data on a landscape, e.g. climatic, geomorphic, vegetational, faunistic, land use, is not enough to describe properly our subject, even if we add other aspects, like social and economical ones, in a multifunctional vision. On the other hand, the attempts at overcoming these difficulties by borrowing ecological methods from other disciplines to study the environment at a landscape scale (Vos and Stortelder 1992) has intrinsic limits in the theory of the integrative levels of biological organisation, as the ecosystemic level is a lower, simpler level of organisation.

Odum (1971) noted that an evident principle of functional integration subsists inside the biological spectrum: for this reason the results obtained at a certain level of organisation help us in the study of a higher level, but they can never completely explain the phenomena occurring at that level. This is only a corollary of the most important Emerging Properties Principle (Weiss 1969; Lorenz 1978): an organic system is much more than the sum of its parts (see Sect. 2.1); so, it is sufficient to change the structural assemblage of the elements of a system to obtain a behaviour that is not only different, but sometimes unpredictable.

It is compulsory to analyse the properties of a specific level of biological organisation, studying the intrinsic behaviour of the level itself, because the information coming from lower levels is at the most partial. Let us consider the law of gravitation, as Gell-Mann (1994) pointed out, which is a wonderful example of the simplicity of the natural principles at the basis of all phenomena we may observe. This law is able to describe processes of aggregation of the matter which led, during the evolution of the Universe, to the assembling of galaxies, and then of stars and planets, the Earth included. Since the time of their constitution, these celestial bodies have shown characters of complexity, diversity and individuality. But the law of gravitation is unable to explain these characters.

The Emerging Properties Principle assumed new meaning when complex adaptive systems (i.e. self-organising systems) appeared. On the Earth, this development is linked up with the origin of life and the process of biological evolution, which produced the surprising variety of species, the behaviour of organisms in the ecological systems, the ability of learning and thinking in animals and humans, the evolution of human society (including the economic-technologic aspect) and the formation and the evolution of landscapes.

The common character of all evolutionary processes is that, in each of them, a complex adaptive system acquires information on its environment and its interaction with it; the analysis of this information allows it to identify regularities, which are integrated in a sort of model able to provide a guide to act in the real world. In any case, different schemes exist in competition with each other and the results of the action in the real world are reintroduced in the system, influencing competition among schemes, when the transformation process is accomplished.

The study of complex adaptive systems should be the major task of ecology, but we know how many limits it still presents. Even today, ecology is far from the new vision of science which changed its paradigms in the last decades of the twentieth century. In addition, ecology seems to consider time as symmetric, but the lack of a direction in time (time as an irreversible process) led us to consider Nature as something without history, mechanistic, under the human domain, with man who sees himself as its engineer. What is more, we can note that determinism not only obstructs contact with reality, but it brings human freedom up for discussion too (Lorenz and Popper 1985; Popper 1983), while freedom implies an idea of creativity. That is why it is absolutely necessary to refer to new scientific paradigms (at least new for biology) in the light of which the ecological discipline is forced to renew.

1.5
The Chief Consequences of the New Scientific Paradigms in Ecology

1.5.1
Paradigms in Synthesis

At present, scientists have to avoid two representations of nature which tend to a world of alienation: 1) the *deterministic* one, with no possibility of novelty and creation, 2) the *stochastic* one, which leads to an absurd world with no causality principle and, for instance, without any ability to forecast. Perhaps, the major incentive toward a new conception of nature comes from scientists like Von Bertalanffy (1968), Weiss (1969), Lorenz (1978), Popper (1982) and Prigogine (1996), who observed how nature creates its most fine, sensitive and complex structures through non-reversible processes which are time oriented (time arrow).

This group of scientists asserts that: (1) an organic whole is more complex than the sum of its parts, and (2) the description of the behaviour of a dynamic system presents more solutions than the classical ones. Thus, they reach the conclusion that life is only possible in a Universe far away from equilibrium and that indeterminacy is compatible with reality. The self-organising properties of non-equilibrium dissipative structures and the basic feature of indeterminacy show the real nature of our universe. In Chap. 2, we will see a coherent synthesis of these scientific paradigms; but here it is compulsory to show some of the main consequences on the behaviour of natural systems, on observation and prescriptions and on scale.

1.5.2
Consequence: The Behaviour of Natural Systems

The study of every branch of the real world, which seems to be continuous in itself, begins with a separation from a systemic point of view. Thanks to the formulation of at least the above mentioned important scientific theories during the twentieth century, we know that mechanistic paradigms are no longer sufficient to foresee the behaviour of natural systems in a more general way.

The state of a dynamic system is not exhaustively determined by the knowledge of the co-ordinates and the parameters of its elements. We need to know also the behaviour of the system, the laws guiding its state.

Prigogine (Prigogine et al. 1972; Prigogine and Nicolis 1977; Prigogine 1996) pointed out that we need a new objectivity, different from both the the the classic thermodynamic one, which identifies the knowable with the confirmable, and the dynamic one, which tries to understand the evolution of a system   always assuming stable initial conditions. But we can not design a system by considering its initial conditions in a way such that it should be possible to follow an exact evolution or trajectory, since it is impossible to reach an infinite precision.

We need to replace linear and deterministic processes by non-linear chaotic and cybernetic ones. It is the case of the well known assertion of Prigogine "Order through fluctuation". A system which is far away from thermodynamic equilibrium, and thus unstable, develops by running into several branching points where "it makes a choice" between one of the possible paths (self-organisation of the dissipative structures): near these points, fluctuations (i.e. the variation of disturbances) acquire a primary rule.

So, a structure created by a succession of extended fluctuations may be only understood in relation to the history of its choices: the past has been produced by unpredictable events and must be considered as unique and not repeatable. This assumption points out the importance of history in ecological studies, even in the absence of man, and agrees with the principle of emerging properties. It affirms that through its self-regulative properties the system becomes more than the sum of its parts, not in a quantitative-additive way but in a qualitative-structural one. The emergence of a new probabilistic description shows some properties that are not compatible with a deterministic description.

As suggested by Popper (1990) the "propensities" of events to follow one another allows the description of changes, e.g. development or evolution, in a way that is beyond the capabilities of Newtonian description.

Moreover, if we apply Prigogine's concept of self-organisation of dissipative structures to the human society, it is possible to demonstrate that human activity, which is creative and innovative, is not foreign to nature, but can also be considered as an extension and an intensification of some aspects already present in the physical world, which may now be more easily understood with the discovery of models for processes far from equilibrium.

1.5.3
Consequence: Observation and Prescriptions

Following these new paradigms of science, many problems related to the study of biology, and especially to ecology, can be viewed from a quite different perspective, e.g. the problem of observation and of scale, but also the problems of description, stability, diversity, evolution, etc.

It is crucial to clarify the significance of observation. Remember that Galileian objectivity arose from scientific observations. But after Descartes the understanding of nature and obtaining of certainty have been achieved by separation from the natural world and by its precise measurement. And today a well known statement derived from quantum mechanics warns: if a physicist chooses to suppress the role of the observer in the system, the consequence is incorrect measurements. This is correct, if it is not brought to the extreme.

In quantum mechanics there are indeterminacy and duality: the wave function of Schroedinger (that describes a potentiality and is reversible) must be reduced to a real case through a measurement, an irreversible operation. Responsible for this transformation is the observer who takes the measurement. Thus, an observer who changes a process by measuring it, becomes responsible for the realisation of a potentiality of nature. But how is human action (the observation) responsible for a process? This is in contrast with common sense! When the measurement (through a physical device or our senses), following the new laws of dynamics (e.g. the non-equilibrium thermodynamic), breaks temporal symmetry, it does not need any duality. The direction of time is in common to the measurement apparatus and to the observer.

On the other hand, the reduction of an object of knowledge to an instrumental relation or quantifiable value is in contrast with the new paradigms of science. These paradigms imply that an understanding of the process of knowing (i.e. epistemology) has to be included in the description of natural phenomena, which the observer is an inseparable part of. This vision is also contained in the natural theory of knowledge (Lorenz 1978), in which is affirmed the validity and the reliability of the abstracting capacity of human perception. For these reasons observation is at the top of the clinical method of diagnosis.

In landscape ecology, a correct observation can underline the need to expose different possible scenarios and to realise the most desirable one. Even in this aspect landscape ecology can be compared to medicine (Ingegnoli 1993, 2001; Naveh 2000). We have to diagnose the health of whole landscapes, anticipating their fate and the risk involved in their further misuse and alteration. Our obligation is to try and prescribe the best remedies for their planning, management, conservation and restoration.

1.5.4
Consequence: The Problem of Scale

After the demonstration of Weiss (1969), that the information about the whole, about the collective, is larger than the sum of information about its parts, we are now able to understand the contradiction contained in considering something scaled from a small space-time dimension to a wide persistent one but analysing the same characters - that is what happens to the concept of ecosystem. Note that, even if we would like to consider an ecosystem only as a criterion for observation (sensu Allen and Hoekstra 1992), the contradiction persists, because even the observation method must be adequate to the scale.

Remember that the scale is a necessity for perception and knowledge, it is not a principle *per se*. The concept of scale means relative size, extent, arrangement in steps or degrees; that is, a set of related dimensions. The scale derives from the principle of relative invariance of a system as a whole, in comparison with the greater variability of its constituents, and the consequent hierarchic organisation of an ordered wholeness in a space-time dimension.

This hierarchy depends on the irreversibility of time, which gives a sequence of processes: the flux of time is not homogeneous, thus processes occur at different times, thus they are scaled (Fig. 1.4).

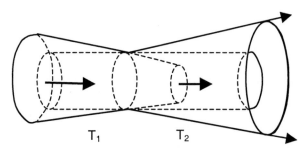

T_1 T_2

Fig. 1.4. The variation (or the constancy) of the characters of a system through time (T as irreversible) leads to different levels of processes, both in space and time. In fact, entropic time (i.e. the rate of energy dissipation) has very different results in our universe. This figure represents two processes with two different times, the first (T_1) concentrating with time, the second (T_2) dilating time

As observed by Prigogine (1996), entropic time, indicating the evolution of the system, has very different results in our universe, it does not follow our watches! Even Gould (1994) underlined the diversity of temporal scales during the evolutionary process. In landscape ecology this observation is more evident in comparison with other fields of biology. A landscape is formed by elements which follow different time-space scales, e.g. prairie vegetation - characterised by a small tessera and few years of life - vs a mature forest - characterised by many large tesserae and several centuries of life; or even a temporary pond vs a ten thousand year old lake. Sanderson and Harris (2000) note that thinking of processes as verbs (e.g. to move, to disperse), we see that the outcomes of processes are scale-dependent, but the processes themselves are scale-invariant. This is not always true, because some processes appear (even as verbs) only in a particular range of scale: e.g. all living systems are able to reproduce but not to be born. Anyway, it is theoretically possible to recognise two process behaviours in relationship to the time arrow: variant and invariant (see Chap. 2).

1.6
Revision of the Hierarchy of Levels of Life Organisation

1.6.1
The Present Situation

As shown, Western science has begun only recently (last two decades) to be able to study the entire biological spectrum (*sensu* E.P. Odum), i.e. the hierarchic levels of life organisation. From the end of the nineteenth century, the emergence of ecology has been due to an awareness of the levels of: community, ecosystem, biome, and, a few years later, the genome and the biomolecular level of protoplasm. The two most recent levels are the landscape and the biosphere. But scientists are still divided.

Many of them consider the biological spectrum not applicable because those levels do not correspond to levels which result from observations at different scales (Allen and Hoextra 1992). On the other hand, there is no doubt that the exception proves the rule, and in fact everybody knows the significance of an individual organism and of a population. And we know that speech is not possible without names for its elements. In fact, Sanderson and Harris (2000) wrote: "Though we and others do not subscribe to the theory that nature is necessarily hierarchically ordered, a simple hierarchy will suffice for the purpose of describing where entities fit into our schema of ecological systems."

What prevents consideration of this conventional hierarchy (from cell to biosphere) corresponding to levels defined by a scale of observed natural systems is probably a sort of residual reductionism, which does not consider time irreversibility and other related aspects: it considers an incorrect sequence of levels and an incorrect relationship of ecological characters with the scales. Anyway, a deeper analysis of the levels of the biological spectrum reveals some problems. We have to agree with Allen and Hoextra, but their observation is not enough to eliminate the question, conversely, it is a call for a revision. That is what we are proposing hereafter.

1.6.2
The Hierarchy of Levels of Life Organisation

Organism and population generally correspond to quite definite ranges of scale, but only if we consider respectively their vital space *per* individual and their minimum habitat, because any ecological system must include both a biological element and its environment.

An *organism* is defined by a genetic integrity, a physiological autonomy, a discrete bodily form and its own environment; it may vary from millimetres and a few hours to several metres and several centuries.

A *population* is an integration of organisms: it may be delimited by many criteria, but the genetic similarity and the difference in habitats and geographic discontinuity are generally good boundaries; its range, considering its minimum habitat, may vary from some square mm and a few days to many square km and several millennia.

Also the upper part of the biological spectrum generally corresponds to quite definite ranges of scale: *ecoregions* (better than biomes) and the *ecosphere* (better than biosphere). On a regional scale (Bailey 1996) other characters appear which are impossible to study in an ecosystem or a landscape, such as soil ordination, vegetation belt formation, faunal changes, biological gradients in rivers, climatic role in ecological changes, geomorphic transformation, etc. According to Forman, a mix of local ecosystems (i.e. biogeocoenosis) or land-use types is repeated over the land, forming a landscape, which is the basic element in a region of the next broader scale, region composed of a non-repetitive, high-contrast, coarse-grained pattern of landscapes.

An *ecoregion* may range from several millennia and about ten thousand sq. km to some millions of years and about a million sq. km. The *ecosphere* obviously includes the entire planet spatio-temporal scale.

Major problems arise in the middle part of the biological spectrum: *communities*, *ecosystems* and *landscapes*. For many scientists they are levels usable almost at any scale (Allen and Hoekstra 1992; Wiens and Moss 1999) or at least they occupy the same wide range of scales. This is in contrast with the Emerging Properties Principle. The differences among them seems to be only the criteria of observation. But they are presented as three different levels: this is another contradiction.

If the range of scales was the same, they should converge coherently, therefore going toward the integration of the complex behaviour of an ecobiota in a given area so as to produce a cohesive and multifaceted whole, which manifests properties of self-organisation. Now we may ask: can the integration of an ecobiota have a range of validity from the population level to the ecoregion? Certainly not, since the range of scale is too large. The Emerging Properties Principle has to integrate two types of environmental systems:

1. The different components of the eco-mosaic, such as plants, animals, men, soil, climate and their influence, in a geographic site (first level of integration: species-locality).
2. The different eco-units which form the ecological mosaic in a geographic land (second level of integration: ecobiota-territory).

Therefore, it seems probable that we have to consider two hierarchic levels in the middle biological spectrum: (1) the ecobiota, composed of the community, the ecosystem and the microchore (i.e. the spatial contiguity characters), which we will name *ecocoenotope*, and (2) the *landscape*, formed by a system of interacting ecocoenotopes.

1.6.3
Proper and Exportable Characters

There is a possible objection: some characters of community and ecosystem are available also at landscape level. No doubt, and even the inverse is true. But also some characters from other levels may be common to a wide part of the biological spectrum. By the way, it is only reductionist thinking which pretends to separate all the characters related to each level. We can note that each level presents *proper* characters and *exportable* ones. It is again a consequence of the balance between realism and indeterminacy in nature. The integrability needs correlation among the diverse systems, and so they need common characters and distinct ones.

Each system which presents *proper* characters is an entity, and we can find properties characterising cell, organism, population, ecocoenotope (or ecobiota), landscape, ecoregion, ecosphere (or biosphere). But pay attention: we can not pretend to have fixed boundaries related to each level, in a deterministic criterion. Any complex adaptive system (or self-organising system) has a structure, difficult to focalise and to represent.

Every biological system shows a structure composed by well defined functional sets in a context of variable substrates, starting from cells up to ecosystems, landscapes, regions and the ecosphere. As expressed by Sanderson and Harris (2000), scaling up an ecosystem does not produce a landscape.

As we try to synthesise in Fig. 1.5, passing from one level to the next level of organisation, the latter presents new, proper characters even if the exportable ones remain the same. There is a biunique interference among these characters, which may be visualised by hypothetical fields representing their potential relationships.

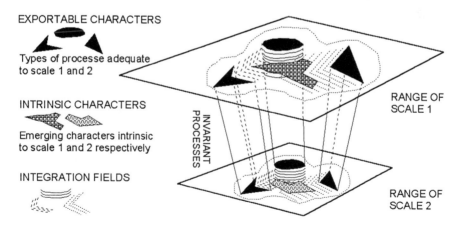

Fig. 1.5. Differences between exportable and proper characters of two hypothetical systems of life organisation. Passing from the first to the next level of organisation, the latter presents new proper characters even if the exportable ones remain the same. There is a biunique interference among these characters, which may be visualised by hypothetical fields representing their potential   relationships

This is in accordance with the Principle of Functional Integration inside the "biological spectrum", noted by Odum (1971). This is above all in accordance with the self-regulative properties of a system that becomes more than its parts, not in a quantitative-summative way but in a qualitative-structural way.

There is no doubt that the use of the exportable characters of a landscape, the chorological ones, is useful in different branches of ecology, because the spatial aspect may aid in studying many levels of biological organisation. Note that landscape ecology focused the attention on scale in ecology, but the landscape itself is not a concept valid at any scale.

A synthesis of proper and exportable characters is, for instance (Table 1.2):

Table 1.2. Schematic representation of the main characters of the life systems having a strong ecological interest

Biological levels and range of scales	Main proper characters (Ecological chapters)	Exportable characters
Organism $S = 10^{-2}$- 10^6 m^2 $T = 10^{-3}$- 10^3 years B = multicellular E = vital space	Genetic integrity, phenotypic growth, discrete bodily form, physiological autonomy, metabolism, ethology, etc. (Auto-ecology)	Basic bio-systemic: (structure, dynamics, reproduction, maintenance, etc.)
Population $S = 10^0$- 10^9 m^2 $T = 10^{-1}$- 10^3 years B = organisms E = minimal habitat	Genetic similarity, ecological density, age distribution, birth/death ratio, logistic growth, social behaviour, etc. (Population ecology)	Carrying capacity, habitat, etc.
Ecocoenotope $S = 10^2$- 10^8 m^2 $T = 10^0$- 10^4 years B = species/environment E = site	Dominant/rare species, niche, succession, trophic web, speciation, competition, foraging, etc. (Community and ecosystem ecology)	Energy flux, biodiversity, disturbances incorporation, etc.
Landscape $S = 10^6$- 10^{10} m^2 $T = 10^2$- 10^5 years B = ecocoenotopes E = land	Permeant populations, source-sink dynamics, ecotope role, landscape apparatuses, transformation control, etc. (Landscape ecology)	Spatial contiguity characters, context conditioning, etc.
Ecoregion $S = 10^{10}$- 10^{12} m^2 $T = 10^3$- 10^6 years B = landscapes E = region	Biogeographical processes, fluvial basin ecology, regional geomorphic processes, zonal climate characters, etc. (Eco-geography)	Land compensation, biome characterisation, etc.
Ecosphere $S = 10^{13}$- 10^{14} m^2 $T = 10^7$- 10^9 years B = ecoregions E = world	Atmospheric and oceanic bio-equilibrium, thermal balance and vegetation, organic limestone and plate tectonics, etc. (Global ecology)	Biogeochemical cycles, climatic cycles, etc.

Main dimensions: S space, T time, B biotic components, E environmental components

In conclusion, the biological organisation depends on the thermodynamic of dissipative structures (Prigogine 1972); the hierarchic evolution of biological systems (O'Neill et al.1986), their historical dimensions, and what we have described up to now is linked to these theories. Related to the Emerging Properties Principle, we have just seen how to distinguish the levels of biological organisation.

1.7
Toward a Widening Foundation

1.7.1
The Basis for a More Unified Discipline

We noted that to unify landscape ecology as an unitary discipline requires the recognition of what is important about landscapes. There is no doubt that these problems require a widening foundation of landscape ecology in order to reach a manifold but unique definition of landscape and to recognise what is important about landscapes. We need to focalise the main consequences of the new scientific paradigms and the revision of the hierarchy of levels of life organisation previously discussed. In this framework, the main basis for a more advanced ecological integration should be:

1. *To specify definitively that the landscape*, as a system of ecocoenotopes, *is a proper biological system.* The observation that man has a spiritual life does not contradict the definition of man as a biological organism. Similarly, the observation that a landscape is a multifunctional system does not contradict its definition as a proper biological level.

2. To emphasise that even *culture does not implicate the subjection of nature to the dominance of man.* We may demonstrate that in many cases cultural changes of landscapes express natural needs. Remember that the distinction between a living being and his environment has no substantial significance; besides, as clearly expressed by Ingegnoli (1999b) and by Sanderson and Harris (2000), landscapes have properties, therefore they are entities. For example, landscapes can support species that require forage in a grassland ecosystem and refuge in a forest ecosystem, or can direct ecosystem transformation to maintain a steady state of the landscape itself.

3. Given that a landscape is much more than a set of spatially heterogeneous characters, *to specify the intrinsic biological characters* of the landscape (structural and dynamical) as opposed to the *exportable* characters (mainly chorological). In doing this, we reconcile the conventional hierarchy (from cell to biosphere) to the non-deterministic levels defined by the scale of observed natural   systems.

4. Remembering the expressions of Forman and Naveh, who mentioned the "interweaving of the ecosystems in a landscape", *to refer to a new, more complex structural model* based on the concept of tissue, which could be named *ecotissue* (Ingegnoli, 1993, 1999b). Actually, all the other disciplinary models present their spatial schemes based on the concept of mosaic. But in this way it is impossible to fully integrate their differences. The multifunctionality of landscapes needs more complex structural models.

5. *To consider landscape ecology as a discipline like medicine*, biologically based but transdisciplinary. In fact, since the landscape is a biological level, it is the physiology(ecology)/pathology ratio which permits a clinical diagnosis of a landscape, after a good anamnesis. No doubt that landscape ecology has its own predictive theory, e.g. the landscape dynamic as opposed to the ecocoenotope one requires a new landscape vegetation dynamic. Nevertheless, it is necessary for landscape ecology to develop not as a simple predictive science, but as a prescriptive one - just like medicine, and for similar reasons inclusive in the new scientific paradigms.

1.7.2
The Core Characters of the New Disciplinary Model

If we compare the new disciplinary model derived from a widening foundation of landscape ecology with the previous four models (see Table 1.1), we have to consider:

- *Definition of the landscape:* System of biogeocoenosis as a specific biological level. System of ecocoenotopes repeated over the land, different from the scale of the region.
- *Structural scheme of the landscape:* Hierarchical multidimensional tissue (ecotissue).
- *Essential characters of the landscape:* Intrinsic biological characters of the level (structural and dynamic) and exportable characters (spatially dependent processes).
- *Scale:* From ecotope to region.
- *Main contributions in landscape models and indices:* Landscape apparatuses, biological territorial capacity indexes, standard habitat per capita indexes, transformation deficit, integrated vegetation evaluation, natural role and influence of an ecotope.
- *Disciplinary field:* Ecological, unified. Transdisciplinary.
- *Main epistemological basis:* Natural philosophy of knowledge (Lorenz, Popper, Prigogine,…).

In this framework, it is possible to utilise the contributions in landscape models and indices given by the main disciplinary models. Moreover, we have to note that a unified discipline of landscape ecology may arrive to change many principles of traditional ecology, first of all regarding vegetation science, going and leading to unified ecology.

2 Some Concepts on a General Living Systems Theory

2.1
Living Systems and Complexity

2.1.1
Main Characters of Living Systems

Man is an organism, thus the concept of life is spontaneously transferred to all organisms. If we want to define, even in a synthetic item, the attributes of life, we may observe that it is a very complex self-organising system, which needs to exchange material and energy with the outside, is able to perceive and process information, to reproduce, to have an history and to participate in the evolution. According to Naveh and Lieberman (1984), in an evolutionary view, processes and structure become complementary aspects of the same evolving whole.

Therefore it must be evidenced that life can not exist without the presence of its environment. They are a unique system, because the conditions permitting life are necessary both inside and outside an adapted distinguishable entity, like an organism. That is why the concept of life is not limited to a single organism or to a group of species, and therefore we describe life organisation in hierarchic levels, as we have shown in Chap. 1.

Works such as *The Selfish Gene* by Dawkins (1976) are fascinating but misleading regarding the concept of life, because even if the genetic messages of life were formed by clusters of selfish genes, natural evolution would force these genes to co-operate in complex adaptive systems, becoming other levels of life. Thus life created a large proliferation of forms and processes in order to utilise the world around it and to acquire more and more information and skill.

Furthermore, the world around life is made also by life itself; thus the integration reaches again new levels. This is another reason why biological levels can not be limited to cell, organism, population, communities and their life support systems (as geographic zones): life also includes ecological systems such as ecocoenotopes, landscapes, ecoregions, and the entire ecosphere.

We talked about the large delay between the definition of landscape by von Humboldt (1846) and the emergence of landscape ecology as a discipline (1950) (see Sect. 1.2.1). Another large delay is registered in the history of science dealing with the observations of Poincaré (1906) on the problem of the sensible dependence of a deterministic system from initial conditions, which led to the chaos paradigm (Ruelle and Takens 1971). In both cases, the problem derived from the study of some characters of complex systems, and the limitations were due to the reductionistic view of science, which impedes the global vision of events.

2.1.2
Studying Complexity

The complexity of ecological systems is difficult to investigate, especially if we consider landscape ecology in the same way as medicine, because clinical analysis, diagnosis and methods need a rich theory for application. Complexity may be defined as the attribute of a system (natural or cultural) which contains information that is hard to understand (Ruelle 1991). Complexity does not depend on the number of components of a system, but especially on the type of interaction among them. Weinberg and Weinberg (1979) classifies three types of complexity:

1. *Organised simplicity:* systems formed by a few components with simple interactions, which mathematics can formalise (e.g. Lotka-Volterra equation).
2. *Unorganised complexity*: systems formed by a very high number of components, with casual interactions. They may be formalised by statistical analysis and deterministic chaos (e.g. turbulence theory).
3. *Organised complexity*: systems formed by a medium number of components, with organic interactions. They can not be completely formalised by mathematics (e.g. biological-ecological systems).

Biology, thus ecology, is interested in all three types of complexity, but crucial problems generally deal with the third type. In biology, experimental data can not be very precise, even if collected over a long period; moreover, it is practically impossible to get equations of the temporal evolution of an ecological system. Actually, it must be remembered that a biological system is able to learn (it is an adaptive system), since it is continuously changing.

On the other hand, the mathematical theory shows even the theoretical impossibility of dealing with a perfect formalisation of the third type of complexity: once some inference rules and a number of axioms are fixed, we may find exact assertions which are impossible to demonstrate as being true or false. This is a basic theorem of Goedel (1931).

Anyway, remember that it is not enough to characterise living systems simply as open, adaptive, non-equilibrium and learning systems. They are even self-transcendent, capable not only of representing and realising themselves, but also of transforming themselves. Thus, in the study of these systems it is not generally possible to follow simple and univocal cause-effect connections. A good example is the insoluble linkage between physiology and pathology, as the removal of a disturbance needs the understanding of the normal process, which is rather understood through a disturbance (Lorenz 1978). It is not possible to analyse global entities (i.e. an organic whole) studying only their parts. As suggested by the new scientific paradigms (see Sect. 1.5), it is necessary to study ecological systems with the help of system theory and its corollaries on irreversibility, information, etc.

2.2
The Systemic Paradigm

2.2.1
About the Definition and Significance of a System

The concept of system was introduced (end of nineteenth century) into scientific fields by thermodynamics, the theory dealing with entropy and irreversibility at a macroscopic level. A set of elements closely interacting forms a system. The totality of relations among the components and their states constitutes the structure of the system. Because of its relations a system is a whole.

Thus, a system is always more than the sum of its elements. This fact is very important, being the basis of the holistic axiom. Holism is the view that the entirety of a complex system, such as an organism or a landscape, is functionally greater than the sum of its parts.

The Emerging Properties Principle affirms that some of the characters of a whole (i.e. a system) are determined by the properties of its elements, but other characters of the system are the consequence of "the way in which the elements organise themselves". That is: the whole is greater than its components, as the epistemological school of *Gestalt* (perception of the form) proclaimed in the first half of the twentieth century.

The first part of the mentioned properties is quite easy to understand. For example, when fungi (i.e. mycorrhizae) colonise the roots of trees, the fungi-root system is able to extract mineral nutrients from the soil more efficiently than roots alone; or when a fluvial ecosystem is combined with vegetated corridors, forming a riparian landscape, the cleaning capacity of the system is greatly enhanced. Similar mutual relationships are common in nature, and also in a well-ordered human society.

The second part is less intuitive and needs to be illustrated: when in the development of the mammalian nervous system there are some mutant factors changing the disposition of the same elements, the resulting system changes its functions dramatically (e.g. in the cerebral cortex) (Fig.2.1A); when in the management of an agricultural landscape, we change the disposition of the same elements (e.g. fields and hedgerows), the functioning (and productivity) of the ecological mosaic modifies drastically (Fig.2.1B) (note that, if repeated at a larger scale, this process could cause desertification).

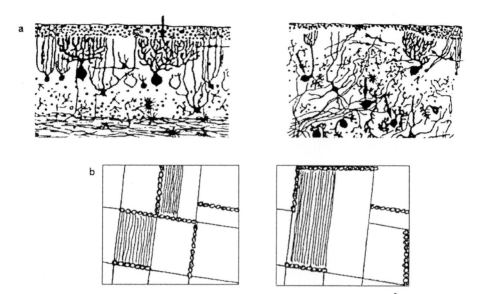

Fig. 2.1A,B. Some examples of the Emerging Properties Principle. **A** Cerebellar cortex disease causing severe damage to its functions by changing only the disposition of its cells. **B** Changing the disposition of the same elements in an agricultural landscape means changing its ecological parameters (wind effects, humidity, animal home range, etc.)

2.2.2
Hierarchical Systems

A researcher studying a certain level of interest among natural systems, such as a patch of trees, has to inquire at a more synthetic scale if he wants to know the significance and the constraints of this patch, for instance in what kind of vegetational landscape is it growing, what are the climatic constraints, etc.; then he has to investigate on even a more detailed scale, e.g. single trees, if he wants to know the components of his plant association and its reason for existence.

This is a typical hierarchic sequence, and to understand our complex world it is helpful to think in terms of level of organisational hierarchies, as we have shown in Chap. 1. Note that the rates of processes in a hierarchic system are asymmetrical: symmetry may exist only for a small (transitory) step of the entire gradient of linkages. In fact, the central concept of the theory is that the organisation of a system results from differences in process rates.

Levels in the hierarchy are isolated from each other because they operate at distinctly different rates. Boundaries, or surfaces, separate the set of processes from components in the rest of the system. Koestler named these surfaces "holons". The surface of a holon may be tangible and visible, as the skin of an organism or the boundary of a lake, or intangible, as in the case of populations. Anyway, processes are held within surfaces; in fact the limits of a process define where the surface shall be. Note that if processes would have not been structured in a hierarchic sense, we could not understand how biological structures may appear concrete, although the substance of the structure is in constant flux. According to Allen and Hoekstra (1992), we can recognise five criteria ordering higher and lower levels:

- *Bond strength*: the higher the level the weaker is the strength of the bonds that connect entities at that level. In fact, a greater organisational structure is formed when the components are able to more highly integrate, therefore needing less strength of connection.
- *Relative frequency*: higher levels behave at a lower frequency, that is they have a longer return time. Nature appears regular only if what behaves at a lower frequency is defined as occupying an upper level.
- *Contest*: low frequency behaviour permits the upper level to be the context and the constraint of the lower level.
- *Containment*: in non-nested hierarchy, the higher level is a distinct entity and does not contain the lower levels. So, if a system like this changes its hierarchical criterion, it changes its structure (e.g. food chains).
- *Constraint*: upper levels have lower constraints, but generally a constraint should not be seen as an active condition; it is a passive one. Constraints are important, because they permit forecasting in ecological systems.

The hierarchical characters of systems underline that observations pertaining to a high level of organisation ignore the distinctions at the surfaces of entities of far lower levels. Nevertheless, in ecology, a highly deterministic methodology often employs levels of organisation that are too low. Thus, a low process speed corresponds to the behaviour of a high level in a system and vice-versa (Fig.2.2).

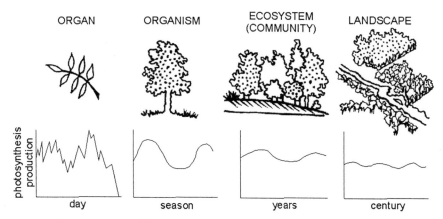

Fig. 2.2 Hierarchical levels of ecological systems. Note that a low process speed corresponds to the behaviour of a high level in a system, and vice-versa, e.g. photosynthesis in a forest patch and in a single tree.

Thus the inferior level components explain the origin of the level of interest (i.e. allow an inner description), while a superior level system explains the significance of the level of interest (i.e. allow the characters derived from transferred conditions to be understood). As suggested by O'Neill et. al. (1986), one of the most important consequences of the hierarchical structure of a system is the concept of *constraint*. It is more correct than the concept of limiting factor and it shows the behaviour of an ecological system as limited: (a) by the behaviours of its components and (b) by environmental bonds imposed by superior levels of organisation. Remember, there is a linkage between constraint and information. Regarding a landscape, for instance, we can find a range of conditions representing its constraint field, which provides information on the limits of the landscape dynamics.

2.2.3
Dynamic Systems

A system is defined as *dynamic* if it is able to evolve over time. A system is a physical entity upon which some actions may be exerted on through an input a and from which, a reaction, an output y, is derived. If we want to correlate inputs (i.e. causes) and outputs (effects), we note that at the instant t, the value of the output y can not be determined by the value of the input a in the same instant t.

The output in fact depends on the *history* of the system (i.e. on the sequence of its dynamic states) and its inner characters. Therefore, we must admit the existence of a third element, the *state*, which includes information on the past, present and potential evolution of the system. The value $x(t)$ assumed by the state at the instant t must be sufficient to determine the value of the output in the same instant. This statement is particularly important (see Sect. 2.3.6).

Knowing the values of $x(t_1)$ and $a(t_1,t_2)$, the state (then of the output) in the instant t_2 can be calculated. A dynamic system may be defined by six sets of variables (Fig. 2.3), correlated by two functions:

- Input parametric variables $a(t) \in A$ and input functions $a(.) \in \Omega$
- Output parametric variables $y(t) \in Y$ and output functions $y(.) \in \Gamma$
- State variables $x(t) \in X$
- Time variables $t \in T$
- Function of the state transformation $\varphi[t, t_0, x(t_0), a(.)]$ from which
- $x(t) = \varphi[t, t_0, a(.)]$
- Function of the output transformation $\eta[t, x(t)]$ from which
- $y(t) = \eta[t, x(t)]$

The couple state-time (x, t) has great importance because the set X, T is the set of events, the history of the system. Once an instant t, an initial state $x(t_0)$, an input function $a(.)$ are fixed, the transition function $\phi[., t, x(t), a(.)]$ is univocally determined, and named "movement".

$$S = (T, A, \Omega, X, Y, \Gamma, \varphi, \eta)$$

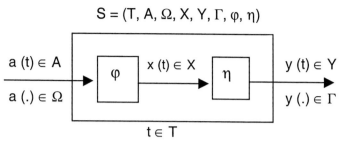

Fig. 2.3 Theoretical representation of a dynamic system, defined with six sets of variables

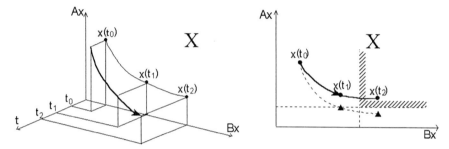

Fig. 2.4 Movement and trajectory of a dynamic system. *X* represents the field of the states of the system (state of the phases). *AX* axis and *BX* axis are, for example, a set of structural variables and a set of functional variables respectively. The trajectory is the projection of the movement of that X field. Note the drawing of a delimited set (*right*), useful to control the trajectory

Figure 2.4 shows the main elements of a dynamic system. The space containing the points which corresponds to the states of the system is called the *space of the phases*. The projection of movement on the vertical plane represents the trajectory of the system. This function is particularly useful in building models, as we will see in other applications in the next chapters.

Each dynamic system may be characterised by a kinetic energy, which depends on the speed of its components, and by a potential energy, which depends on the interactions among the components. These interactions can interfere with the movement of the system or even bring the trajectory to a collapse.

We may have hypothetical systems composed of elements without interactions, in which future and past could be interchangeable. But in a world isomorphic to a set of bodies, subject to no interaction, there would not be place for the arrow of time, nor for self organisation, neither for life. Fortunately, a real system like this does not exist, as observed by Poincaré (1906).

In the general case, the particles of a system (described from a statistical point of view) interact. This fact leads to the presence of resonance among the degrees of freedom of the system which determines conditions of non-integrability in a mathematical sense, as discovered by Poincaré. Remember that mathematical integration signifies the zero setting of potential energy of a system, and therefore non-interacting parts. By contrast, in engineering systems the integration is the arrangement of components in a system so that they function together in an efficient and logical way. In fact, each degree of freedom of a dynamic system is characterised by a frequency, and Poincaré proved that resonance is an insuperable obstacle to mathematical kind of integration. The KAM theory (Kolmogorov, Arnol'd and Moser) studies the influence of resonance on the trajectories of a system; thereby, it studies the transformation of the topology of the space of the phases with increasing energy. Once a critical threshold is reached, the system behaviour becomes chaotic (see Sect. 2.3.5), the trajectories of its components diverge and we may even observe diffusion (an irreversible phenomenon).

Remember that resonance is a phenomenon exhibited by a physical system upon which an external periodic driving force acts, in which the resulting amplitude of oscillation of the system becomes large when the frequency of the driving force approaches a natural free oscillation frequency of the system. Therefore, resonance leads to the coupling of events (like in music).

Moreover, we have to distinguish between an individual level (trajectory) and a statistical *intrinsic* level (sets). For example, the existence of the phase transitions show an emergent property, not compatible with individual description: the single particles are not solid or liquid, but the gaseous, liquid, solid states are properties intrinsic to the system.

2.2.4
Some Notes on Entropy and Time

Two well known general principles (laws) are at the base of thermodynamics:

1. Energy and matter can not be created or destroyed but only transformed.
2. Spontaneous transformations occur along a direction of irreversibility.

The concept of entropy has been introduced to indicate that direction: *entropy* is a function of the state of a system and its value rises when a system spontaneously transforms itself (without outside interference). If S is entropy, t is time, d is the derivative, we will have:

$$dS/dt > 0$$

We can also affirm that entropy measures the degree of change in a system; that is, the degree of mixing up of the components of a system in random way. A system tends toward the most probable state. Thus, in statistical mechanics the meaning of the second law has been broadened, and entropy has become the measure of structural *disorder*. In this sense the concept of entropy is opposite to the one of *order*, reached by a decrease of entropy, negative entropy or *negentropy*. As expressed by Boltzmann for the phase-space distribution of gas molecules in physical systems at different times

$$S = k \ln D$$

where: S = entropy, k = Boltzmann's constant, D = thermodynamic probability, that is, the number of possible microstates.

The first equation implies an isolated system, but we have to deal essentially with closed and open systems, depending on the exchange with the environment: only energy or energy and matter. These exchanges (fluxes) are spontaneously possible only from more to less concentrated matter and from higher to lower energy. A flux is established only over a threshold value. These processes of exchange are irreversible, so open systems are connected to their environment through energy (or matter) fluxes which eliminate in an irreversible way the gradients established by previous differences of symmetry.

In reality, we have to note that the production of entropy always contains two dialectic elements: the creation of disorder but also a creation of order. These two elements are linked together, as we will see, from dissipation. When there is a breaking in the temporal symmetry, the past and the future start to perform different roles in the statistic formulation, as instability destroys the equivalence between the individual level and the statistical level, and consequently the probabilities assume an intrinsic meaning (see Sect. 2.3.3).

The Principle of Increasing Entropy states a first-after relation, introducing the concept of time. According to Prigogine (1996), *irreversibility* must not be considered as a simple appearance, manifested by the enormously long time of return to an improbable initial state. It is something more essential, without which the emergence of life on Earth is inconceivable. But time is not only irreversibility and evolution. Prigogine suggested (1988) that time is a potential state which becomes topical through a fluctuation phenomenon. Time is before existence.

2.2.5
Dissipative Systems

Photosynthetic processes have the main responsibility of energy transfer in biological systems. Chlorophyllous enzymes produce an endoergonic reaction utilising 673 Kcal/M of reduced CO_2, as follows:

$$6\,CO_2 + 12H_2O + (h\nu) \;=\; C_6H_{12}O_6 + 6O_2 + 6H_2O$$

The question is: how is it possible? Ecological systems must be open. Otherwise, the free energy F would not be available: in fact, at the equilibrium state, E should be completely transformed in S_i (internal entropy of the system) and, consequently, F should tend to 0 ($F = E - TS$, where E is the total energy, T is time, and S the entropy).

In an open system we have to consider two fluxes of entropy: d_eS, that is the entropy flux due to the exchanges with the environment, and d_iS, which is the entropy flux due to the irreversible processes within the system. The second term must have a clear positive sign, but the first term does not have a definite sign. So the inequality of Clausius-Carnot becomes:

$$dS = d_eS + d_iS \quad being\ d_iS > 0$$

Let us consider a period in which the system is stationary $(dS = 0)$, thus

$$d_eS + d_iS = 0$$

consequently

$$d_eS < 0 \quad (\ being\ d_eS = - d_iS)$$

So, in an evolutionary process, when the system reaches a new state of lower entropy (new stationary state) $S\ (t_1) < S\ (t_0)$, it is able to maintain it in a similar way by "pumping out" the disorder. But this is possible only in non-equilibrium conditions, in so-called *dissipative* systems: a dissipation of energy into heat is necessary to maintain the system far from equilibrium. Note that dissipation is also necessary to eliminate transitory phenomena.

2.2.6
Stability of a Dissipative System

An energy dissipation, which allows work to be done, has to be coupled, for example, with the transformation of a system from state A_0 to state A_1. The process able to perform this transformation is an example of an *operator* (Op), a rule of action on a given function. If we express it in the form $A_1 = (Op)\ A_0$, the complete transformation process is

$$A_1 = [(Op)\ A_0] \cup (e_w \rightarrow e_d)$$

where: e_w = available energy, e_d = dissipated energy. Generally, an operator transforms a function into another one. For example, if $Op = d/d_x$, consequently x^2 will be transformed into $2x$. But every operator may find functions which remain invariant to its actions. So, e^{kx} is invariant to the derivation and is only multiplied per k. Similar functions are called *auto-functions* of the operator, and the constants, which they are multiplied by, are called *auto-values*. A theorem states that an operator can be expressed through its auto-functions and its auto-values.

If a state of a system becomes an auto-function for a certain operator, the system does not undergo further changes. This state is called a *fixed point* of the system, and it may represent a stationary state. An attractor basin is generally associated with this particular state. But, what is an attractor?

An *attractor* is a state of a system toward which the system itself is directed starting from initial positions. It can be viewed as a geometrical object toward which the trajectory of a dynamic system, represented by a curve in phase space, converges in the course of time. The attractor can be single or manifold. In some conditions, for instance in far from equilibrium systems, there could be a set of attractors. In non-equilibrium thermodynamic, an attractor is a state corresponding to a local minimum of *dP/dt*. Therefore the *stability* of a system depends on the variations of the internal entropy d_iS. Its derivative

$$d_{i2}S/dt_2 = dP/dt$$

expresses the production *P* of the entropy *S* in excess. Consequently, we may distinguish two components of *dP/dt*:

1. d_xP/dt , due to the variation of *X* (*t*), the thermodynamic state
2. d_jP/dt , due to the variation of the fluxes, depending on the organisation

If we assume *dP/dt* as locally linear, a system is able to maintain a stationary state only if d_xP/dt is decreasing or constant: therefore the stationary states correspond to a local minimum of *dP/dt* and can be named attractors.

2.3
Complex Systems and Self-Organisation

2.3.1
Cybernetics and Self-Organising Systems

The science of the behaviour of interacting systems is called *cybernetics*: it is interested in mutual causal effects. We may have principally two interacting conditions: (1) information coupling and (2) feedback coupling.

1. The effective *connection* between two sub-systems S_1 and S_2 is realised by signals that cause a change in system behaviour. This change is conditioned by a bridge created by a common pool of signs Z_1 and Z_2 . For the regulation of biological or technological processes this information can be carried out both in an energetic and/or a material way. This means the *structuring* of energy with the help of information.
2. The output values again affect the input values of the system, creating feedback loops. They may be positive or negative. In *positive feedback* couplings the effect and the reaction act in the same direction, amplifying each other (e.g. exponential growth of a population). In *negative feedback* couplings the effect and the reaction are opposed, controlling each other.

As expressed by Naveh and Lieberman (1994), a good example of negative feedback is the present tendency to lower population growth rates because of density stresses. Increases in fertility and life expectancy, *A*, coupled with depopulation of agricultural landscapes, C, has led to greater population densities, especially in the larger cities, *B*; but in return, the increased density stresses *C*, which adversely affects fertility and health, slowing down the rates of population increase and densities.

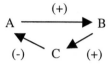

In living systems the capacity to maintain a dynamic equilibrium as a whole is called *homeostasis*. It is ensured by a great number of closely interrelating cybernetic feedback mechanisms, hierarchically ordered. These biological and ecological processes of auto-regulation can be active also at a landscape level.

Complex interaction systems in which cycling, structuring and auto-regulation are realised from the inside may be called *self-organising* systems. Their dynamics can be synthetically expressed by:

$$X_t = f(X_0, t, \lambda) \tag{1}$$

where: X_t is the state of a system at time t, X_0 is the state of the system at time 0, λ is a specific parameter for the examined system indicating the acquisition of energy and matter from outside. Depending on the parameter λ and its values, X may tend toward a stationary state or a chaotic one.

2.3.2
Some Notes on Information Theory

According to Shannon and Weaver (1949) *information, I*, can be defined as the measure of the uncertainty of an event and of its removal by a message. The message is intended in a broad sense and it is independent from the mechanism of its transmission. If the amount of information is expressed by the number of choices between two equi-probable possibilities, the measure of these binary decisions is in "bits". For example, if we have to choose the right habitat from a set of 16, the probability P_0 of choosing the right one before receiving information is *1:16*, or *0.063*; but after receiving exact information P becomes *1* (certainty). Using a binary decision method, we may measure I:

$$I = log_2 P/P_0 \quad thus \quad I = log_2 16 = 4 \ bits$$

Information has been considered an equivalent of negentropy, because its measure implies the removal of uncertainty and randomness and increases order and organisation. Thus, remembering the concept of negentropy, we can define also the *content* of information of a message, I, as the difference between two entropies:

$$I = S_{(Q/X_q)} - S_{(Q/X'_q)}$$

where: Q is a defined problem, X_q is the understanding before a message, and X'_q after the message. For instance, the ecosphere receives an energetic flux of *1.2 Gcal /year* per square metre (medium), which promotes a decrease of entropy on

Earth of about 10^{38} bit s^{-1} before being irradiated into space as black body radiation.

The concept of negentropy allowed Shannon and Weaver to use the Boltzmann entropy function as a measure of information. So the information content of, e.g. a plant community can be expressed by the diversity index H' (average mutual information) in a set of plant species:

$$H' = - \sum_{i=1}^{n} log_2 P_i \quad bit$$

However, information is more than a physical dimension expressed by this formula. Remember that a message with a high information content is a message selected from a large class of admitted ones or a message which contains a lot of chance.

This case may correspond only partially to disposable information, the rest being noise, often without interest. The example of the quantity of information in works of art (e.g, paintings) is clear and important (Ruelle 1991): a high content of information does not correspond generally to a high quality of art and significance. The quantification of autocatalytic processes (see Sect. 2.3.3) needs the study of information, or constraint (the terms are interchangeable in information theory), and we can thus use the measure of average mutual information (AMI).

2.3.3
Autocatalytic Systems

A new perspective on causality was expressed by Karl Popper (1990): indeterminacy is a basic feature of the real nature of our universe. Thereby, we need to generalise our notion of force in a non-deterministic way as the "propensities" for events to follow one another. These propensities are related (not necessarily identical) to conditional probabilities when forces deals with determinacy: if A then B, $p(B|A) = 1$. But under more general conditions, dealing with indeterminacy, if A then probably B, so $p(B|A) < 1$; furthermore, $p(C|A)$, $p(D|A)$, etc. > 0.

Moreover, (a) propensities may change with time and (b) only forces exist in isolation while propensities do not. In particular, propensities exist in proximity to and interact with other propensities. The end result is development or evolution. Changes of this nature are beyond the capabilities of Newtonian description.

Ulanowicz (2000) proposed a way to quantify these propensities underlying the concept of *mutualism*: A and B interact in positive feedbacks (+,+). When mutualism exists among more than two processes, the resulting set of interactions has been defined as *autocatalysis*.

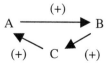

Thus process *A* has a propensity to enhance process *B* which exerts a propensity for *C* and similarly back to *A*. Indirectly, the action of A increases itself, whence the "autocatalysis". Note that the temptation to identify this process as a mechanism, as in the field of chemistry, is highly inappropriate. The behaviours of autocatalysis under open conditions are non-mechanical, because they can exhibit any or all of eight characters which, taken together, distinguish non-Newtonian phenomena (Ulanowicz 1997). Autocatalytic loops have the following properties:

1. They *enhance growth:* an increment in the activity of any member engenders greater activities in all other elements.
2. They *perform a selection pressure* upon its components: if a random change in the behaviour of one member makes it more sensitive to catalysis, then the effects will return to the starting element as a reinforcement.
3. They *break symmetry*: autocatalysis imparts a definite direction to the behaviours of the system.
4. They *enhance centripetality*: the transfer of material and energy between their components and the rest of the world is centripetal.
5. They *induce competition:* if a new element D appears by chance or is more sensitive to catalysis than A, then D will grow to overshadow the role of B in the loop or will displace it.
6. They *are autonomous* from lower level events. The autonomy of a system can not be apparent at all scales: if one's field of view does not include all the members of an autocatalytic loop, the system will appear linear in nature.
7. They *emerge* from a wider vision, once the observer expands the scale of observation enough to encompass all members of the system.
8. They *are formal*: an autocatalytic system is in itself a kinetic form (project).

A convenient example of autocatalysis in ecology is the community of processes connected with the growth of macrophytes of the genus *Utricularia* (Ulanowicz 2000). The growth of leaf and stems is *A*; each leaflet is a good substrate for periphyton (process *B*) and component *C* is the zooplankton feeding on it. Small bladders can suck a copepod (or similar animal) which decomposes and gives nutrients to the plant.

But another example is more common in ecology: the main process of urbanisation of a landscape. A rich city (*A*) in a favourable landscape begins to build industries in its peripheral small towns (*B*). So these towns need more people who emigrate (*C*) from other regions. The emigrants can not be hosted in the small suburban towns, therefore they settle in the city, which is able to build new residential wards, and begins to grow, absorbing its periphery. Then a new cycle starts. To quantify an autocatalytic system, we need the following definitions:

- The node *0* represents the *setting input*.
- The node *n+1* represents the *flux output*.
- The node *n+2* represents the *dissipation* part output.
- T_{ij} is the *generic flux* going out from node *i* and entering in node *j*.

The total system throughput will appear as: $T_{..} = \Sigma_{i,j}\, T_{ij}$

where the dotted subscript indicates that the particular subscript has been summed over all components, from 0 to $+2$. Any increase in the level of system activity will be reflected as a rise in $T_{..}$.

Calling S_i the measure of the indeterminacy of event i, we can write it in the classical form of information theory: $S_i = - k \log p\,(A_i\,)$, where $p\,(A_i\,)$ is the probability of event A_i happening, k is a constant, and S_i is the *a priori* indeterminacy associated with i. When a component B_j precedes A_i any constraint that it exerts upon the latter will be reflected by a change in probability that A_i will occur. Thereby, this is a conditional probability of A_i , given B_j . The indeterminacy has been diminished to

$$S_{ij} = - k \log p\,(A_i\,|\,B_j)$$

where S_{ij} is the a posteriori indeterminacy of A_i given B_j. The information (reduction in indeterminacy) will be consequently:

$$S_i - S_{ij} = - k \log\,[p\,(A_i\,|\,B_j)\,/\,p\,(A_i\,)]$$

Baye's Theorem allows one to calculate $p\,(A_i\,|\,B_j)$ as $p\,(A_i\,,B_j)\,/\,p\,(B_j)$, where $p\,(A_i,B_j)$ is the joint probability that A_i and B_j occur in combination. Hence:

$$S_i - S_{ij} = - k \log\,[p\,(A_i,B_j)\,/\,p\,(A_i\,)\,p\,(B_j)]$$

Moreover, because A_i and B_j are an arbitrary pair of events, it is possible to calculate the average amount of constraint that all system elements exert upon each other. It's sufficient to multiply the precedent equation for each combination i and j by the probability that A_i and B_j co-occur and to sum over all combinations of i and j. The resulting "average mutual constraint" *(AMC)* is:

$$AMC = k\, \Sigma_{i,j}\, p(A_i,B_j)\, \log\,[p\,(A_i,B_j)\,/\,p\,(A_i\,)\,p\,(B_j)]$$

To make this equation operational it remains to estimate the three probabilities in terms of measured quantities. The probability $p\,(A_i\,)$ of a quantum of flux to leave node i at time t is $\cong T_i\,/\,T_{..}$. Similarly $p\,(B_j) \cong T_j\,/\,T_{..}$, and $p\,(A_i,B_j) \cong T_{ij}\,/\,T_{..}$. Therefore, remembering that constraint and information are interchangeable, we will have:

$$AMI = k\, \Sigma_{i,j}\, (T_{ij}\,/\,T_{..}\,)\, \log\,(T_i\,/\,T_{..}\,/\,T_j\,/\,T_{..}\,)$$

Applying this equation to any variety of flow network configurations, one obtains the measures of the intensive change in kinetic structure, as in Fig. 2.5.

We have to distinguish between extensive and intensive properties in a system: extensive ones pertain to the size of the system, whereas intensive attributes refer to those qualities that are structural. Thus, growth enhancement is extensive; the remaining properties are intensive and serve to prune from the structure those pathways that less effectively participate in autocatalysis (Fig. 2.5). Both properties, however, are strongly influenced by the process of autocatalysis.

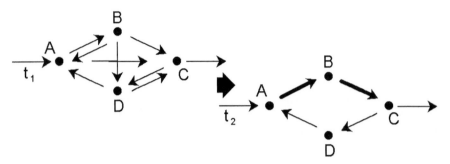

Fig. 2.5 Schematic effects that autocatalysis exerts on network pattern systems

Therefore we may use the scalar factor k to impart physical dimensions to our measure of constraint. Setting $k = T_{..}$ in the last equation gives:

$$A = \Sigma_{i,j}\ T_{ij}\ log\ (T_{ij}\ T_{..} / T_i\ T_j\ )$$

where the scaled index A is renamed the system "ascendency". It can be also desultory, and does not concern the role of biomass or stocks, so we need a correction:

$$A = \Sigma_{i,j}\ T_{ij}\ log\ (T_{ij}\ B^2_{..} / B_i\ B_j\ T_{..}\ )$$

where B_i is the *biomass* of component i. The flow T_{ij} could represent the movement of a given amount of a species from spatial position i to location j and B_i could represent the density of the given population at location i. Therefore A in this case applies to the migration of a given population over the landscape.

2.3.4
Some Notes on the Chaos Paradigm

A concise definition of *chaos* is very difficult. The most traditional could be: the sensible dependence of a system on initial conditions, leading to a sort of "deterministic noise". Systems of this sort constitute the typical problem Poincaré referred to, asserting (1908) that case and determinism are made compatible by a long-term unpredictability. But scientists tried to study questions like these (e.g. turbulence) by observing the "modes", that is the periodic oscillation of fluids. Finally in 1971, Ruelle and Takens proved that by exciting a fluid with increasing forces from outside, a continuous spectrum of frequencies, and not a gradual intensification of their modes, suddenly appears. The work of Ruelle was based on "strange attractors", which are objects of non-entire dimension, or "fractals", in particular the Lorenz attractor (1963).

We have to note that the uncertainty given by chaos does not depend on complexity: even a simple deterministic system can be chaotic. A system is chaotic when it amplifies initial conditions, thus magnifying small differences, for instance between two trajectories. If the distance between the trajectories is d at the beginning, it becomes $10d$ after a characteristic time T (intrinsic to the system), and these differences amplify exponentially at times proportional to 2T, 3T, 6T, etc. (Ekeland 1995).

The amplification of small deviations shows the presence of casualty. A basic principle of determinism states that two identical initial position will produce the same trajectory. But it is impossible to give to a physical system the same precedent position, therefore after nine powers of ten the difference of only one atom reaches one metre!

Remembering the concept of closure, a structure can be reproduced when the generative rule is known, therefore a short description is enough. This is normal for each organised system, while random systems need a total description. Obviously, if the shortest algorithmic description of a system coincides with the description of the entire system (i.e. with the longest one), the system is chaotic.

It is impossible to shorten the description of a chaotic system because of its unpredictable behaviour due to branching possibilities of evolution, thus to a manifold of attractors. These behaviours may also assume regular (repetitive) forms, or patterned chaos, for instance, in clouds or waves.

2.3.5
Complex Systems between Order and Chaos

To study the behaviour of complex adaptive systems from a mathematical point of view, scientists like Stuart Kauffman (1993, 1995) have recourse to Boolean stochastic webs. In these webs each variable can assume a definite state, active or inactive, and it is regulated by other variables, which may be considered as entries. Rules of logic commutation define the activities of a variable as reaction to inputs.

For example, the function *OR* (disjunction) activates the output variable if at least one of the entrance variables is active. By contrast, the function *AND* (conjunction) activates the output variable only if all its entrances are active. If the entries of each binary element are K we will have 2^k possible combinations of inputs; having to specify a signal of activity or not, note that the Boolean functions for each element are *2 raised to 2^k*.

In an autonomous Boolean web the K entries per element N come only from inside the system; entries and Boolean function are assigned by chance. Giving values to N and K it is possible to distinguish a family of webs with the same local characters. A stochastic web is chosen random in this family. Each combination of activity/inactivity of the binary elements forms a state of the web. A system passes through different states in a so-called trajectory.

The Boolean stochastic web has a finite number of states, hence the system may return to a previously assumed state. Having a deterministic behaviour, the system will repeat the same succession of states, in cycles. These cycles are called dynamic attractors of the web. The set of states related to a given cycle is called the "attraction basin" of that cycle.

Thus, a web reaches one of its attractors, but, if disturbed, its trajectory may change. A web may show diverse perturbations, e.g. minimum or structural. What is important to underline is that by changing the parameters of a Boolean system its behaviour is modified and it can change from chaos to order.

The $K = N$ webs express a maximum of sensitivity to initial conditions because each following state is random, and enlarging the number of elements N, the length of state cycles increases exponentially. Thus the web presents a chaotic behaviour. Despite this chaotic behaviour, $K = N$ systems express an unthinkable first principle of order because the medium number of possible state cycles (and attraction basins) is very small. This number is equal to N/e: therefore for a system of 120 elements – hence of 2^{120} states – (a normal magnitude for a landscape) the behaviour regimes are only *44*, two-thirds of which pertains to small attractor basins.

Chaotic aspects of Boolean webs appear also when K decreases until *3*. When $K = 2$ the characters of these systems change suddenly and a spontaneous collective order appears. In the case of $K = 1$, the structure degenerates into separate feedback loops. In $K = 2$ webs, the number of different cycles and their length are reduced until the value of the square root of N. In our example of $N = 120$, the cycles number will be *11*. Moreover, these systems are near insensible to perturbations.

According to Kauffman, the reason for the emergence of order in Boolean webs depends on the formation of a frozen nucleus, a connected weft of elements blocked in an active or inactive state. Variation of activities can not propagate among these nuclei, thus the system becomes ordinate, because modifications of its behaviours have to remain small and localised. But even in highly connected webs, when commutation rules are non-symmetric, order may emerge.

In summary, highly chaotic webs are so disordered that the control of complex behaviours is impossible, while highly ordered webs are so rigid that it is impossible to express a complex behaviour. But if frozen components begin to melt, it is possible to have more complex dynamic behaviours leading to a complex co-ordination of activities within the system. Thus, the maximum complexity is reached in a "liquid" transition between solid and gaseous states, where the best capacity of evolution is expressed. For instance, it is possible to see a similar situation in DNA and its capacity to remain ordered but also to change by mutations.

The threshold between order and chaos seems to be an essential requisite of complex adaptive self-organising systems. As these systems are dissipative too, an order through fluctuations is effective in working between the above mentioned conditions.

2.3.6
Order Through Fluctuations: an Answer for Complex Organised Systems

When an intense energy flux passes through a system, generally some dissipative structures appear, characterised by high instability. The self-organised living systems are able to capture this kind of energy and to utilise it to produce new structures: "order through fluctuation" writes Prigogine. He showed (1972) that even simple physical systems present processes of order. For example, the variation of heating under a fluid may change its structure and typical convection cells will begin to emerge. Or, in a chemical solution in water, we can pass from a disordered solvent to a more ordered one by adding some benzoate: this polar molecule gives a greater order to the H_2O dipoles, diminishing the entropy of the system.

According again to Prigogine, attention should be paid to the concentration of the intermediate product B in a chemical reaction: during the course of the reaction, going further on the stable thermodynamic branch, the intermediate product enters a field of instability with the appearance of subsequent bifurcations (Fig. 2.6).

When a system arrives at a branching point, disturbances, like fluctuations, become important, allowing the system to choose one of the two branches of new relative stability. The evolution of this kind of system has an historic element in itself. For example, if one observes the system in the position d_2, it means that the system passed through the states b_1 and c_1.

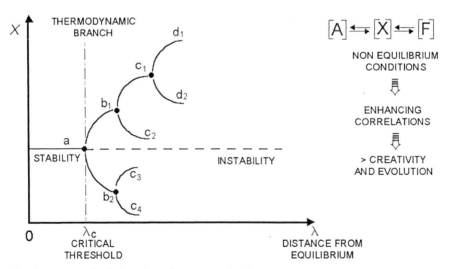

Fig. 2.6 Successive bifurcations in a non-equilibrium system: going further on the stable thermodynamic branch, the intermediate product enters a field of instability with the appearance of subsequent bifurcations. Thus, disturbances become important, allowing the system to choose one of the two branches of new relative stability

The fluctuation-dissipation sequence may be viewed as a feedback process. We can observe that macro-fluctuations produce instabilities that move the system to new temporary organisational states. These new states permit increased dissipation and move the system toward new thresholds. New fluctuations can move the system beyond these thresholds, permitting it to reach a better organised new stationary state: then the process is able to repeat itself.

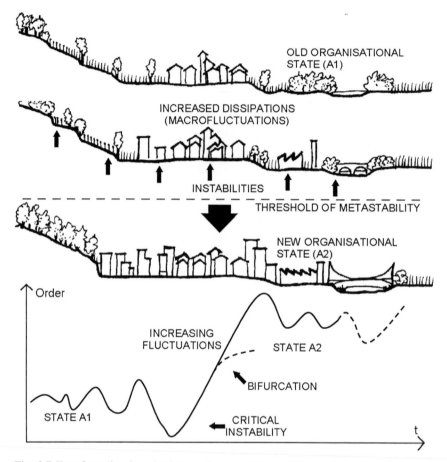

Fig. 2.7 Transformation in a landscape. From a state *A1* of lower order through increasing dissipation, a system reaches a critical threshold and, after a branching point, it arrives at the state *A2* of higher order. The old organisational state is an agricultural landscape; an increased flux of energy (e.g. agricultural improvement and social-economic richness) produces macro fluctuations of the local organisation and then some instabilities (i.e. land abandonment, use of the fluvial valley, building of the first industries, etc.). These instabilities lead to an increased dissipation of energy, the system becomes more difficult to maintain: when a threshold is reached, characterised by the prevailing of metropolitan structures over the previous rural ones, a new organisational state results, that needs a different kind of management.

We can see a theoretical representation of this process in the lower part of Fig. 2.7: from a state *A1* of lower order, a system, through increasing dissipation, reaches a critical threshold and, after a branching point, arrives at the state *A2* of higher order. The related typical case of landscape transformation as an "order through fluctuation process" is shown in the upper part of Fig. 2.7. The old organisational state is represented by an agricultural landscape at the end of the past century in Europe; an increased flux of energy, due to agricultural improvement and social-economic richness, produces macro fluctuations of the local organisation and then some instabilities (i.e. land abandonment, use of the fluvial valley, building of the first industries, etc.). These instabilities lead to an increased dissipation of energy, with the system becoming more difficult to maintain. When a threshold is reached, characterised by the prevailing of metropolitan structures over the previous rural ones, a new organisational state results, that needs a different kind of management.

In conclusion, we underline again that disturbances, like fluctuations, are important for a system at a branching point, enabling it to choose one of the two branches of new relative stability. Under these conditions, mutual relations of large range occur among the components. The matter acquires new properties, a new sensitivity of matter to itself and to its environment takes place, associated with dissipative and not reversible processes.

A far from equilibrium system is able to self-organise through intrinsic probabilities, exploring its structure and realising one among the possible structures, but not a random one. For instance, in protein synthesis, the possible chains are so many, but in the cell only a few chains are realised: they are attractors.

2.4
Metastability and Disturbances

2.4.1
The Concept of Metastability

The concept of metastability is not a compromise between a form of stability and one of instability. When a system is oscillating around a steady attractor, but may even move toward another attractor, then we have metastability (Godron 1984; Naveh and Lieberman 1984; Forman and Godron 1986).

The significance of metastability in ecological systems (e.g. a landscape) is based on their capacity to maintain themselves within a circumscribed set of conditions, with the possibility to reach other conditions if their constraints envelope is going to change.

Higher or lower metastability depends on the distance from the position of maximum stability and on the height of the thresholds of local (far from equilibrium) stability. Ecological systems with low metastability have a low resistance, but a high resilience to disturbances. By contrast, high metastability systems have high resistance to disturbances. For example, a prairie patch has a higher resilience than a forest one. The concept of metastability is derived from the paragraph on non-equilibrium thermodynamics, and allows the traditional concept of ecological equilibrium to be update: "equilibrium" does not stay around 0, but it identifies various stationary or equilibrium states far from 0. A system reaches a new organisation after instabilities and the passage to a new metastable level.

A quite simple model of a metastable system can be plotted based on a potential function. This function y may represent the potential given by a state function x and a parameter z, for instance the U shaped surface in Fig. 2.8.

The continuous line represents the manifold, that is the attractor set. The return speed to the attractor is considered higher than the change speed of the potential, which could modify how the attractor intervenes on the state of the system. In the figure, a system, $A1$, can be displaced by disturbances in $A0$. Then usually it returns to $A1$ before the possibility of a change in the entire potential. If this potential should change (dotted line), $A0$ reaches $A2$, which may present a bifurcation.

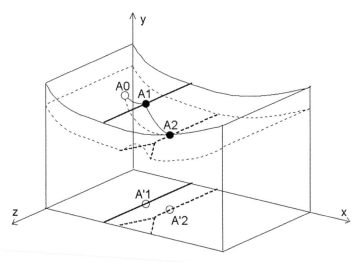

Fig. 2.8 Example of a metastable system in theory. The function y represents the potential given by a state function x and a parameter z, for instance the U shaped surface. The continuous line represents the attractor set; the return speed to the attractor is considered higher than the change speed of the potential function (i.e. the environmental conditions), which could modify the attractor set intervening on the state of the system. The system $A1$ may be displaced by disturbances in $A0$: then usually it returns to $A1$ before the possibility of a change in the entire potential function; but if this potential function changes (*dotted line*), $A0$ may meet a bifurcation, going to $A2$

The system dynamic is guided, in the long-term, by these slower speeds, which may modify the attractor in a way that, after a perturbation, the return line will not be *A0-A1*, but *A0-A2*. In this case *A2* may coincide with a bifurcation, from which new changes of the parameter, even small, can cause great transformations in the system. A familiar example is an ecological system regarding a fluvial trait. After a perturbation a new evident metastability threshold may be reached. As we will see in the chapter on landscape transformations, every change is generally related to this kind of processes.

2.4.2
The Importance of Disturbances in Ecological Systems

Even the concept of disturbance is linked to the non-equilibrium thermodynamics and all the other characters emerging from the synthesis reported in this chapter. Once a branching field is encountered, a self-organising system needs some disturbances in order to "make a choice". Remembering the hierarchic theory of systems, we know that some limitations on the dynamics of an ecological system come from inferior levels of scale and are due to the biological potential of its components. Other limits are imposed by superior levels as environmental constraints.

A range of conditions emerges for every kind of ecological system, for instance a landscape, and can be expressed as the constraints field or optimum set of existence. It represents the metastability level of the system and it is expressed in the space of phases of a dynamic system. This field is related to the range of disturbances which characterise a landscape.

Natural disturbance regimes can be measured in relation to surface S, intensity I, and frequency F. If S and I are dominant, the effect will be very strong, sometime destructive (e.g. a storm); if S, I, F, are of the same magnitude, the disturbance is of local effect, as the fall of a tree; if F is dominant, the scale is generally small and the effect well controlled even by a small ecosystem.

Human disturbances may be added to the previous ones: they occur in a system always subject to a natural disturbance regime. These anthropic effects have to be distinguished into two groups: the maintenance activities, generally on a natural basis, and the technological activities, which can be in contrast with an ecological principle.

Note that in many cases the majority of disturbances can be *incorporated* into ecological systems. The mentioned constraint field of an ecological system is based on a resistance strategy to a current regime of perturbations. Therefore, we can speak of disturbance incorporation when the system organisation exerts control over some environmental aspects that is impossible to control at a lower level of organisation. This process may limit possible alterations to its stationary state; meanwhile it may utilise perturbations as structuring forces. For these reasons the incorporation of disturbances has been proposed as a stabilising process of ecological systems (O'Neill et. al. 1986).

As we will discuss in later chapters (see Sect. 4.8.3), disturbances have to be reconsidered by many traditional ecologists. For example, in some cases the imperative of an E.I.S. (environmental impact statement) to minimise every kind of perturbation is incorrect environmental policy. Even fires (Fig.2.9) have to be regarded in less dramatic way. The large fires in Yellowstone National Park (1988) taught us that this kind of landscape has been structuring itself in the presence of fires for thousand of years, and a too strong limitation of them by man can cause two negative aspects: (a) the rising up of exceptionally strong fires, and (b) the decrease of biodiversity.

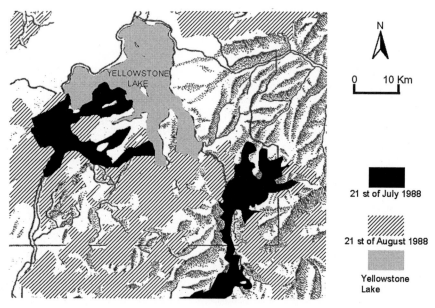

Fig. 2.9 Part of the great burning of the Yellowstone National Park (USA) of 1988. This area measures more than 4000 km^2; from about 30 to 35 % of these forests were burned between August and November 13, when a dense snowfall stopped all the last fires. Note that in each burned area many vegetation patches were only partially destroyed or even untouched.

3 Landscape Structure

3.1
Structural Models

3.1.1
Mosaic and Variegation Models

Describing and studying a landscape can be extremely complex. The increasing attention that has been paid to the spatial structure of the environment coincides with historical progress in ecology, too.

Prior to 1960, general ecology usually emphasised spatial homogeneity. The reference was a sort of traditional *ecosystemic* model, in which - given a scale of interest - only a homogeneous spatial unit was considered, for example, a wooded area in an agricultural environment (Fig. 3.1**A**). This unit should have been studied as a community (e.g. vegetational association) or as an ecosystem (e.g. a watershed). Outside this spatial unit, there was an indistinct (heterogeneous) environment. A more complete model considered the boundaries of that ecological unit: the edge belts, or the ecotonal belts if an environmental gradient separated the ecosystem from ones nearby.

Because theories assuming homogeneity developed early in the history of ecology, they had a powerful and persistent effect on how ecological systems were viewed. That is why E. P. Odum, in his famous book *Fundamentals of Ecology* (1971), mentioned the landscape, but he was not able to properly represent it, even when writing on land use and the necessity of ecological planning, "undoubtedly the most important application of environmental science". Nevertheless Odum mentioned the book of Ian McHarg, *Design with Nature* (1969), which showed how the natural landscape can provide guidelines for quality urban development. Mc Harg, a landscape architect, was able to represent the natural landscape as a patchy environment, probably on the basis of the vegetational maps made by botanists of the school of Braun-Blanquet.

Consequently, we passed to a new model of representing the landscape which has been fully elaborated by many authors, e.g. Forman and Godron, Haber et. al., at the beginning of the 1980s. This second model could be named *mosaic* model, and has two aspects (Fig. 3.1**B**): the "patch-matrix" and the "ecological mosaic".

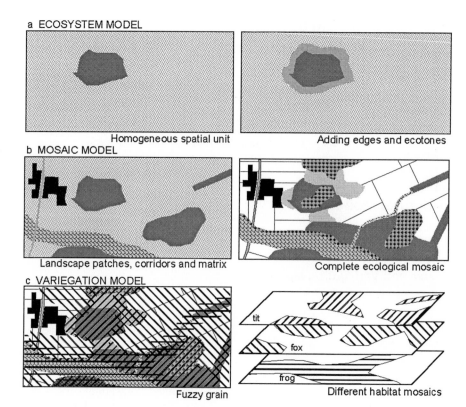

Fig. 3.1A, B, C.　Main structural models of the landscape. **A** ecosystemic model. **B** mosaic model. **C** variegation model. Each model may be represented in two versions, the second more detailed

The first highlights the patches formed by the types of ecological communities, at a given scale of interest, and the corridors composed of natural or human elements, within a sort of landscape matrix: this is the main type of element which characterises the landscape. The second emphasises the entire landscape mosaic, the elements of which we could name *tesserae* and "ecotopes": these are the delimitations of the principal types of ecosystems (i.e. biogeocenosis or, better, ecocoenotopes) and they constitute a sort of geographic map, apparently similar to the land use maps of the human territory, but with an ecological sense. These maps are more similar to the vegetational maps.

The concept of human fragmentation of the landscape implies that vegetation remnants are surrounded by unfavourable environments (e.g. a boreal forest patch in an arable fields mosaic). Consequently, as pointed out by Ingham and Samways (1996), organisms may not be able to move across or inhabit the fragments. Much ecological research has been carried out on the basis of this model. However, the form and extent of the change in landscape varies, and the mosaic model may not always be applicable.

Many zoologists have observed that relatively few organisms perceive the landscape in a similar biological way, because of the existence of stenotopic and euritopic taxa. Moreover, the biology of an adult insect, for instance, may differ from that of the larvae. Therefore a species may be dependent on two or more landscape elements for its life cycle, or may live in one area but forage in another one. The landscape - thus - may act as a filter, separating out the species (Wood and Samways 1991), or even becomes a species-specific environment (Farina 1993). So the landscape seems to disappear, evanescing into a sort of fuzzy-edged mosaic. Thus, a new landscape model emerged, the *variegation* model (Fig. 3.1**C**), because the ecological mosaic is composed of tesserae having a variable conformation (e.g. fuzzy-edged boundaries), in other words, it is an overlapping series of different patch-matrix mosaics (proportionate to the main groups of steno-euri-topic habitats).

In summary, we have passed from the ecosystemic model, undoubtedly too simplified and unrealistic, to two divergent new models: the mosaic one and the variegation one, each of which registers the real situation, but in a typical complementary contrast (sensu Bhor and Heisenberg) resulting in a typical uncertainty. On the other hand, many researchers knew that reality needs both the mosaic and the variegation model, e.g. when studying a mosaic of forested landscape and its bird communities.

The main limitation of the mosaic model is that differently scaled entities may be forced together in a rigid mosaic; other serious problems concern the variegation model too. As pointed out by Allen and Hoekstra (1992), if organisms of different species on the same spot of ground respond to different scaled patches of landscape, then the concept of "organism" becomes inconsistent as a scaled unit. Alternatively, if we use the organism as the normalising framework, then the landscape becomes a curved space, because different types of organisms view it at their own scale of reproduction, dispersal, movement, or home range.

This seems to be absurd. The theory of complementarity may lead to a misunderstanding of the problem of observation (see Sect. 1.5.3). As sustained by Popper (1990) and Prigogine (1996), the reality per se remains measurable. Therefore we need to change our criteria if we want to add another step towards a true descriptive model of the landscape.

In fact, some observations regarding the variegation model are not completely true. No doubt that an euritopic species like *Orthoptera* or *Odonata* perceives the landscape differently from a stenotopic species like *Hemiptera* or *Carabidae (Coleoptera)*. An euritopic species forages indifferently in a patch of meadow or in the closest patch of trees, but its physiologic and behavioural responses will not be the same for each patch, because the microclimate (e.g. light, humidity) changes, the quality of food (e.g. amino acids, lipids, carbohydrates of leaves) changes, many predators change, etc.

Therefore, we may find diverse hierarchies of ecological factors, some of them depending on the mosaic model, others on the variegation one. Moreover, many animal species are definable as "permeants" (*sensu* Shelford in Odum 1971), simply to indicate their use of many types of ecosystems. These euritopic animals, which move, feed, protect, drink, nurse, with periodic rhythms, have a different perception than stenotopic ones, and they are even able to recognise the landscape as a complex mosaic in detail, from macro- to micro- scales (e.g. apes, wolfs, gooses, bees, etc.). Some of them are able to change directly some parts of the landscape, for example, beavers, but also ibexes, elephants, termites, etc.

3.1.2
The Ecotissue Model

As just discussed, the descriptive models of the landscape are mainly: the "mosaic" model and the "variegation" model (Fig. 3.1). These models represent two completely different outlooks, one static and one dynamic. But it is not completely correct to compare a mosaic of juxtaposed tesserae of landscape elements (mosaic model) with a mosaic of species-specific tesserae and variable geometry (variegation model). In reality, as we know, each biological system shows a structure made by well defined functional groups in a context of substrates changeable in space and time (or variegated configurations over a fixed weft, and whose elements are not only juxtaposed, but also overlapped and intersected). Both Forman and Godron (1986) and Naveh and Lieberman (1984) talked about the interweaving of ecosystems in a landscape.

As a consequence, it is better to introduce the concept of ecological tissue (from the Latin *textus* or *textilis*; in English: textile) or *ecotissue*, which is a complex multidimensional structure represented by a basic mosaic and a hierarchic succession of correlated mosaics and attributes:

$$Ect = Mm \cup(\cap) M' \cup(\cap) M'' \cup(\cap) Mk$$

where: Ect = ecotissue, Mm = basic mosaic, M' ... Mk = correlated mosaics and the set-theoretical $\cup(\cap)$ symbols of union and of intersection may be interchangeable or may even coexist, since they are often heterogeneous groups.

The basic mosaic is generally formed by the vegetational coenosis because control of the flux of energy and matter and the capacity to create the proper environment pertain to it. This fact is in accordance with non-equilibrium thermodynamics. Whereas an energy concentration (i.e. photosynthetic plants) produces structure and organisation in a landscape matrix with increasing entropy, the order through fluctuation process creates a patch, which acquires a specific landscape role. This may be the principal way by which ecological systems become heterogeneous (Ingegnoli 1980, 1999b; Forman and Moore 1991).

Anyway, all the other mosaics are correlated to the basic mosaic, and compatible with its main scale. Trying to detect and gather information on organisms and communities outside this main scale (see Sect. 1.5.4.) of interest is generally non-sense, because of the hierarchic organisation theory.

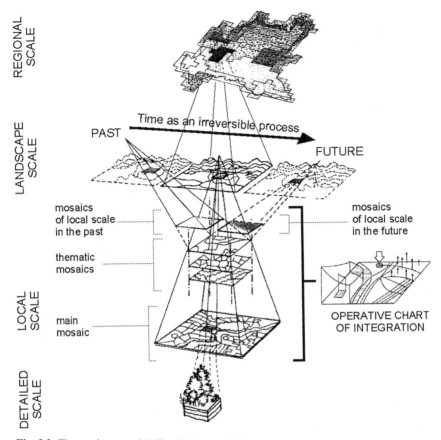

REGIONAL SCALE

LANDSCAPE SCALE

LOCAL SCALE

DETAILED SCALE

Time as an irreversible process

PAST

FUTURE

mosaics of local scale in the past

thematic mosaics

main mosaic

mosaics of local scale in the future

OPERATIVE CHART OF INTEGRATION

Fig. 3.2 The ecotissue model. The basic mosaic is generally the vegetational one. The complex structure of a landscape has to integrate diverse components: temporal, spatial, thematic. An operative chart of integration is therefore necessary to elaborate plans. Note that the integrations are intrinsic, that means they have to follow integration functions derived from the intrinsic characters of that level of life organisation

We can study a self-organising system only through projections and sections of an hyper-space, so that it is possible to gather information to be integrated, step by step, in a hierarchic correlation with the basic mosaic.

It is necessary to consider constraints at different scales and intersections with specific thematic mosaics, so that they are compatible with the characters pertaining to the intrinsic level of life organisation of a landscape (Fig. 3.2).

The ecotissue gives the right importance to the landscape and integrates three fundamental dimensions of the landscape:

1. A range of spatial scales, from regional to local
2. A set of thematic mosaics, and correlated arguments

3. A range of temporal scales, which permits the evolutionary dynamic of the landscape to be forecasted and reconstructed.

This result may be configured in an operative chart of integration.

As we will see in the section on methodology, the importance of the concept of ecotissue can be found especially in the integration of elements and processes in a landscape. Actually the study of a multifunctional landscape is usually led by a certain number of different mosaics (geomorphologic, vegetational, zoological, agronomic, of land use, of human needs, etc.) which have to be integrated for a certain purpose. But this kind of integration is a sort of a posteriori process, obtained with a traditional multidisciplinary criterion, because the landscape is ultimately viewed as a support for biological and human systems. It results from an integration like this.

By contrast, if a landscape is defined as a living system, we must refer to a more complex structural model in which the integration is made *intrinsically*. This means that the mentioned thematic mosaics have to be related to the structure and behaviour which characterise that landscape. A clinical and pathological methodology is applicable only if we analyse the state of an organism knowing something about its anatomy and physiology. Similarly, we have to analyse the state of a landscape following the knowledge we have about its structure and functions.

3.2
Landscape Elements

3.2.1
The Landscape Elements in Our Discipline

A landscape exhibits the same three fundamentals characters of all living systems, as an organism or an agricultural system, and represents a challenging research frontier. These three characters are structure, function and transformation:

1. *Structure*: The spatial relationships among the distinctive ecosystems (better: ecocoenotopes) or "elements" forming the landscape.
2. *Function*: The interactions among the spatial elements, that is, the flow of energy, materials, and species among the component ecosystems, and the intrinsic behaviour of its complex mosaic.
3. *Transformation:* The evolution and alteration in the structure and function of the complex mosaic over time.

If we define the landscape as a system of ecosystems, each type of landscape may be related to a particular configuration of interacting ecosystems, that is, with a specific structural pattern. But these patterns share a similar structural model. Carl Stainitz, landscape planning professor at Harvard University, presenting the book of Forman and Godron on landscape ecology (1986), underlined the theoretical importance of the observation that all landscapes, from the wilderness to the central city and from the natural to the developed, share a similar structural model. We know that the reason for this behaviour lies in the fact that the landscape is a specific biological level of the ecological hierarchy (see Sect. 1.6.2), having its proper structure and functions. So we propose the ecotissue model as its synthetic appropriate reference.

In general, given a particular range of scales, the structural model of a landscape is concerned with the selection of elements, representing the minimum part that the landscape may be divided into. Then, it is necessary to analyse the configurations of these elements in a hierarchical sense, enhancing the ecological mosaics formed by them. Other larger units, forming a principal landscape, can be detected.

Before expressing our view on landscape components in coherence with the ecotissue model, let us review the main traditional schools.

In the vision of Forman and Godron (1981, 1986), the basic elements of the landscape are more generically named patches, or corridors (if linear). These patches are intended as ecosystems or biogeocenosis in the sense of the German-Russian school. They describe the patches in relationship to their shape, size, biotic type, number and configuration. Systems of patches form an ecological mosaic, in which we may detect the landscape matrix, as the most extensive and connected landscape element type present having a functional importance. Thus, at any spatio-temporal scale it is possible to analyse the landscape's "overall structure". These authors described its heterogeneity (macro and micro), the contrast among types of patches, and the grain or medium size of the components.

Naveh and Lieberman (1984, 1994) considered the landscape to be composed of ecotopes. The smallest, homogeneous and mappable landscape unit is the ecotope: an holon composed of viable pieces of the biosphere, the pedosphere, the geosphere, the lithosphere, the hydrosphere and the atmosphere. If small, it can be called tessera. The ecological holarchy of concrete systems, with the ecosphere as the larger system and global landscape of the total human ecosystem, integrates ecotopes of the biosphere (open landscapes) and of the technosphere (built-up landscapes) through the geosphere.

Haber (1989, 1990) asserted that, for a landscape ecology approach, it is useful to transform the functional model of an ecosystem into an ecotope model, the ecotope (Fig. 3.3) being the spatial representation of the ecosystem. Where the local physical and chemical composition of the site changes, another ecotope will begin, thus defining the ecotope boundary. Landscapes are composed of ecotopes or ecotope types and they can be assembled into regional natural land units (RNUs from German *naturraumliche Einheiten*, Finke 1986). These RNUs are determined by common geomorphological properties and regional climate, and they have their characteristic landscapes.

Zonneveld (1989, 1995), described three principal dimensions connected with landscape studies: topologic, chorological, geospheric. The topologic one emphasises ecological understanding of a relatively homogeneous area (ecotope or land unit), involving co-operation among disciplines like soil science, vegetation and zoology. The chorological dimension studies the relationships among the ecotopes in a landscape mosaic, without neglecting topologic knowledge. The geospheric dimension regards the upper scales, that is, the relationships with the regional and global (biosphere) structure. Zonneveld defines a land unit as independent of scale. The land units may be arranged hierarchically into: (a) the land facet (microchore), (b) the land system (mesochore), (c) the main landscape (macrochore), (d) the larger landscape (megachore) in transition to the geospheric dimension.

In summary, an elementary unit that is still a holistic unit at the proper range of scales of a landscape may be defined as a landscape element. As we have seen, usually ecologists name the elements as *tesserae,* or ecotopes, or land units, or patches, but many other expressions can be found, such as habitats, biotopes, geotopes, etc. It may be that scale and culture have a strong effect on this problem. Anyway, the definitions of these landscape elements seem to be somewhat confused. For instance, Zonneveld (1995) lists the following synonyms for ecotope: site, tessera, Fliese, land cell, facies. We need some attempts at clarification.

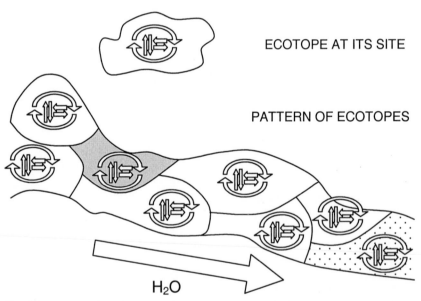

Fig. 3.3 Representation of a pattern of ecotopes along a mountain slope (following the definition of Haber 1990), where the downward flowing water creates a dependence of lowland on upland ecotopes. In our vision, flowing water and morphology differentiate the landscape functional role of each ecotope (e.g., a protective role for the *grey* ecotope or a recervoir role for the *dotted* one)

3.2.2
The Hierarchy of Pattern Components of Landscape

Note that, generally, an ecotope, or a patch, is defined with spatial and ecosystemic characters, rarely with other biological attributes, and this is not sufficient. Remembering the proper and exportable characters of life organisation levels (see Sect. 1.6.3.), we need to define a landscape element within this perspective.

For example, all living systemic units have to maintain, even at the landscape level, some specific characters available for every life unit, like their own structure, layer of delimitation and filter, self-reproduction system, development phases, specific physiognomy. Biotic types of landscape elements depend, among other things, on their type of ecocoenotope structure (i.e. producers, consumers, decomposers, etc.), and it is possible to analyse the energy flux among ecocoenotopes, exchange of species, etc. The local physical and chemical composition of the site is necessary to define an element boundary. And dominant/rare species, niches, etc. are also needed. Landscape characters are obviously the most important as are permeant populations, source-sink dynamics, elementary unit role, landscape apparatuses, transformation control, spatial contiguity and context conditioning.

Therefore, in order to define an acceptable hierarchical pattern of landscape elements, in coherence with the ecotissue model, let us recall the definition of landscape: a landscape is a system of ecocoenotopes in a recognisable configuration, the structural model of which is the ecotissue, a multidimensional structure with a basic mosaic and a set of other mosaics and attributes hierarchically correlated.

The smallest landscape unitary element that can be considered to be a landscape in the sense of the definition is the minimum system of ecocoenotopes; thus, not less than two elements have to be intended as the first-level unit of integration, i.e. species-locality. Following Troll (1963) and Neef (1967) we may call it *ecotope*.

The components of an ecotope, which represent the ecocoenotopes, are the parts of land which are uniform throughout their extent in landform, soil and vegetation: the ecosystem characters (functional model) integrated with the spatial characters of the community. This multifunctional entity can be called *tessera* when referred to landscape ecology.

A hierarchy of pattern components of landscape can be proposed observing that the definition of landscape leads to a field of existence varying from a system of two-three tesserae to the ecoregional level. The spatial resolution in studying this large field has to be enhanced, as in other biological disciplines (Fig. 3.4). For instance, if we have to study a cell we know that at a particular resolution scale it is possible to see only the cell as a whole (e.g. 150/1), but at a more detailed scale we may see the nucleus (e.g. 1000/1) or even the Golgi apparatus, and finally, with an electronic microscope also the mitochondria (e.g. 20000/1). Or, by contrast, reducing the scale (e.g. 50/1) we may see cell types in their histologic tissue, and so on. A similar concept is applicable to the landscape. A vegetated tessera of prairie can measure a few square metres and can be mapped on a 1/200 scale, a forested tessera can be observed on a scale 1/1000, a mesoscale ecotope can be surveyed on a 1/5000 map, a simple landscape unit   from a scale of 1/10000 to 1/25000, and so on. An ecoregional unit can seldom be mapped under a scale of 1/ 250000-1/500000.

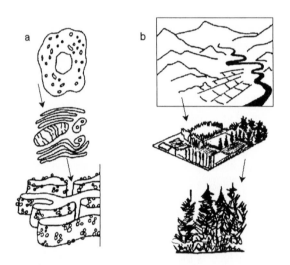

Fig. 3.4A, B. Scale dependence of structure in physiological and ecological fields. Examples of two scale sequences. **A** From a cell to a mitochondrion. **B** From a landscape unit to a tessera. The spatial resolution in studying a landscape has to be enhanced, as in other biological disciplines

At this point we have to note that in landscape ecology it is possible to find two different criteria in studying landscape structure.

The first criterion is independent of scale. The concept of patch/corridor (Forman and Godron 1981, 1986) or the concept of land unit (Zonneveld 1989, 1995). For instance, the definition of land unit is: a tract of land that is ecologically relatively homogeneous at the scale level concerned. In the case where a hierarchical arrangement is needed, one should use by approximation the terms ecotope, micro-, meso-, macro- and megachore.

The second criterion is consistent with the above mentioned observation on scale dependent elements. We prefer this last criterion, because we are convinced that:

- We can not pretend to have boundaries related to each levels of life organisation in a deterministic criterion: but there is no doubt that every biological system shows a structure formed by well defined functional sets in a context of variable substrates, starting from cells up to ecosystems (ecocoenotopes), landscapes, regions and the ecosphere.

- The described concept of ecotissue needs to be referred to a hierarchical pattern of landscape components.
- Many patches may be considered relatively independent, but many others not, having an explicit strong linkage among them.
- In any case, it is better to refer to vegetational *tesserae* and ecotopes than refer, using informatics tools, to manifold clusters of various types of patches, the most part of which have no biological significance.

Consequently we propose as pattern components of a landscape: tessera, ecotope, simple landscape unit, complex landscape unit, landscape. This is similar to the Australian (D.O.S.) system: site, land facet, simple land system, complex land system, main landscape.

Moreover, the second criterion does not reject the first: it is possible to use "by approximation" the first criterion with the second. For example, we may speak of a patch of tesserae of the same typology, or a simple landscape unit may be viewed as a patch at a very synthetic scale (e.g. 1/200000).

3.2.3
Tesserae and Ecotopes

The minimum landscape unit, in the sense we have just described, is the ecotope, (Fig. 3.5), which is the minimum system formed by two or more ecocoenotopes. As expressed by Leser (1978) and Pignatti (1994) an ecotope is formed by interdependent tesserae, for instance a small trait of a river with its bed, its banks and their vegetation complex.

In order to analyse an ecotope, Vos and Stortelder (1992) suggested surveying the so-called physiotopes (geomorphologic and microclimatic characterised localities) and their biotopes (community-ecosystem characters). But this is not sufficient. Recurrence (topographical) and genetic (geomorphological) relations, in fact, are not always enough to locate an ecotope. That is why Ingegnoli (1993) proposed a survey also of the functional role in the landscape unit: for example, along a river we may have groups of tesserae more applicable to contain a flood.

Moreover, remember that the main characters of an ecotope can be numerous and they are to be analysed on three scales: ecotope scale, simple landscape unit scale, tessera scale. They may be:

- *Basic mosaic characters*: ecological state of vegetation and its medium biomass; biodiversity degree; not-incorporated disturbances; land use functions; natural and human habitats; grain size; connectivity with the outside mosaic; potential transformability; landscape pathologies.
- *Function characters, shared to upper scales:* physiotope participation; role in hydrography; landscape apparatuses; role in regional landscape system; importance of ecotope for heurytopic animals; contribution in structural orientation of the landscape; presence of technological webs.
- *Lower scale characters*: see tessera

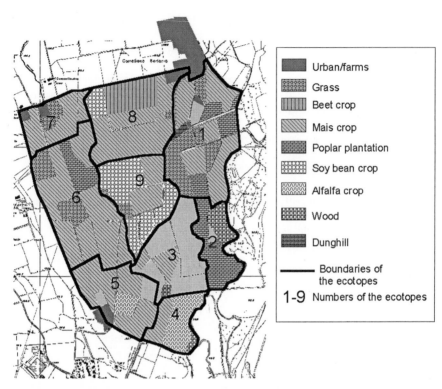

Fig. 3.5 An example of ecotope and tesserae, from the regional natural Adda River Park (Lombardy, Italy). When mapping the ecotopes, note the differences existing between them, which can be composed even of tesserae of different type and function (as an agricultural ecotope with various cultivation tesserae, one or more wooded tesserae, a farm tessera, etc.) and the patches of tesserae (all tesserae with the same land use, which are not ecotopes)

In this framework, if we need to locate a tessera a proposed series of analysis should be done as follows:

1. Plant association or sub-association
2. Other vegetational characters: growth phase and form, structure, biomass, etc.
3. Influence of eco-mosaic
4. Geomorphologic characters
5. Geophysical characters
6. Natural boundaries (or human)
7. Land use management
8. Faunal habitat

The development phases of vegetation are particularly noteworthy, because this characterisation of a tessera regards other factors too, such as the animal distribution in an ecological mosaic, for instance in agriculture or forested ecotopes, even if the vegetational association is the same. For this purpose (Fig. 3.6) we can see the four typical phases (A,B,C,D) in a forested ecotope.

- The innovation phase is A, lens-shaped, in which the information of the propagule bank is operational. The highness is very limited in contrast to the density of organisms.
- The aggradation phase is B: the vertical structure is now differentiating, low trees survive, and the rhyzosphere is deeper.
- The biostatic phase is C: trees have built structural ensembles that organise forest architecture and all other component behaviour. The canopy is again closed.
- The degradation phase is D: trees of the present decay, and forest architecture breaks down to liberate many other forest components. For instance, light and organic matter activate part of the propagule bank to prepare a following innovation phase.

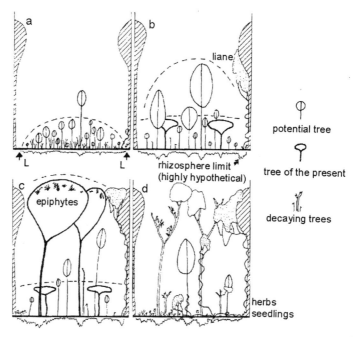

Fig. 3.6 Vegetation phases of growth in a forested tessera (from Oldeman 1990, re-drawn). The innovation phase is A; the aggradation phase is B; the biostatic phase is C; the degradation phase is D

3.3
Spatial Configurations

3.3.1
Patches and Corridors

Working on ecotopes, it may be useful to consider them as patches and corridors, depending on the shape, and -as always- on the scale. The patch is defined as a non-linear surface area differing in appearance and/or in substance from its surroundings. The corridor is a narrow strip of land that differs from the surrounding elements of the landscape. Many observations have been made on patches and corridors (e.g. Forman and Godron 1986) and it is better to try and synthesise them more in detail. It is possible to classify many types of patches in relationship to their origin :

1. Disturbance (isolated, chronical, cyclic)
2. Remnant (associated or not with a disturbance)
3. Environmental resource (biotic or abiotic or both)
4. Human activity (direct or indirect)
5. Colonisation (natural or human)
6. Ephemeral patches (natural or human)

Generally patches are formed by some environmental change and each patch is transformed in time. The dynamics of the species is involved in this process, particularly extinction and immigration in relationship to the different origins of the patches. The persistence of a patch depends on the presence of a range of perturbations, and it is proportional to its stability, essentially for homeorhetic reasons.

Since the concept of landscape element is linked to a certain range of scales, we note that it is hierarchic. Many patches may be considered relatively independent, but many others are not, having an explicitly strong linkage. For instance, some forested patches in a sub-rural landscape could be independent; by contrast, on the same landscape, gardens and villas or ponds and reeds are linked.

The extension of a patch does not influence the flux of energy and nutrients of its ecosystem: there is a proportionality between the surface and these processes. The extension of a patch may influence its biomass, because of the difference between the border edge and the interior part of the patch. The influence of the borders of a patch is proportionally the inverse of its extension. In these marginal belts of a patch, biomass is denser than the interior, where its species have less diversity. Anyway, large patches seems to have more species (especially animal) than small ones. A patch may be viewed as an isle, for instance a forested patch surrounded by a cultivated mosaic; therefore the species/area diversity generally follows what community ecology has expressed as:

$$S = f[(+) H_{habitat}, (-) Ds, (+)A, (+)Is, (+) age]$$

where : (+) = positive relation, (-) = negative relation, H = diversity, Ds = disturbances, A = area, Is = isolation, S = number of species. In these terrestrial islands we have to observe that the species diversity S is positively correlated with the surrounding mosaic and its heterogeneity and negatively correlated with the margin discontinuity of the patch. In a case like this, disturbances may have positive relationships.

The shape of a patch normally has ecological significance not only as a relationship between edge-interior (or littoral-lake), but even when the diameters are different or the shape is annular or dendritic. The shape affects the interaction of the patch perimeter with the landscape mosaic, the movement of species through the mosaic, the functional exchange with a patch archipelago, the type and number of species, and the role as an ecological attractor in the landscape. In the case of a peninsular shape we note particular effects, like the diminishing of the species toward the top, the changing of environmental gradient and a so-called interdigitation process.

The ecological influence of the shape of a patch may be observed also in a three-dimensional sense, as in marginal belts. A strong difference in light, temperature, humidity and species changes the structure of patch margins both in shape (e.g. more branched trees) and regarding the presence of more shrubs and herbs. Even in the case of patches of direct urbanisation (built-up areas) we have a margin effect, if the environmental relationships are not degraded.

According to Forman and Godron (1986) corridors may also be classified in relationship to their origin, but the main characters of corridors are the following :

1. Stream corridors (which are formed by a river or a canal)
2. Line corridors (patch boundaries, roads, paths, hedgerows, etc.)
3. Strip corridors (wider bands with interior species)

To study the structure of corridors it is necessary to analyse: curvilinearity, nodes, breaks, connectivity, and their transverse section. Corridors are generally composed of multiple elements (Fig. 3.7).

Stream corridors are characterised by the presence of the river bed and its banks, but often by a strip of forest, which controls water and nutrients fluxes and allows the movement of species. That is why it is necessary to preserve river corridors with all their structure, at least on one side. In a stream corridor, species change as the stream order increases and in relationship to the heterogeneity of the landscape mosaic.

Linear corridors show many types: hedgerows, dikes, roads, railroads, power lines, narrow ridges. They may act as through ways or barriers, depending on the landscape mosaic. Their micro-environment is particular, because of the effect of the winds, the light and the interactions with other contrasting elements of the mosaic.

Strip corridors may be of the same type as linear corridors, but they contain an interior environment. Bird communities of a forest differ generally with the bird community of a corridor, because of the length gradient. Also the plant species of a corridor may differ from those of the nearby forests, especially if the landscape matrix is very transformed by the human population.

Hedgerow corridors are particularly important since they are able to characterise agricultural landscapes. They may be planted or remnant, and may have a large heterogeneity of trees and shrubs. Complex corridors were found in many regions of Europe, with a structure of a lane and two parallel hedgerows, often with a ditch and a bank. In the last 3-4 decades, the ratio of hedgerow per unit area is decreasing, as is the number of trees /100 metres, e.g. from 5 to 2 in central France (Burel and Baudrie 1999).

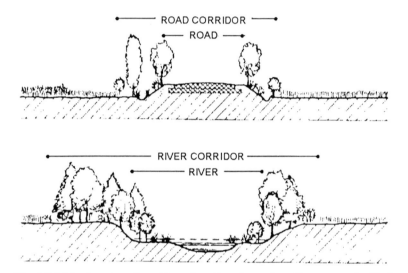

Fig. 3.7 Main characters of corridors. A section of a road corridor compared with the section of a river corridor

3.3.2
Landscape Matrix

The definition of landscape matrix is linked to the concept of the ecological characterisation of a mosaic, therefore a matrix is the most extensive and most connected element type present in a landscape, and plays the dominant role in its functioning. It is possible to distinguish among three types of matrices :

1. Continuous, with a single dominant element type
2. Discontinuous, with a few co-dominant element types
3. Web-shaped, with connected corridors of prevailing functions

Forman and Godron (1986) cited three criteria available to distinguish a landscape matrix: relative area, connectivity, control over dynamic.

The relative area criterion pertains to the existence of the first or the second type of matrix. One type of element is considerably more extensive than the others when it covers in a connected way more than 50% of a landscape. But this element must have a non-degraded ecological state. The percentage may be less than 50% if additional characters indicate the strong ecological importance of the type of element : in cases like these, the highest relative frequency of the element in the mosaic is sufficient. A discontinuous type of matrix may be found.

The connectivity criterion identifies a matrix by evaluating the degree of connectivity. The matrix is the more connected landscape element type present. In fact, when one landscape element is completely connected and encircles most of the others, it has to be considered as a matrix.

The control over dynamic criterion considers the matrix as the system of elements exerting the greatest control over landscape dynamics. This is because any future landscape is determined by the type of elements able to control the transformation of the entire mosaic. Nonetheless, it is necessary to utilise all three methods of evaluation of the landscape matrix.

Even if the matrix is classified as continuous it contains some heterogeneity. Analysing a matrix becomes important to measure its *porosity*, that is, to measure the density of patches in a landscape. The type of patches to measure depends on the case study, for instance forested patches in an agricultural landscape (Fig. 3.8) or built-up patches in an urbanised one.

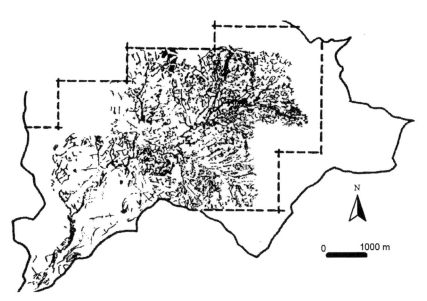

Fig. 3.8 Main characters of the landscape matrix: example of porosity of tree vegetation in a Mediterranean landscape unit (Magazzolo River, Sicily, Italy). Note the loss of structure in the central and south-western parts, due to human actions and clay instability of the hills.

Note that porosity is a concept independent from connectivity. The number of patches per unit of surface may have some significance, but it is necessary to consider the type of elements. In Fig. 3.8 we note that the porosity (wooded parches) may help to identify the degradation of some part of this landscape, when the resulting structure differs from its surrounding.

3.4
Functional Configurations

3.4.1
Landscape Apparatus

The definition of landscape apparatus concerns functional systems of tesserae and ecotopes which form specific configurations in the complex mosaic (i.e. ecotissue) of a landscape. They differ substantially from the landscape units, because they are not districts or sub-landscapes, but complex configurations of patches that may even be non-connected. These apparatuses are distinguished by a specific landscape function (and /or its range of sub-functions), not only by many local characters.

If we want to refer to the ecotissue model, it is necessary to put in evidence the distributive modalities of specific systems of elements, choosing criteria of landscape functionality. For instance knowledge of the dynamic phases of each ecotope does not depend on the ecological state of patches and corridors, but especially on the spatial pattern of the landscape.

A first well known, general but important landscape function is demonstrated by the survey of human habitat (HH) versus natural and semi-natural habitat (NH).

Ecologically speaking, the HH can not be the entire territorial (geographical) surface: it is limited to the human ecotopes and landscape units (e.g. urban, industrial, and rural areas) and to the semi-human ones (e.g. semi-agricultural, plantations, ponds, managed woods). The HH can be defined as areas where human populations live or manage permanently, limiting or strongly influencing the self-regulation capability of natural systems.

The NH are the natural ecotopes and landscape units, with dominance of natural components and biological processes, without direct human influence and capable of normal self-regulation. Note that even near-natural ecosystems (i.e. little changed after human abandonment) are NH. Remember that in landscape ecology the management role of human populations, if not directed against nature, may be considered as semi-natural.

Following the ecotissue model, the presence of HH and NH is generally mixed: for instance, it is difficult to find a completely HH tessera or ecotope. In many margins of human tesserae and ecotopes we may find natural species. Therefore even in urbanised landscapes generally it is possible to find NH patches and corridors. On the other hand, in natural landscapes many semi-managed patches may be found. Hence, the mosaic of NH and HH in a landscape is different from the land use mosaic (Fig.3.9).

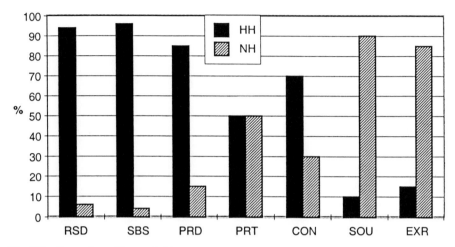

Fig. 3.9 Proportion of human habitat (*HH*) vs natural habitats (*NH*) in tesserae and ecotopes with the following functions: residential (*RSD*), subsidiary (*SBS*), productive (*PRD*), protective (*PRT*), connective (*CON*), source (*SOU*), excretory (*EXR*). Values typical for Central Europe

Rarely these other functions were linked to some basic concept of landscape ecology, such as to its structure and dynamic. In reality, it is possible to distinguish many types of landscape apparatuses:

- GEO = geologic (emerging geotopes or element dominated by geomorphic processes)
- CON = connective (elements with important connective functions in the mosaic)
- STB = stabilising (elements with high metastability)
- RSL = resilient (elements with high recover capacity)
- EXR = excretory (the fluvial web as landscape catabolite processing)
- SOU = source (ecological sources as centres of community expansion)
- DIS = disturbance (elements with a range of non-incorporating disturbances)
- CHG = changing (elements with high capacity of transformation)
- PRT = protective (elements which protect other elements or parts of the mosaic)
- PRD = productive (elements with high production of biomass)
- SBS = subsidiary (systems of human energetic and work resources)
- RSD = residential (systems of human residence and dependent functions)

Note that many landscape functions are typical of human habitats (e.g. productive, residential, subsidiary), others are typical of natural habitats (e.g. source, stabilising, geologic), and others may be in common (e.g. protective, resilient, connective). Note also that even human habitat apparatuses are intended in an ecological sense, not in a geographical or urban planning sense (Fig. 3.10).

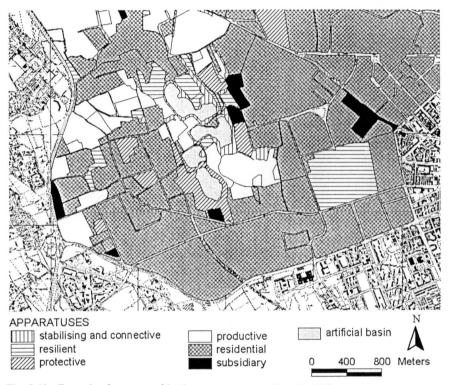

APPARATUSES

			N
stabilising and connective	productive	artificial basin	
resilient	residential		
protective	subsidiary	0 400 800 Meters	

Fig. 3.10 Example of a survey of landscape apparatuses. Note the differences between human and natural apparatuses and the presence of common types of landscape functions (e.g. protective patches or corridors)

The residential apparatus is characterised by settlement and service functions of human populations, transmission of traditions and culture, and administrative centres. Small orchards, small gardens and urban street vegetation are components. Dense and dispersed urbanisation patches generally form its typical parts. The historical heritage of old towns and cities (and roads) has to be studied in order to know the process of formation of this apparatus into a landscape.

The productive apparatus is mainly part of the human landscape, even if in a natural mosaic there are often more productive patches (e.g. density of food resources). The difference between the economic and urban definition of productivity is marked, because in ecology only communities with a high dominance of autotrophic species are truly productive. The degree of ecosystem stage in which respiration is not high plays an important role. Agricultural systems form the most typical productive patches, together with farms.

The protective apparatus is found both in natural and anthropic landscapes. Many ecosystems acquire the role of protection in their own mosaic, for instance resisting slope disturbances at the base of a mountain, and permitting the formation of other types of more sensitive ecosystems. In human landscapes, the hedgerow network in a field mosaic or the gardens around villas are typical protective elements.

The excretory apparatus is composed of the network of stream corridors (much more than the hydrologic network) of which are important the carrying potentiality, the buffer capacity, the cleaning capacity, the structural fitness with the mosaic, etc. Linking this web to landscape principles is not a matter of technical necessity, but is, rather, inherent to the landscape functioning. All the operations regarding the control and monitoring of water resources can thereby reach an higher level of preservation.

The stabilising apparatus is composed of elements of high metastability, for instance mature forested patches, and plays a regulatory role in the mosaic. The resistance stability predominates in those elements. Generally mature fitted vegetation is particularly important, but even other ecosystemic stages may be crucial in difficult environmental conditions.

The source apparatus is intended as in source-sink theory, but is related to the potential expansion of centres of community: populations are rarely concerned. The connections with sink patches are very important and have to be differentiated as barriers, porous, open. Sink patches may also be differentiated as receptive or not. It is necessary in evidence supply, disposal, resistance and retention per patches.

3.4.2
Landscape Units

The landscape unit, or geo-bio-district, is intended as a sub-landscape, that is, a part of a landscape which assumes particular characters or even functions in relationship to the entire landscape. In the old structural classification of Neef (1967), following Troll (1963), this concept is similar to the micro-mesochore and macrochore, which are horizontal arrangements of ecotopes. It is also similar to the D.O.S. Australian classification, as already discussed.

A (simple) landscape unit can be defined as an interacting disposition of recurrent and genetic (*sensu* geomorphology) ecotopes which assume a particular significance (function) in its own landscape. For example, the ecotopes around the lake of Doberdò, in the Karst landscape, or a suburban park in a Berlin urban landscape. In the first case the formation of a karst lake produced a ring of geomorphically characterised slopes converging toward the lake; in the second case an abandoned field area among new expansion districts assumed a recreational function.

A complex landscape unit can be defined as an interacting disposition of a cluster of simple landscape units. It could represent a good proportion of an entire landscape, for example, a good trait of an alpine valley in a Dolomite landscape.

The structure of a landscape unit is not always immediately recognisable, because it is not always a simple arrangement of ecotopes, even if it forms a connected patch of ecotopes. An analysis of the range of functions which form the geo-bio-district is consequently necessary. A landscape unit (Fig. 3.11) is formed by the emergence of many and various functional components: (a) geomorphologic characterisations, (b) ways and areas linked to the movement of permeant animals and/or men, (c) bio-functional effects produced by a community over other communities, (d) changing capacity of some ecotopes, (e) mesoclimate effects, etc.

Note that we may define other types of landscape units, not always directly linked to geomorphic substrate and ecologically related functions, but dependent on particular research needs. In this case, remembering the concept of ecotissue, the sum of the operational landscape units' surface may be greater than the area of the entire landscape. This fact is verifiable especially surveying animal habitats in a natural landscape, or human districts in urbanised landscapes. It is not ever a question of necessary approximation.

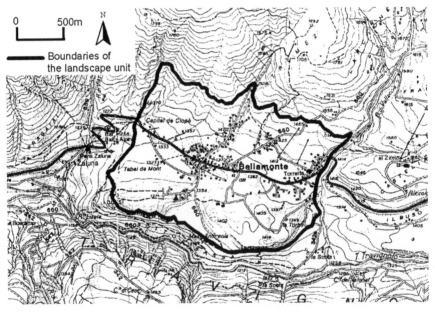

Fig. 3.11 Example of a landscape unit in the alpine valley of Fiemme (Trentino, Italy). The geomorphological characters are the most important in a mountain territory in the definition of a landscape unit, but they are not the unique: also vegetational and human ones have to be followed

Thus, in summary, it is possible to distinguish many types of landscape units, most of which can be integrated into significant sub-landscapes. We may rank for instance the following :

- Subdivisions of a landscape in local sub-patterns
- Districts with a diverse range of naturalness
- Zoning of different criteria of human management
- Zoning of preservation value areas
- Pattern of specific agricultural and urban functions
- Districts of historical and geographic characterisation

Even the degree of alteration or some negative dynamics (e.g. degradation) may be utilised in order to distinguish landscape units. Anyway, note that in case of choosing an operative (merely working based) landscape unit, it is always necessary to aim for a better ecologically localised one.

3.5
Landscape Classification

3.5.1
Landscapes

As we discussed, it is impossible to study an ecosystem (better: ecocoenotope) below the scale of the community, and without knowing its component populations. When, enlarging the scale, other elements are added outside the structure and functions of an ecosystem, it is necessary to define a landscape. For instance, when a land mosaic is reached, new structures and new processes appear: ecotonal webs, connectivity, porosity, landscape matrices, landscape apparatuses, landscape dynamics, new strategies of metastability, etc. Other characters appear at a regional scale, impossible to study properly in a landscape, such as soil order, forest formation, fauna changes, river bio-gradients, climatic role in ecological differentiation, etc.

In fact, not only many characters but also many dominant controlling factors in an ecological system change with the enlargement of the scale: as suggested by the hierarchy theory. That is why it is necessary to define a landscape as being generally limited to a specific range of spatio-temporal scales. It is not a question of a human scale perception, because even in a natural hierarchy of ecological systems every researcher by changing some range of scales can find different behaviours.

Note that at geospheric dimension, domains and regions were traditionally studied in biogeography, before the foundation of ecology, and now they comprise global and regional ecology, together with the biosphere. Sometimes regions (ecoregions) are studied also in landscape ecology (Forman 1995), but this is a normal overlapping margin between ecological fields, necessary when a framework is needed to analyse a landscape.

Now it is possible to ask how to classify landscapes. Today there is no complete taxonomy of the landscape, but only some ordination methods. We can distinguish four principal methods : (a) dominance of man-made artefacts, (b) phytosociologic criteria, (c) hierarchic factors criteria, (d) integrated landscape apparatuses.

3.5.2
Dominance of Man-Made Artefacts

One of the most classical ordinations of landscape types has been proposed by Naveh and Lieberman (1984), according to the definition of Total Human Ecosystem (ecosphere = biosphere + technosphere). Energy, matter, information inputs from bio- and techno-ecosystems are involved. On the x axis is reported the modification, conversion, replacement of natural bio-ecosystems. On the y axis (negative verse) is reported the degree of dominance of man-made artefacts. There are seven major resulting types of landscape, disposed on the diagonal of the figure and divided into open and built landscapes (last three):

- Natural landscape (natural bio-ecosystems)
- Semi-natural landscape (natural bio- and rural techno-ecosystems)
- Semi-agricultural landscape (natural bio-, agricultural bio- and rural techno-ecosystems)
- Agricultural landscape (agricultural bio-, rural techno- and urban techno-ecosystems)
- Rural landscape (rural techno- and urban techno- ecosystems)
- Suburban landscape (rural techno- and urban techno- ecosystems)
- Urban-industrial landscape (urban techno-ecosystems)

Ranked according to decreasing naturalness or increasing artificiality is the ordination of Wolfgang Haber (1990). There are five landscape types, divided into bio-ecosystems and techno-ecosystems (the last) :

- Natural (without direct human influence and self-regulated)
- Near-natural (influenced by humans but little changed after human abandonment and capable of self-regulation)
- Semi-natural (resulting from the use of natural and near-natural landscapes, changing significantly after human abandonment, requiring some management)
- Anthropogenic-biotic (intentionally created by humans, fully dependent on management)
- Techno (technical systems with dominance of artefacts for industrial, economic and cultural activities, dependent on human management and on surrounding ecosystems)

3.5.3
Phytosociological Criteria

Landscapes are classified on the basis of a survey of its vegetation complexes. A vegetation complex is a group of vegetational associations viewed as tesserae of a wider specific ecotope. This survey is divided into two phases : (I) registration of all the plant associations pertaining to the ecotope, in percentage, (II) registration of all dominant elements of the surrounding landscape, but only their presence. Note that the landscape elements are the most important in classifying the type of landscape (Pignatti 1994).

Another method derives from the so-called synphytosociological and geo-synphytosociological analysis (Tuexen 1978; Géhu 1988; Rivas-Martinez 1987): it is possible to identify the relationships between the plant communities of serial and chain types, which appear inside the same dynamic succession, evolutionary or regressive, or between series in the same territory. It is possible also that vegetation integrates various aspects of the environmental system. This method is applicable even to identify sub-regions (landscape systems) at a wider scale.

Vegetational associations interact dynamically and manifest steps of a regressive or evolutionary process (the so-called vegetation series or sygmeta).

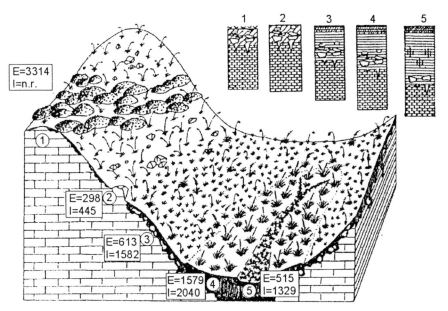

Fig. 3.12 Small landscape unit of the sub-alpine area of the Gran Sasso mountain (Italy) in which a geosygmaetum is shown, composed of diverse vegetational associations linked to different soil types and biomass classes: 1. *Daphno oleoidis-Juniperetum alpinae*, 2. *Seslerietum apenninae*, 3. *Luzulo italicae-Festucetum microphyllae* sub-ass. *caricetosum kitaibellianae*, 4. *Luzulo italicae-Festucetum microphyllae*, 5. *Taraxaco apennini-Trifolietum thalii* (from Biondi et al. 1992)

In a typical South-European example, an abandoned pasture association (e.g. *Centaureo bracteate-Brometum erecti*) will be changed into a shrubby one (e.g. *Juniperus communis-Pyracanthetum coccineae*) which in the future will evolve into a forest (e.g. *Aceri obtusati-Quercetum cerridis*).

In landscape phytosociology the vegetation series (*sygmetum*) assumes the same role as an association in classical (Braun-Blanquet) phytosociology. The sygmetum can be distinguished into climatic or edaphic series, depending on the water support: rain only or rain and soil origin. Note that among neighbouring associations a dynamic relation is not always possible, for instance because of the diverse soil potentiality. Their relationship will be only topographic or by soil chain. Therefore, the last level of classification emerges from the integration of diverse vegetation series, in what is called geosynphytosociology or landscape chain phytosociology. This analysis allows recognition of a geo-series in an homogeneous landscape unit, such as a valley or a mountain or a trait of a coast. A good example (Fig. 3.12) is the study of small landscape units in a sub-alpine area of the Gran Sasso mountain (Central Italy) by Biondi et al. (1992).

We should note that this so-called sigma-syntaxonomy method is not generally accepted (Naveh and Lieberman 1994; Zonneveld 1995).

3.5.4
Hierarchic Factors Criteria

As sustained by Forman and Godron (1986) an important criterion of ordination is based on the assumption that it is necessary to create valid attributes in a proper hierarchy. No natural typology gives all attributes equal priority, because of its latent hierarchy: in fact it is usually followed a descending hierarchy in five levels :

1. Zonal climates, principal climates of the biosphere, e.g. humid temperate
2. Climatic regions, specific sub-climates, e.g. Mediterranean ecoregion
3. Vegetational belts, bioclimatic units, e.g. sclerophyll vegetation (macchia)
4. Geomorphic units, soil structure in local areas, e.g. red ferallitic soils in terraced slopes
5. Human influences, agricultural fields, roads and settlements, e.g. scattered orchards, vineyards and villages

After this first ordering it is possible to identify five distinct configurations of landscape on the basis of their structural pattern. Six types of landscape spatial pattern have been described by Forman (1995): (a) large patch, (b) small patch, (c) dendritic, (d) rectilinear, (e) checkerboard, (f) interdigitated. Moreover, a landscape may be locally:

- *Regular*, with uniform distance among the elements, even if the ecotopes are diversified, e.g. regular fields, large and small patches, webs of small corridors, etc.

- *Aggregated*, with specific clustered elements predominating, e.g. large patches of fields, series of different sub-parallel corridors, etc.
- *Linear*, oriented along a particular direction, following a valley, a big river, ridges, or, in the case of human landscape, a main road or a railway, etc.
- *Spatially linked*, with characteristic association of elements, e.g. fields and hedgerows, moss and small lakes, or villas and gardens, etc.

Note, that it is necessary to avoid considering a landscape as a sum of the above attributes, remembering instead that the landscape is an integrated system.

3.5.5
Integrated Landscape Apparatuses Criteria

Another possibility in ordering landscapes derives from their principal configurations of elements. Remember that a landscape apparatus concern functional systems of elements which form specific configurations in the complex mosaic (i.e. ecotissue) of a landscape.

If we consider the nine most characterising landscape apparatuses and combine them with different proportions, it will be possible to note the emergence of at least 16 types of landscapes (Table 3.1). With criteria like this it is possible to classify most of the landscape types of the world, because they are compatible with the previous hierarchic ordination. In fact for each ecoregion (see the next section) of the ecosphere we may check the combination of the landscape apparatuses, and finally, if the case, add some elements of the landscape spatial pattern.

Table 3.1. Main composition of landscape apparatuses forming landscape types

Geo	Exr	Stb	Rsl	Con	Prt	Prd	Rsd	Sus	Landscape Types
++	-	-	+-	-	-	-	-	-	Desert
++	-	-	++	-	+-	-	-	-	Semi-desert
-	+-	-	++	++	+-	+-	-	-	Prairie
+-	+-	+-	++	+-	++	+-	-	-	Shrub-prairie
++	+-	+-	+-	+-	+-	+-	-	-	Shrubby
+-	++	++	++	+-	++	++	-	-	Open forested
-	+-	++	+-	++	++	+-	-	-	Closed forested
+-	+-	++	++	+-	+-	++	-	+-	Semi-natural, > bmass
+-	+-	+-	+-	+-	+-	++	-	+-	Semi-natural, < bmass
+-	++	+-	++	++	++	++	+-	-	Cultivated, protective
+-	+-	-	++	-	+-	++	+-	-	Cultivated, productive
+-	+-	-	+-	-	+-	++	+-	+-	Rural
+-	+-	-	+-	-	+-	+-	+-	+-	Sub-urban, rural
+-	+-	-	-	-	+-	-	+-	++	Sub-urban, industrial
+-	++	-	-	+-	++	-	++	+-	Urban, open
-	+-	-	-	-	+-	-	++	++	Urban, closed

GEO geologic, *EXR* excretory, *STB* stabilising, *RSL* resilient, *CON* connective, *PRT* protective, *PRD* productive, *RDS* residential, *SUS* subsidiary, ++ full presence, +- partial presence, -- absence, *bmass* biomass

3.6
Landscape Systems and Ecoregions

3.6.1
Ecoregions

The hierarchy theory and the emergent property principle enhance the importance of the context in ecological analysis: that is why we need to study regions. In fact, a region is defined as an ecological system composed of connected landscapes. Bailey (1996) prefers to name those systems ecoregions. Regions occur in a wide range of scales, like landscapes, and they stand frequently in contrast with one another, while long-distance linkages connect them.

The eco-climatic zones of the Earth are the basis of the criteria used in delineating ecoregion levels. Koppen used the composition and distribution of vegetation in his search for significant regional climatic boundaries, as did many other authors (Tricart and Cailleux 1972 ; Bailey and Cushwa 1981).

Where disturbances and secondary successional stages make regional boundary placement difficult, Bailey considered the patterns displayed on soil maps of broad regions, such as the FAO-UNESCO World Soil Map, because soils tend to be more stable than vegetation.

The ecoregions delineated by Bailey (1996) are grouped in four domains, as shown in Table 3.2: (a) polar, where the frost action primarily determines plant development and soil formation, (b) humid temperate, with a variable importance of winter frost, (c) dry, which comprises arid and semi-arid areas of middle and adjacent latitudes, (d) humid tropical, with persistent high moisture and high temperature.

3.6.2
Landscape Systems

The 30 macroscale regions listed in Table 3.2 are well defined due to climatic zones and macroscale vegetation. At mesoscale, regions are more difficult to define, because they are mostly dependent on landform differentiation and mesoscale vegetation. Anyway, at this meso-scale we can not speak properly of landscape: rather, we can define landscape systems. To have an idea of landscape systems, it is necessary to refer to a subcontinental example, like the Italian peninsula or the British isles in Europe.

A landscape system is therefore defined as an arrangement of landscapes which have significant geographical and vegetational features in a certain type of ecoregion (*sensu* Bailey).

The Italian peninsula has been known for centuries as one of the most rich in landscape types in the entire world. It can be divided into 4 ecoregions:

I-   Marine regime mountains
II- Marine division
III- Mediterranean regime mountains
IV- Mediterranean division

Table 3.2 Ecoregions of the continents (from R.G. Bailey 1996)

Rank n°	Ecoregions	Km^2	%
100	Polar domain	38 038 000	26.00
110	Ice cap division	12 823 000	8.77
110M	Ice cap regime mts	1 346 000	0.92
120	Tundra division	4 123 000	2.82
120M	Tundra regime mts	1 675 000	1.14
130	Subartic division	12 259 000	8.38
130M	Subartic regime mts	5 812 000	3.97
200	Humid temperate domain	22 455 000	15.35
210	Warm continental division	2 187 000	1.49
210M	Warm continental regime mts	1 135 000	0.78
220	Hot continental division	1 670 000	1.14
220M	Hot continental regime mts	485 000	0.33
230	Subtropical division	3 568 000	2.44
230M	Subtropical regime mts	1 543 000	1.05
240	Marine division	1 347 000	0.92
240M	Marine regime mts	2 194 000	1.50
250	Prairie division	4 419 000	3.02
250M	Prairie regime mts	1 256 000	0.88
260	Mediterranean division	1 090 000	0.75
260M	Mediterranean regime mts	1 561 000	1.07
300	Dry domain	46 806 000	32.00
310	Tropical/Subtropical steppe division	9 838 000	6.73
310M	Tropical/Subtropical steppe regime mts	4 555 000	3.11
320	Tropical/Subtropical desert division	17 267 000	11.80
320M	Tropical/Subtropical desert regime mts	3 199 000	2.19
330	Temperate steppe division	1 790 000	1.22
330M	Temperate steppe regime mts	1 066 000	0.73
340	Temperate desert division	5 488 000	3.75
340M	Temperate desert regime mts	613 000	0.42
400	Humid tropical domain	38 973 000	26.64
410	Savanna division	20 641 000	14.11
410M	Savanna regime mts	4 488 000	3.07
420	Rainforest division	10 413 000	7.11
420M	Rainforest regime mts	3 440 000	2.35

mts mountains

Following geomorphologic characters, regional climatic gradients, vegetation belt differences, human settlement types and agricultural landscape differences, it is possible to describe and rank the landscape systems of Italy (Fig. 3.13; Table 3.3).

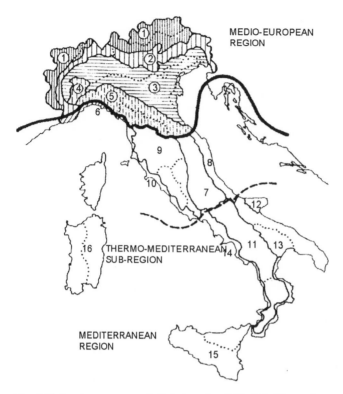

Fig. 3.13 Landscape systems in Italy (see also Table 3.3). The southern part of Central Europe is limited by the Tuscan Apennine; the meso-Mediterranean zone is limited by the *dotted line* between the Gargano and south Lazio

Table 3.3 Ecoregions and landscape systems in the Italian Peninsula

Zones	Landscape systems	Identification number	Sub-divisions
Central Europe	Intra-alpine sub-continental	1	A, B, C
Alps	Pre-alpine southern	2	A, B, C
Central Europe	Po valley plan	3	A, B, C
Po Plains	Hilly Monferrato	4	-
Central Europe-	Apennine northern	5	A, B, C
Mediterranean	Central Apennine	7	A, B
Appennine	Gargano peninsula	12	-
	Southern Apennine	11	A, B
Meso-	Ligurian coast	6	-
Mediterranean	Central Adriatic coast	8	-
	Tirrenic hills	9	A, B, C
	Tuscany-Lazio coast	10	-
Thermo-	Apulia Tablelands	13	A, B
Mediterranean	Neapolitan-Calabrian coast	14	A, B
	Sicilian insular	15	A, B

4 Landscape Dynamic

4.1
Complex Articulation

Landscape ecologists have studied many dynamic aspects of landscapes, most of them concerning: the role of edges, the effects of the relationships between the exterior and the interior part of a patch, the fluxes across the network of edges, the influence of mosaic structure on organism dispersal, the influence of mosaic configurations on biogeochemical fluxes, gene flow and differentiation of population in spatial heterogeneity, the connectivity of patches, the heterogeneity of disturbances, movement and corridors, etc. In many cases these arguments have been well studied, but in the absence of a true disciplinary framework.

The text of Forman and Godron (1986) has remained till now one of the principal references in studying landscape dynamics. It is divided into six groups of arguments: (1) natural processes in landscape development, (2) the human role in landscape development, (3) flows between adjacent landscape elements, (4) animal and plant movement across a landscape, (5) landscape functioning and (6) landscape change. Major transformation processes are divided into: (1) specific geomorphic processes (long time periods), (2) methods of organism colonisation (any temporal scale), (3) local disturbances of single ecosystems (short time periods).

We know that a landscape shows all the principal characters of any level of life organisation: structure, movement, reproduction, metastability, etc. Now, we have to observe that it is necessary to study the processes related to each character, even for structural components. In doing that, a complex articulation of landscape dynamics emerges, never completely examined up to now. So we prefer to suggest a more articulated sequence.

First of all, it is possible to distinguish between: (a) general dynamic processes and (b) biosystemic processes. These two groups of processes have to interact but, in a so complex and articulate field, it may be convenient in some cases to separate landscape functions from transformations.

Landscape functions may emerge from the interaction among climate, soil, erosion-deposition cycles, vegetation, animals, human population (i.e. general processes) and landscape structure processes, landscape delimitation, landscape information, movements, growth and reproduction, etc. (i.e. biosystemic processes). Landscape transformations may emerge from the interaction among the functional processes and: metastability and maintenance, ecological succession, alteration and pathology processes.

We expose this framework of articulation in Table 4.1. Rather than synthesise what is described in the table, it could be interesting to divide the study of landscape dynamic into two chapters, because the questions concerning landscape transformation, alteration and pathology need to be separate as in medicine (see Sect. 1.7.1).

It is extremely important to underline that generally the landscape components have a complex functional behaviour, therefore the processes ranked in the table have only a didactic and disciplinary significance. For instance, a corridor (e.g. an hedgerow) may have a structural function, but also an ecotonal one, or a movement one, etc.

Table 4.1 Main articulation of landscape dynamic processes

Principal arguments	Following subdivisions
LANDSCAPE FUNCTIONS	
General processes	
Geobiologic processes	Main interactions
	Geomorphologic processes and soil formation
Biosystemic processes	
Structurally related processes	Functions of landscape elements
	Fragmentation and connectivity processes
Delimitation processes	Boundary processes
	Ecotonal network functions
Movement processes	Geoclimatic movements
	Biological movements
Information processes	Genetic processes at landscape level
	Human coevolution and control
Reproduction processes	Self reproduction processes
	Colonisation processes
LANDSCAPE STABILITY	
Metastability processes	Biological territorial capacity
	Landscape metastability
	Disturbance influences
LANDSCAPE TRANSFORMATIONS	
Transformation processes	Transformations and successions
	Basic processes
Pathology	Alteration processes
	Pathology and consequences

Note : Landscape transformations are discussed in Chap. 5

4.2
General Processes

4.2.1
Main Interactions

The most important interactions among the geophysical, climatic and biologic components of landscapes can be visualised in Fig. 4.1 in a very synthetic scheme. The main sequence of general processes may be described in six phases:

1. The *climate* influences directly the morphology of a landscape (erosion, transport, sedimentation, etc.), the soil formation, and the vegetation.
2. The *vegetation* is influenced by climate, but plays a basic role in soil development and protection, as in the formation of ecotope mosaics and in the maintenance of local microclimates.
3. The *soil* interacts in return on the vegetation.
4. Climate, vegetation and soil constitute the *environmental conditions* for the establishment of animals.
5. The *animals,* in return, through their behaviour modify both soil and vegetation (pollination, grazing, seed dispersion, etc.) and complete the landscape formation.
6. With the evolution of *man*, inventive interactions integrate the precedents creating new landscapes. Feedback may influence almost all processes, but the possibility of a managerial control derives from these interactions.

These phases enhance the related processes belonging to three groups: geophysics-climatic (1); adaptive-evolutionary (2, 3, 4, 5); inventive-cultural (6). Major interactions among these three groups of processes take place not only on a macro-scale, but they are transmitted to various levels on lower scales too, even if the influences of each sub-process do not reach every level of the organisation, as stated by the hierarchic theory.

An in-depth treatment of general processes can be found in every text on general ecology. Here it is sufficient to remember the importance of the open thermodynamic dissipative process, starting from solar irradiation, which is absorbed in partial and differential ways into the atmosphere, generating wind circulation and all the climatic phenomena. Again the irradiation permits plant photosynthesis, heating and transpiration, and water extraction from the soil. Vegetal and animal communities form the soil, stabilise it and participate actively in its deposition. Contributions to soil formation derive even from alluvial sedimentation and from endogenous phenomena. We have to note that landscapes appeared in the ecosphere during evolution, in parallel to the establishment of life forms on the Earth's surface (mainland), starting from the Cambrian, as indicated in Table 4.2. See the strong increase of landscape heterogeneity in the Palaeocene.

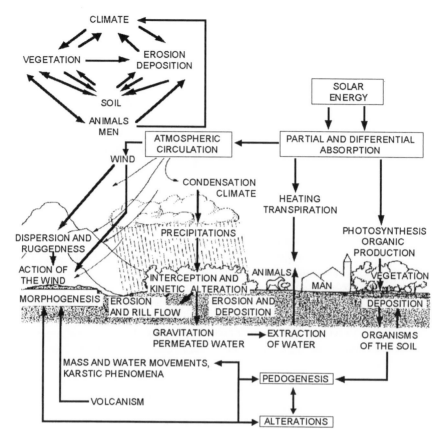

Fig. 4.1 Synthetic scheme of main general processes describing the interactions among the geophysical, climatic and biologic components of landscapes

Table 4.2. Principal types of landscape systems and their appearance in the ecosphere

Types of landscape systems	Time by present (years)
Submersed	550 million
Semi-aquatic	450 "
Semi-desert	420 "
Paleo-forested	350 "
Forested	130 "
Meadow	50 "
Woody-sub-anthropic	0.5 "
Semi-agricultural	50 thousand
Agricultural-rural	10 "
Urban	5 "
Sub-urban	0.2 "

4.2.2
Geomorphic Processes and Soil Formation

Geomorphic phenomena have two important functions in a landscape: (1) the generation of fluxes of matter and the consequent modification of the regional surface, which signifies energetic ties for a landscape, (2) the derivation of characters from the transport of eroded matter. As anticipated in Sect. 1.2.1, geomorphology is seen as a factor regulating many others landscape functions, so assuming a strong ecological significance. On the other hand, soil formation acts through four groups of processes:

1. *Humification processes*: formation of organic-mineral complexes, generally in well drained soils with short cycles which have been altered bio-chemically in temperate and cold regions (e.g. brown soils)
2. *Seasonal conditioning*: formation of vertisoils by strong seasonal contrasts (wetting-drying) which alters the humification and the formation of distinct horizons in cold and continental regions
3. *Geochemical alteration*: total hydrolysis in soils without organic soluble elements, which eliminates silicon and bases and accumulates aluminium and iron compounds through long cycles in hot climates, warm regions
4. *Station conditioning*: production of physical and chemical environments mainly by water excess, which form oxidation-reduction processes and salt, e.g. pseudogley soils

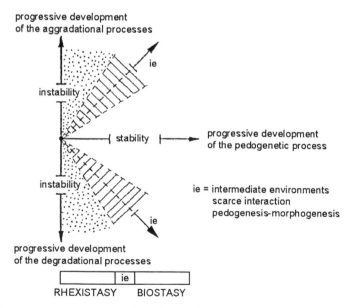

Fig. 4.2 Schematic relations among pedogenesis (soil formation) and morphogenesis. Stable, intermediate and unstable environments are possible

Tricart and Kilian (1979) noted that *pedogenesis (soil formation)* and *morphogenesis* are not cyclic, as proposed by the rhexistatic theory of Erhart (1956), but they are contemporaneous and interfere with each other. Therefore, it is necessary to distinguish between localised and diffuse processes, both on a spatial and a temporal scale, and especially to locate one of the three types of geo-dynamic environment: (1) stable, (2) intermediate and (3) unstable (Fig.4.2).

(1) Stable geodynamic environment: slow evolution, tendency to a steady state, possible in regions with a weak internal geodynamic activity and a weak intensity of external processes. The balance is favourable to soil formation. Two forms exist:
(a) Long-term stability, with surface evolution quite absent
(b) Almost recent stability, remnant forms due to Quaternary oscillations, where sometimes windy actions may be present
Here it is important to maintain a vegetational cover near to maturity, because these formations may be the basis of new soil transformation.

(2) Intermediate environment: contemporary interference between pedogenesis and morphogenesis, with a weak balance in favour of one of them. Morphogenesis assumes a great importance and permits two cases to be distinguished:
(a) Diffuse pluvial erosion, surficial creep, with a possible development of soil base, and
(b) Diffuse superficial mass movements, soil-flux type, able to notch all the soil layers.
Sometimes a partial deforestation may arrest the glides, because it diminishes the load and the water absorption, especially on clay soils.

(3) Unstable environments: predominance of morphogenesis. Two main causes:
(a) Aggressive bio-climatic conditions, with irregular intense variations, transmitting a great amount of energy, and
(b) Impervious reliefs, high inclinations, intense internal geodynamic, no remnants possible
Soil formation can not develop, thus hetero-dynamic mosaics appear, with litho- and rhego-soils. Sometimes reforestation is not able to stop mass movements: lands are not cultivable, and become marginal.

The soil is a natural environment, in continuous formation and evolution, which is extended to the superficial layer of the lithosphere. Soil evolution, in its history, follows three processes (Duchaufour 1984):
(1) Rock decomposition and alteration, leading to formation an altered complex, (2) increase of the amount of organic matter due to colonising vegetation, with an equilibrium phase between fresh inputs and mineralised outputs, (3) transport of soluble or colloid elements through the profile, due to water flux, forming impoverished horizons A and enriched ones B.
A typical chain of soil evolution is shown in Fig. 4.3, regarding brown soils on a slime-sandy substrate (following Duchaufour 1984).

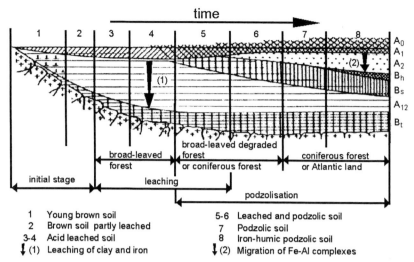

Fig. 4.3 Evolution and degradation of a typical brown soil in Europe. Soil evolution phases (1 to 8) show two main chemical processes (leaching and podzolisation) and three main types of vegetation which are adapted to this soil chain. When cultivated in the last phases, this acid soil is re-saturated by manure, eliminating its absorbent complex (degradation). (From Duchaufour 1984)

Some basics of soil dynamics, i.e. the zonality law, have been enunciated along general lines beginning with Dokuchaev, beginning with 1898. In every climatic zone, soils develop by different rocky substrates and are colonised by different ecological communities. When the ecosystems develop toward a steady state, soil and vegetation originating from different substrates converge toward a uniform ecological system. This assertion involves interesting implications that we shall analyse in Sect. 5.1. Even if the zonality law states that the climate is the first of all ecological variables, the local characters of the geosphere maintain an important role, particularly in human landscapes.

Therefore, in a region or in a landscape, some units dependent on geomorphology appear, called ecological sectors (Godron 1984). These units locate diverse soil mosaics within an uniform environment, and this may be very important in landscape analysis (e.g. for landscape unit delimitation). Concerning this, we have to underline that the integration of pedology (soil science) in the field of landscape ecology is often more difficult than that of geomorphology. In fact, a typical pedogenetic environment does not always produce the same soils everywhere, and this is in apparent contrast with the zonality law; in any case it remains much more interesting than the analytical knowledge of any type of soil located in a given landscape.

A specific field of study includes the geomorphologic processes pertaining to water, which are too complex and detailed to be reported here: but we can not avoid referring to at least fluvial characters, because rivers are generally the most important geobiologic elements in a region. Water courses due to water runoff are organised in corridors with a diverse drainage density and spatial distribution. These corridors result from the basic process of erosion and are regulated by diverse hydrologic regimes: glacial, snow-pluvial, pluvial.

Many scientists have proposed a river zoning based on ecological and hydrological characters. The better known are: Hypocrenon (spring brooks), Epirhytron (mountains streams), Hyporhytron (valley torrents), Epipotamon (hilly rivers), Metapotamon (plan rivers), Hypopotamon (final rivers).

Anyway, we have to observe that in landscape ecology rivers assume diverse, more complex functions. They are:

1. Ecological systems with an accelerated landscape dynamic, with a structure in continuous transformation, which many adapted ecosystems are connected to
2. Excretory apparatuses of the catabolism of surrounding ecotopes, with a prevalent function of buffer and filtration and consequent cleaning
3. Complex corridors, dependent on the scale both in function and structure, available even for a protective role and for landscape movements
4. Polarised ecological systems regarding the ecological mosaics, with a function of biological conservation, thus particularly important in human landscapes

In karst zones and some deserts with very high percolation rates, river corridors may be not present or result only temporarily. Nonetheless, water circulation remains an important landscape factor, as we can see in talwegs or huadys.
Furthermore, human colonisation has a strong influence on water courses and its consequences on landscapes.

4.3
Processes Related to Landscape Structure

4.3.1
Functions of Landscape Elements

Some of these functions have been shown (see Sect. 3.3.1) to explain structure elements. We may study the functions of the patches in a landscape, considering their shape in relationship to a flow of species or energy: the influence of the shape is generally evident also in relation to its boundaries and edges. A well known process, interdigitation, is dependent on the intersection of two patch types with folded boundaries. The species distribution reaches a maximum along the barycentric line of interdigitation, while the species of the two patches decrease starting from the same line.
Synthesising the principal characters of a patch, as orientation, interaction with adjacent ecosystems, etc., it is possible to describe a principle of patch shape

(Forman 1995): to accomplish several key functions, an ecologically optimum patch shape usually has a large core with some curvilinear boundaries and narrow lobes and depends on the orientation angle relative to surrounding flows.

The characters of the patches show peculiar variations as a function of the different types of landscapes. This process was plotted by Forman and Godron (1986) in relationship to the degree of human influence on a landscape (Fig. 4.4A). For instance, remnant patches are generally present in all types of landscape with a maximum in suburban. Plotting similar variations in more articulated landscape types, and referring them to the landscape apparatus types, we may observe other processes, which can be useful to compare structural patterns with a local case study (Fig. 4.4B).

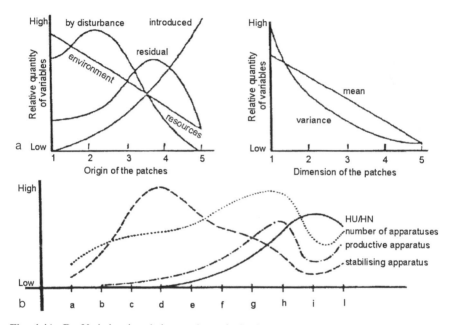

Fig. 4.4A, B Variation in relation to the main landscape types. **A** Of landscape elements. Landscape types: (1) natural, (2) forested, (3) cultivated, (4) suburban, (5) urban. **B** Of apparatuses. Landscape types: (a) desert, (b) semi-desert, (c) open forested, (d) forested, (e) shrub-prairie, (f) semi-natural, (g) cultivated, (h) rural, (i) urban, (l) suburban

Note that it is impossible to distinguish in a complete sense the structure and the function in a natural system. That is why, studying landscape structure, many patterns emerge from the functional role of its components, such as in a landscape apparatus (see Sect. 3.4.1). Remember, for instance, that the excretory apparatus is composed of the web of stream corridors, in which the structural fitness with the mosaic, carrying capacity, buffer capacity, cleaning potential, etc., all play a role. Corridors can be seen as a conduit for movements, or a habitat for particular species and a barrier for other species, or even as a source for an ecological mosaic. A corridor may be also an element of a connective landscape apparatus (e.g. hedgerow) or an element of a subsidiary one, in case of a human landscape (e.g. highway). The density of species in a corridor is a function of its width. The number of species (diversity) shows two aspects: the first concerns the edge species and follows an asymptotic curve (i.e. after a critical width the number of species does not increase); the second concerns the interior species and follows a semi-exponential curve in proportion to the width of the corridor. The hedgerow width in temperate landscapes can host both types of species when it reaches at least 10-12 m of width (Forman and Godron 1986).

The interactions among ecotopes in a landscape are very wide. The grain and the size of the patches forming a mosaic express this reciprocal influence. The high level of ecological complexity, biodiversity and naturalness of some patches may give also a dynamic character to the landscape. A coarse grained landscape, containing fine grained areas, represents an optimum to provide for large ecological benefits of the patch, multi-habitat species including humans, and a breadth of environmental resources and conditions (Forman 1995).

The complementarity among ecotopes or tesserae may create a polarity in an ecological mosaic, which changes the hierarchical spatial functions of the landscape. We may find many types of *polarity*, caused by the dominance of some groups of patches over the entire landscape mosaic, e.g. urbanised ecotopes in an agricultural mosaic, or humid ecotopes in an arid natural mosaic. These dominant patches acquire the role of attractors and direct many landscape patterns, becoming dense multifunctional areas (Fig. 4.5).

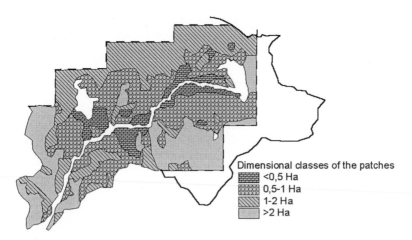

Dimensional classes of the patches
- <0,5 Ha
- 0,5-1 Ha
- 1-2 Ha
- >2 Ha

Fig. 4.5 Examples of the polarisation process and the emergence of influence fields. The grain of the patches near the Magazzolo river (Sicily, Italy) shows the role of attractor of the river, which polarised patch distribution: the patches became larger, going far from its axis (see Fig. 3.8)

These functions may be put in evidence by the emergence of *influence fields*: the intensity of this influence varies with the distance from the patch and it is proportional to the degree of its dominant role. As we will see in the second section, it may be useful to analyse some field of influence, for instance to control the role of a biological conservation area (e.g. a regional park).

The functional impacts due to strong contrasts in the grain of a landscape resemble these processes, depending on structural patterns. This condition is particularly evident in human habitats, but it can be verified also in natural landscapes. An example is the adjoining presence of humid and arid tesserae characterising some highly disturbed landscapes.

4.3.2
Fragmentation and Connectivity Processes

The breaking up of an habitat, patch or land type into smaller parcels and/or the dissection of them with roads or similar is usually called fragmentation. Normally fragmentation is considered as a negative process in landscape ecology, because: (1) it generates the loss of species (especially animals) more dependent on the size of patches, (2) it greatly disturbs species with a high standard habitat (low density species), (3) it produces an increase of alien species, especially in human landscapes, (4) it leads more isolated populations toward an inbreeding depression.

According to Forman et al. (1976) the demonstration of the species loss due to fragmentation was done by observing the species number in a large forested patch, which is higher than in the same area composed of a set of smaller patches. Let us present a typical example. Given a forested patch of about 15 ha, we consider a depth-of-edge influence (DEI) limited to the tree height (20 m) and measure its interior-margin area in parallel with its fragmentation due to corridors of a width of 20 m (see Fig.4.6 and Table 4.3). The effects are very strong. After the opening of the first corridor, the loss of only 6.7% of its area leads to an increase of the edge species of 3.3%, to a decrease of the interior species of 10% and of the greatest patch interior of 34.8%. After a third corridor and the loss of only another 14.1% of the forested area, the former forested patch changes its character. In fact,

Table 4.3. Effects of fragmentation on a medium size forested patch

Patch divisions	Patch No.	Forested area (%)	Corridor area (%)	Greatest interior area (%)	Interior species (%)	Edge species (%)
Entire patch	1	100	0	82.5	82.5	17.5
1st corridor	2	93.3	6.7	47.7	72.5	20.8
2nd corridor	3	90.6	9.4	43.0	65.1	25.5
3rd corridor	5	79.2	20.8	22.8	50.4	28.8
4th corridor	6	76.6	23.4	22.8	44.8	31.8

The *greatest interior area* identifies the part of the initial patch which remains untouched after the cuttings and the dimensions of which are so large as to be able to preserve interior species.

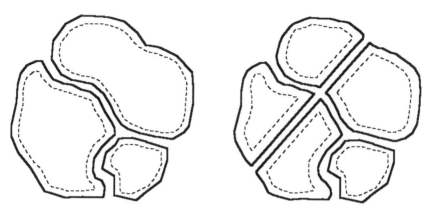

Fig. 4.6 Example of the effect of fragmentation on forested patches. Within an already fragmented former patch *(left)*, the introduction of a new corridor *(right)* leads to an increasing of the number of the small patches and to a fall of the number of the interior species

the greatest interior area loses another 25%, the presence of interior species decreases from 72.5 to 50.4% and one fifth of the total area becomes open to the entry of allochthonous species.

Nevertheless, sometimes fragmentation can be also a positive process, for instance when a rural landscape matrix is fragmented by forest corridors, or when a forest landscape is partially opened by pastures. For example, the enrichment of habitat produced by opening small grazing pastures in a wide forested landscape in Folgaria (Trient) creates a more favourable habitat for capercaillies (*Tetrao urogallus*).

Connection is a topological concept. The mathematical meaning of connectivity is topological: a space is completely connected if it is not divided into two open sets. The word connection implies the organic union of elements forming an environmental ecological unity, e.g. an ecocoenotope, or a specific configuration of the landscape, definable as a specific landscape apparatus. In a given landscape it is possible to recognise a connective apparatus (Fig. 4.7) composed of the elements which have the capacity to connect different patches as an expression of their prevailing landscape function. A clear correlation between connected elements of the same biological type has been demonstrated by many scientists (Burel and Boudrie 1999). Note that the concept of permeability is linked to the concept of connectivity, thus to the dynamic of movement. The permeability may obviously be different for each animal group and also for man.

Corridors are considered as the principal elements of connectivity in a landscape. Fluvial corridors, canopy roads, windbreaks, greenbelts and hedgerows all play a role in the rural landscape, just as equestrian trails, wooded avenues and bicycle routes can be of value in urban areas. All represent linear connectors that permeate the landscape and play a role in an interconnected habitat system.

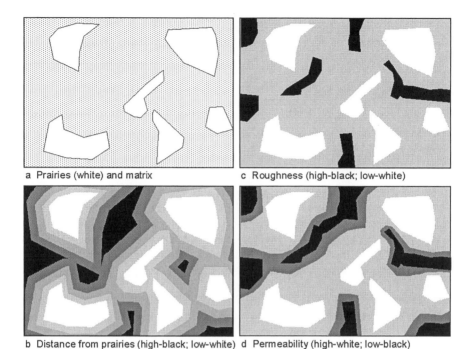

a Prairies (white) and matrix   c Roughness (high-black; low-white)

b Distance from prairies (high-black; low-white) d Permeability (high-white; low-black)

Fig. 4.7 Example of a connective landscape apparatus and its main dynamic. (From Burel and Boudrie 1999, re-drawn). Prairies are shown, as are the distances from them; to the *right* the roughness and the permeability of the landscape unit.

Anyway, we have to consider at least two other elements : (1) the ecotones between different ecosystems or ecotopes, with their peculiar structures and multiple functions, as buffer and filtering areas, (2) the connective tesserae among an ecological mosaic (ecotissue), with quite simple structures and the function of allowing the dynamics of other patches.

The evaluation (see Chap. 7) of connectivity needs the definition of an optimum set of parameters, because the higher thresholds limit the change of the type of landscape, while the lower ones avoid a negative circuitry.

4.3.3
Complementation, Resistance and Strategic Points

The relationships among patches in a landscape mosaic include the study of the heterogeneity of a landscape unit, its topological relations, the distribution of organisms and the proper character of the ecotissue.

Occurrence of subpopulations of a species in a set of habitat patches depends on the diversity of habitats. If the habitat quality is constant, both patch size and degree of isolation affect the distribution pattern. If the habitat quality is divided into optimal, sub-optimal and marginal, the occupation will be strictly hierarchical, beginning from the optimal tesserae. Dispersal may be two-way among optimal patches, but among optimal and marginal ones the flow is unidirectional.

When a species needs the resources existing in patches of, e.g. two different types, its density is higher in landscapes in which these two types of patches are both near and present: this process is called *complementation*. If a favourable patch is situated in a landscape unit gifted with additional resources, it usually supports a larger population, and we may speak of *supplementation* (Dunning 1992).

Note that a fine-grained landscape has predominantly generalist species, since specialist ones, requiring a large patch of a single type of land use, can not survive. A fine-grained landscape is monotonous, although site diversity is high: species that survive need only to move short distances.

All these qualities pertain to the landscape matrix. Landscape *resistance* was described (Forman and Godron 1986) as the effect of structural characteristics of a landscape impeding the rate of flow of objects (species, energy and material). This resistance is mainly dependent on the type and the frequency of boundaries; that is why Forman proposed a boundary crossing frequency, the number of boundaries per unit length of route, as a valid measure of landscape resistance. Different groups of animals are sensitive to different levels of landscape resistance, obviously. Built areas and busy roads are considered to increase resistance, while woody vegetation decreases resistance.

The multifunctional processes of a landscape matrix often result in the presence of *strategic points*, small areas in which more than two landscape functions may interact and/or resistance structures leave a passage, characterising patterns and movements into a landscape unit. It could be, for instance, the case of a small pond in a semi-arid land, used as a drinking pool and meeting point, or a remnant forest patch in a suburban landscape, used as a refuge for many animal populations and floristic wild species. In altered landscapes, strategic points assume a greater presence and can be used for rehabilitation processes.

4.3.4
Landscape Types and Ecological Population Density

The principal landscape functions typical of human habitats (productive, residential, subsidiary), of natural habitats (source, stabilising, geologic) and in common (protective, resilient, connective) have been already presented (see Sect. 3.4.1).

Another important process, particularly typical of human landscapes, is the aggregation of different types of patches which follow a positive feedback loop linking the landscape patterns and the *ecological density* of population. This density is referred to a population (human or animal) in relation to its real habitat, not to a generic territory usually intended in a geographical sense. The inverse of the ecological density gives the needs of space (habitat per capita, see Sect. 8.4.1) of an organism (both human and animal): thus, for permeant species, the need of space is intended as a collection of adapted resource patches within a landscape.

Consequently, it is possible to recognise landscape unit types in relation to population density. For example, in natural landscapes a researcher (but also a hunter) knows where it is highly probable to find a group of moose, observing the assemblage of most favourable habitat patches, such as reeds, wetland, willow bushes, forest margins, etc., which form their most typical landscape unit. Similarly, any traveller knows where to find the highest density of population in a new country, visiting the largest towns. It is possible to find a relationship between the habitat per capita and the landscape unit structure, especially for "engineer species" (e.g. beavers or elephants) and particularly for humans. In central Europe we may rank seven types of human landscape units as a function of the habitat per capita of population (Table 4.4)

Table 4.4. Relationship between human landscape types and habitat per capita in Central Europe

Human Landscape Types	Habitat per capita $(m^2/inhabitant)$
Urban, dense	100 - 300
Urban, rare	300 - 600
Suburban	500 - 1000
Suburban, rural	700 - 1400
Rural	1300 - 2700
Agricultural	2700 - 12000
Agricultural, semi-natural	10000 - 30000

As we will see (Sect. 8.4), it is possible to check the dynamic of the human habitat per capita through an ecological index of *Standard Habitat* (SH), comparing its changes landscape transformations. This measure can acquire an important diagnostic significance in human landscapes. For example, the SH in Rheinland-Pfalz (1950-96) decreased from 3770 to 2750 $m^2/inhabitant$, while in Lombardy (1951-99) it was a bit stronger: from 2430 to 1700.

4.4
Delimitation Processes

4.4.1
Landscape Boundaries

All landscape components may be distinct from their surrounding through their boundaries. This seems to be true even if each range of scale shows a particular type of boundary and if there is a distinction between a sharp margin and an ecotonal belt. Generally speaking, processes are held within surfaces. In fact, the limits of a process define where the surface shall be. As pointed out by Allen and Hoekstra (1992), the multiplicity of devices that can be used to detect the surface of an entity that is resistant to transformation reflects the multiplicity of the processes responsible for that surface.

A surface can be tangible (e.g. tree bark) or intangible (e.g. thermocline of a lake) or mixed, such as the margin of a patch. A surface, or boundary, disconnects the internal functioning of entities from the outside world, but sometimes this fact may be not sufficient to warrant the correct design of the boundary and the surface remains arbitrary. In some cases not all the boundaries of a single element are definable. Therefore it is necessary to evaluate the boundaries through observations derived from a higher level of scale.

Remembering the principles of hierarchical systems, we note that frequency and constraint are the most important criteria for ordering levels, because upper levels constrain lower levels by behaving at a lower frequency. This conditioning is a passive process, in which upper levels constrain lower ones by doing nothing, or refusing to act. That is why if we look at an ecological system at a too low level of scale, then the scope of observations will probably not extend to the principal constraints.

Thermodynamic theory suggests that an open system with energy input becomes spatially heterogeneous in two ways: (1) through gradual concentration gradients of the existing elements that make the system heterogeneous, but not patchy; (2) through the formation of a mosaic with boundaries. Due to inherent environmental patchiness and non-equilibrium thermodynamics, a landscape follows basically the second way, leaving the first only to partial or transitional phenomena (e.g. population invading an area).

The layers composed of vegetational biomass form the current boundary of an ecotope, but topographic and human barriers also play the same role. In this sense, human actions have had an evolutionary function in defining many boundaries once undetermined in natural ecotopes, therefore giving a higher functionality to landscape mosaics.

The delimitation of a landscape element plays an important and multiple functional role, first of all as filter. The internal structural characters of the boundaries are responsible for its permeability. With a correct theoretical analysis, Forman and Moore (1992) claimed that landscape boundaries might be considered as functional analogues of cellular membranes (Fig. 4.8).

A series of statements of the membrane theory may be identified as pertinent to landscape boundaries : (1) boundaries appear and multiply spontaneously in open systems, (2) intense disturbance destroys boundaries, reducing the amount of variety within a system, (3) a boundary contains the record of interactions across itself; strong interactions produce a rich textured (heterogeneous) boundary, (4) when boundaries move, they extend the space of the system with higher variety and decrease the space of the system with low variety.

Like a cellular membrane, energy and matter are concentrated at the edge of a patch, such as the plant biomass in the mantle of a forest edge. Herbivores and predators are often also concentrated at the edges.

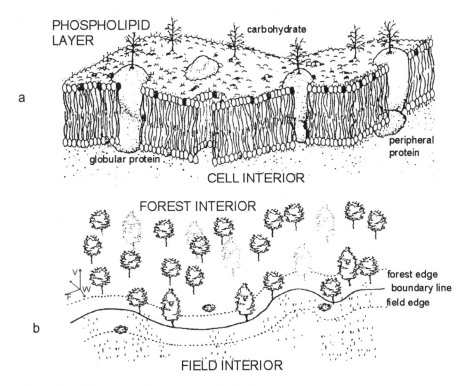

Fig. 4.8A, B. Characters of a boundary of a landscape element: comparison with a cellular membrane (from Forman and Moore 1992, re-drawn). **A** Molecular structure of a cellular membrane. **B** Structure of a forest-field boundary. The boundary zone includes the edges of both sides of a boundary line and its long three-dimensional structure, characterised by width (*W*), verticality (*V*), and form (*F*). Boundary surfaces, such as convex, lobe, and straight, may be in reference to either the forest or the field

Microclimatic effects are relevant in a boundary of different patches, such as from forest to open fields. The well known edge effect has been defined in relation to the increase or the decrease of species in the margin belt of an ecotope: it is necessary to remember that this effect may be positive, void or negative.The width, verticality and form dimensions may determine functional roles of landscape boundaries. The principal functions are: (I) conduit, (II) filter or barrier, (III) source, (IV) sink, (V) habitat. Note that there is an essential affinity between boundaries and corridors, even if spatially the concepts differ and connectivity differs too.

Recent studies (Chen et al. 1992) have shown that the fragmentation of the North American forests, which multiplied the margins of forest patches, influenced these patches up to a distance of 1-3 times the medium height of the trees. Depth-of-edge influence (DEI) is an important issue in the theory and practice, and we note (Table 4.5) that there is no "interior forest" for a patch <10 ha if the tree height is about 36 m (DEI=72), or (same tree height) that we need a patch of 45 ha to have >50% of interior species.

Table 4.5. Amount of edge influence in rectangular forested patches

Patch size (ha) (m x m)	DEI (m)	Interior species (%)	Edge species (%)
5 ha	30	53.3	46.8
(250 x 200)	60	20.8	79.2
10 ha	30	64.6	35.4
(400 x 250)	60	36.4	63.6
20 ha	30	79.9	20.1
(500 x 400)	60	53.2	46.8
45 ha	30	82.1	17.9
(900 x 500)	60	65.9	34.1

DEI Depth-of-edge influence: we can assume about twice the height of the trees

4.4.2
Ecotones

The changes of the boundaries may be abrupt in space and sudden in time, but also gradual and progressive. In this second case we know that it is better to define the concept of *ecotone* as the:

> zone of transition between adjacent ecological systems, having a set of characteristics uniquely defined by space and time scales, and by the strength of the interactions between adjacent ecological systems (di Castri and Hansen 1991).

Ecotones are structural and functional discontinuities within landscapes (Fig. 4.9), having their primary significance in presenting a gradient, thus the separation between two different ecotopes of landscape units is not sharp, but gradual. Being a transitional zone between ecological units, the ecotones represent places where interactions among patches are particularly characterised, and they may modify flows between patches. Moreover, ecotonal network functions are linked with the transmission of information across a landscape. The main effects pertaining to the ecotones are:

- Gradient: progressive reduction of elements
- Zoning: irregular distribution of flux and concentration of elements
- Accumulation: retention of a great quantity of an element
- Barrier: stopping the fluxes
- Margin: maximum local biodiversity (edge effect)
- Neighbouring: influence of the ecotone on adjacent ecosystem
- Transmitter: selection of information directed to nearby ecosystems
- Ecotissue: landscape elements not simply juxtaposed

One of the most important ecotonal networks is given by the set of river corridors which characterise a landscape. The main effects in this case are:

a. Chemical filter, with a retention rate of about 80% (forest) or 8% (meadow), for a 30 m large strip of riparian vegetation. Denitrification is particularly important in forested soils.
b. Biological filter, e.g. when hydrocore plants, such as alders, etc. (*Alnus* spp., *Polygonum* spp., etc.), do not serve as a barrier in relation to the water-land interface, in contrast to the barocore oaks and hazels (*Quercus* spp., *Corylus* spp.).
c. Margin, e.g. when the heterogeneity of the ecotone may favour some animal species, such as the otter (*Lutra lutra*) if there are willows (*Salix* spp.) and oaks or ashes (*Fraxinus* spp.).

An evident ecotonal effect is produced in the peripheral belt between an urbanised landscape and its surrounding agricultural or natural landscapes. In a case like this, we have a large scale delimitation which produces typical gradients. A chain of landscape units became structured by the ecotonal effect, with a series of passages from cultural and technical components to rural and natural ones. If not polluted or too much altered by human infrastructures, these belts can be rich in biodiversity and can be useful for many animal populations (e.g. starlings).

The shifting position of ecotones on a regional scale is a convenient temporal and spatial marker of landscape change. The long-term analysis of historic ecotonal shifts at local, regional, and global scales provides a practical approach to contemporary ecological issues affecting our environment. For example, main climatic changes in the Sub-Atlantic period, such as the warm Roman Climate Optimum (250 BC – AD 250) and the cold Little Ice Age (1580-1870), may be studied by analysing the regional European ecotonal shifts.

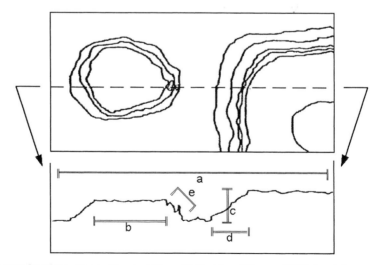

Fig. 4.9. Ecotone functions, plan and transect (from Di Castri and Hansen 1991). Isopleths can represent ecotonal discontinuities within landscapes.

4.5
Main Movement Processes

4.5.1
Geo-climatic Movements

Geo-climatic and biological movements pertaining to a landscape present many processes: it is not possible to discuss all of them. These movements depend on a small number of vectors: wind, water, flying animals, large animals, men and their vehicles. As underlined by Forman, all the ecosystems in a landscape are interrelated, with movement or flow rate of objects dropping sharply with distance, but more gradually for species interactions between ecosystems of the same type.

Vertical energy flow, interconnecting mosaic elements, is given by the different reflections of solar energy (i.e. albedo) by different surfaces, ecosystems and landscapes. Albedo is higher for smooth, light-coloured surfaces, but lower for rough dark surfaces (Fig. 4.10). This process results in a diverse energy budget among local ecosystems even if the incoming energy in that landscape is constant.

The energy absorbed by the tesserae of an ecological mosaic heats them up and provides heat for metabolism, growth, decomposition and especially evapotranspiration too. The vaporisation of water mainly from soil and plant surfaces is called evapotranspiration (ET). Potential monthly ET (=EP) may be calculated by the well known Thornthwaite expression:

$$EP = 0.1645 \ (P/T + 12.2)^{10/9} \quad (cm)$$

where: P is precipitation, T is temperature. EP is proportional to the medium temperature of a landscape and to its cloudiness: for instance we may find an annual EP of 60 cm in the Alpine border, and an EP of 95 cm in Sicily and North

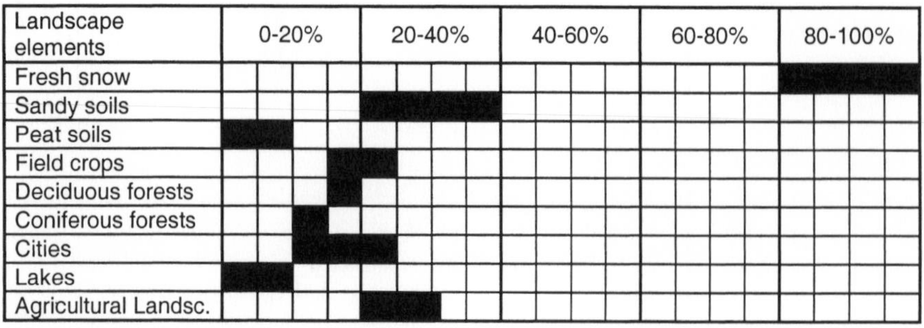

Landscape elements	0-20%	20-40%	40-60%	60-80%	80-100%
Fresh snow					■
Sandy soils		■			
Peat soils	■				
Field crops		■			
Deciduous forests		■			
Coniferous forests		■			
Cities		■			
Lakes	■				
Agricultural Landsc.		■			

Fig. 4.10 Albedo values in %

Tunisia. ET is highest in the windbreaks and lowest on bare soil (a two-fold difference). Among the fields, meadow has the highest ET and wheat the lowest.

The horizontal movement of energy or material carried by air is called advection. The contrasts in albedo and ET create energy gradients in the landscape. A 3-4 °C night temperature difference between forest and adjacent open patches means that, on a windless night, heat and air are moving horizontally to the cool clearing, while the cool patches are radiating heat, moving air masses vertically. These processes carry aerosols, gases, seeds, spores, and other particles (on a calm night).

It could be interesting (Forman and Godron 1986) to compare wind patterns near dense versus porous windbreaks (Fig.4.11). Wind reduction is typically effective only in the 2 m or so next to the ground, and at a distance of about six times the height of the windbreak: in the case of a porous vegetated corridor we have less reduction (40% of wind speed vs 25%) but less turbulence.

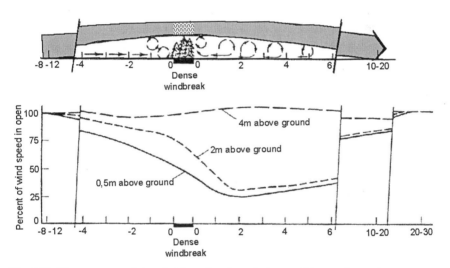

Fig. 4.11 The movement of the wind through a dense hedgerow. Distance from windbreak (*h*) is measured proportionally to the mean height of trees (the number on the horizontal axis times the height of the windbreak). (From Forman and Godron 1986)

Water is transported by the planetary boundary layer (i.e. airflow affected below 1-1.5 km by friction) in the form of vapour, raindrops and snow. Shortly downwind of a source, such as a lake or a forest, is a plume of moist air. That is why in irrigated land surrounding an oasis, ET in the upwind portion causes moister air in the downwind part. In northern Japan (Hokkaido) a 100 m wide "fog-removing forest" collects enough droplets to clear the bottom 300 m of the planetary boundary layer, permitting solar radiation to reach rice paddies downwind. Briefly, water in all its forms, and the hydrologic cycle as a whole, are quite dependent on horizontal and vertical flows in the giant conveyor. Water is collected over a broad area and deposited in a concentrated form in a small area. Wide floods often extend greatly over a landscape.

Water velocity is generally low, but because the land has rarely experienced these floods, the removal of biomass and damage to landscape and human structures may be great. In addition, nutrient rich sediment is spread in patchy deposits, and if the process is not too strong, may be in some cases positive.

Even if this chapter has to be very synthetic, we can not avoid mentioning something on the dynamic of rain shifting. Even an intense rain can be intercepted by the tree canopy up to 30%, while 55% percolates into the forest soil (10% due to retention, 45% to gravity). Only 15% flows over ground! But without vegetation or in the presence of a very degraded plant cover, the rain flow may reach up to 60-70%.

4.5.2
Biological Movements

The movement of species in a landscape has many reasons. Plant movement is linked essentially to dispersal processes (seeds), to the colonisation capacity (pioneer species) and to local expansion due to changes of the surrounding patches. Animal movements are much more complex, being linked to dispersal, home range, foraging, migration, territorial defence, invasion, etc. In many cases movements depend principally on the behaviour of the animals and evolved species may move even for exploration of a territory.

In a system of ecocoenotopes, some tesserae rich in resources and some populations living in favourable ecotopes (i.e. sources) export toward other areas and populations, in some case negative ecotopes (i.e. sinks). This is a corollary of the definition of a biological system as an open and dissipative one. Consequently we may speak of *metaclimax* and *metapopulation*, that is, interconnected sets of sub-units in different phases.

Each subpopulation is connected by the dispersal of individuals in a metapopulation, which changes in size over time and may persist for long periods. Metapopulation dynamics do not concern only movement: here we underline the local extinction and recolonisation process that is facilitated if the landscape resistance is low. The landscape resistance is determined mainly by the arrangement of the spatial elements, especially barriers, conduits, and highly heterogeneous areas. Output and input fluxes of energy and information follow in a landscape the main law of ecology (Zonneveld 1995): not too much, not too little, just enough. The flux is regulated by the following situations:
- Too much input (e.g. overfeeding resulting in poisoning)
- Not enough input (e.g. underfeeding resulting in deficiency)
- Not enough output (e.g. clogging resulting in constipation)
- Too much output (e.g. leakage resulting in depletion)
- Normal flux (e.g. proportional with distance)
- Indifferent flux (e.g. diffuse transitory species passage)
- Oriented exalting flux (e.g. species over-attracted by particular tesserae)

The flux, going from a source toward a sink, may find in the landscape configuration remedies for regulations, such as: supply (underfeeding), resistance (overfeeding), drainage (constipation), retention (depletion), new sources (too great distance), new orientations (few attractors), etc.

It is particularly important to remember the double interaction between the input-output fluxes among the landscape components and their mosaic context characters and behaviour. All ecosystems in a landscape are interrelated, with movement or flow rate of objects dropping sharply with distance, but more gradually for species interactions among ecosystems of the same types. For subpopulations on separate patches, the local extinction rate decreases with greater habitat quality or patch size, and recolonisation increases with corridors, stepping stones, a suitable matrix habitat, or short inter-patch distance (Forman 1995).

The transfer of new genes (alleles) and genetic combinations among subpopulations is called gene flow. Associated with the movements in a metapopulation, we can note and underline an important gene flow that is influenced by the spatial disposition of the patches and corridors of a landscape, and by the size of the sources. For instance, large populations on source patches provide individuals that recolonise empty patches: this is the process that maintains the metapopulation as a whole (Fig. 4.12).

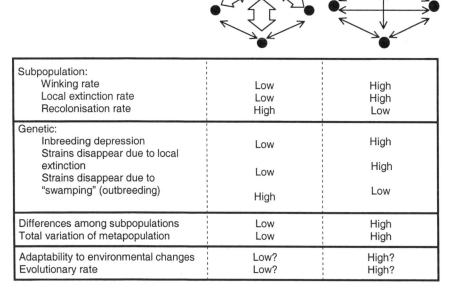

Subpopulation:		
Winking rate	Low	High
Local extinction rate	Low	High
Recolonisation rate	High	Low
Genetic:		
Inbreeding depression	Low	High
Strains disappear due to local extinction	Low	High
Strains disappear due to "swamping" (outbreeding)	High	Low
Differences among subpopulations	Low	High
Total variation of metapopulation	Low	High
Adaptability to environmental changes	Low?	High?
Evolutionary rate	Low?	High?

Fig. 4.12 Expected patterns from gene flow in two metapopulations composed of four subpopulations, living in different cluster of patches. Thickness of arrow represents relative amount of gene flow. (From Forman 1995)

Local extinction is frequent and normal for small subpopulations and a typical process in metapopulation dynamics. If the patches are of the same small size the inbreeding depression is high, the outbreeding low. Furthermore, with a high winking rate (i.e. the rate of local extinction followed by recolonisation), the subpopulations on different patches will be genetically quite different from each other (Forman 1995). Therefore, despite being composed of small subpopulations on separated patches, each threatened by inbreeding depression, the metapopulation as a whole maintains a high level of genetic variation.

Permeant species of animal (i.e. multi-habitat vertebrates) implies a daily use of different habitats, and alternatively different habitats may be used at different stages in the life cycle. Movement is the obliged consequence and in many cases, when these movements are repeated, the pattern of a landscape is structured with paths and places of standstill. Even if vegetation structure is a primary determinant of movement route, chemical signals applied by animals to plants, soil and air play also an important role for many species.

Human activities usually increase the rates of invasion by exotic and alien species, as well as population fluctuation and extinction. This process occurred several thousand years ago in the Mediterranean region, so the present biota in large areas represent those species that could survive there. Human societies need also to introduce their livestock into pastures. Livestock learn movement patterns and are creatures of habit. New herds introduced to a pasture first learn the boundaries, next locate water and then suitable forage. In many cases, livestock needs to be moved far away, e.g. in the mountains in summer and onto a plain in winter. Even if today the transhumance is less frequent, this process has had ecological consequences during the history of many countries and today may continue by road transportation.

The most spectacular movement of animals among different landscapes is probably the migratory flux. In its proper dynamic this process pertains to the ecosphere and regional ecology, but may influence also landscape ecology. One of the main problems is the presence of a series of landscape units (or at least ecotopes) favourable to the migratory routes. When a landscape is altered by human artefacts the migratory movements of many bird species may change or disappear. Nonetheless, the adaptation of some animals is incredible in this sense and in Europe is not infrequent to see a stork nest on iron pylons of electrical plants (e.g. *Ciconia ciconia,* near Milan).

No doubt, the most impressive and important movement of living populations into landscapes is due to human transport systems: from airlines to navigation, and especially road webs and their infrastructures. The multitude of problems linking human transportation and landscape ecology can not be discussed here, and they will be mentioned when opportune. It is meaningful to underline that the entire section of landscape ecology named road ecology is becoming a distinctive discipline (Forman 2002).

Biogeochemical fluxes are also to be mentioned. What is intended in classical ecology as "a cycle" in the reality of a landscape is a chain of cycles. Therefore, the inter-ecosystemic flux plays an important role. For example, the transportation of nitrates through different ecocoenotopes may explain an unsuspected presence of nitrogen concentration in natural non-polluted waters, even underground waters.

4.6
Information Processes

4.6.1
Landscape and Evolution

Inter-ecosystemic fluxes of energy and matter and chains of organisms permit internal webs of communications in an ecotissue. Populations of permeant vertebrates play an important role in managing the system of information, especially humans. This is another evolutionary role of man: building and maintaining a complex system of information into the landscape, which has undoubtedly a stabilising effect (Ingegnoli 1989) in the biological conservation of the entire ecological system.

The informative system of a landscape is essential for the process of coevolution and group selection which may be completed only on a landscape level. The interactions among organisms lead to competition, the interaction organism-environment leads to mutualism, the interaction ecotope-ecotope leads to co-evolution. Gene flow and population differentiation are quite well known to respond to spatial heterogeneity. Note that genetic characters are linked mainly with three levels of the biological spectrum: cell (DNA), population and landscape.

The relation between species diversity and environmental heterogeneity is reciprocal. The process of speciation is endowed with many feedback elements which assure its regulation. Mutations produce diversity, selection eliminates part of the mutants and the remnants are adapted. Random effects contribute to diversify the populations, complementary devices maintain polymorphism and, finally, ecological heterogeneity controls evolution. Therefore, the importance of landscapes is enhanced.

Many animal populations in a new environment tend to a rapid loss of genetic variability. For instance, the heterozygosity rate (i.e. variability) is only 7% for vertebrates, 13% for invertebrates and 17% for plants (Godron 1984).

We note that the equilibrium between evolution through anagenesis (steps along a line) and cladogenesis (steps by forks) is a process more linked to ecological than genetic reasons. In fact, selection is conservative: any species lives in its environment, but explores also marginal environments, and this exploration of biotopes favours polymorphism, because of the interspecific competition.

For instance (Fig. 4.13), the symmetry of the alternative dominance of two species of tit birds (*Parus coeruleus* and *Parus ater*) in different periods is due to competition and changes directly with the transformation of the landscape of Provence (Blondel, 1986).

A reason for polymorphism could be the necessity to decrease the risks of extinction (diverse phenotypes), thus contributing to the stability of the populations, which need coarse grained landscapes and reduced environmental variability. The differences in the mimicry of many butterflies (*Laepidoptera sp.*) may be an example. In any case, the presence of group selection confirms the role of other biological levels, larger than a single species. For instance, the concept of extended phenotype (Dawkins, 1976) sustains that information is controlled even outside the genome, as the relations flower-insect or model-mime.

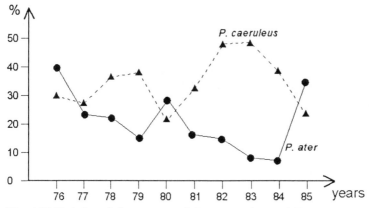

Fig. 4.13 Variation of the presence of *Parus caeruleus* and *Parus ater* on mount Ventoux (Provence), according to Blondel (1986). Note the symmetry due to the competition of the species; the landscape disturbances lead to the alternation of the two species

4.6.2
Information Control

As pointed out by Naveh and Lieberman (1984, 1994), a landscape, being a living system, is self-transcendent. In an evolutionary view, processes and structures become complementary aspects of the same evolving totality. But it is not sufficient to characterise these systems simply as open, adaptive, non-equilibrium or learning systems. They are all that and more. They are self-transcendent, capable not only of representing and realising themselves, but also of transforming themselves, following and generating information and therefore order (see Sect. 2.3.2).

That is why even human landscapes present a modality of transformation led by ecological laws, which may cause in some way even a change in the culture and the ethology of man in order to maintain a metastable equilibrium when landscapes suffer a heavy changing pressure.

Let us examine, for example, the change in garden planning criteria during the industrial revolution. We may observe that when rural landscapes were structured as "gardens" with remnant forest patches and a web of hedgerows in a very heterogeneous field matrix, the theory of garden design was strictly "formal", that is geometric, terraced and with leisure buildings (French gardening); but when rural landscapes, especially around towns, were changed by industries and agriculture began to increase monoculture, a new theory of "English gardening" became dominant, following natural landscape criteria (Fig. 4.14).

Fig. 4.14 An example of the change in garden design during the transformation of the landscapes in Lombardy due to industry and monoculture. Villa Gallarati Scotti, Oreno village, near Milan (Italy). *Above*, the formal garden in the eighteenth century; *below*, a photo of an English park in nineteenth century. Note the island configuration of the park in an agricultural landscape without any more hedgerows and woods

On the other hand, the development of human settlement systems in natural regions and landscapes may be seen as the creation of cybernetic webs applicable to territorial management.

In summary, the beginning of the new era of the Holocene signified principally a change of the rate of information among biological systems, due to the expansion of human populations, both geographically and culturally. The rise of human ecosystems, especially agricultural ones, meant first of all a large scale mutualism among species and a re-structuring of entire landscapes and regions. A second phase was the passage from villages to cities. The presence of a membrane around the most important villages (walls) coincided with the regulation of the exchanges with their landscapes. It was the origin of a differentiated web of circulation, system of transmission and memorisation of information, control of a vast territory, long distance commerce, and specialised districts (Godron 1984).

Consequently, the rise of landscape heterogeneity after the Holocene is incredible, and it is dependent just on information. Moreover, the higher degree of landscape organisation and the lowest instability of ecological systems are proportional (Ingegnoli 1986) with the evolutionary indexes of human population, as D.T.I. (demographic transition index), which considers the growth rate and the longevity of populations. We have to remember that an urban landscape unit may form and develop only in dependence on other landscape units of its territory. When human perturbations do not reach a level of danger (e.g. rupture of cultural-natural control), we could affirm that in general "the evolutionary strategy of biological systems has been to develop human components to reinforce their cybernetic webs through land management".

Countries with higher proportions of urban civilisation show higher ecological metastability. The nature of the present man-environment crisis confirms this observation, being a development syndrome linked with cultural degradation. On the other hand, the ecological evaluation of the negative changing parameters of the environment is another man-nature information flux tending to ecological control through the condemnation of excess deforestation, CO_2 release, human population growth and industrial N fertiliser use.

4.7
Reproductive Processes

4.7.1
Landscape Reproduction

At the organisational level of landscape, reproduction is different from the same processes pertaining to the organism, nevertheless it has to occur. Information to be transmitted has to be circumscribed in time and space, e.g. in a propagule bank.

Table 4.6. Comparison of the main processes of reproduction of an animal population and of a landscape minimum unit

Population renewal	Reproduction process	Ecotope renewal
Predisposed gonads	Reservoir of specialised cells	Propagule bank
Chromosome crossing over	Mutation	Local disturbances
Nest or parental cares	Young protection	Nursery niches
Competition and predation	Selection	Competition and predation
Death, often deferred	Old generation death	Zero event

An outbreak effect must appear, for example, a "zero event" (e.g. fire), allowing the propagulae to substitute previous organic structures (Oldeman 1990).

Through the renewal of the tesserae forming an ecotope, a landscape may reproduce itself, following the typical sequence of reproduction processes (Table 4.6). Remember the process of order trough fluctuations (see Sect. 2.3.6).

The development of an ecological tessera in a forest may be divided into four phases: (A) innovation phase, (B) aggradation phase, (C) biostatic phase, (D) degradation phase (see Sect. 3.2.3 and Fig.3.6). The event of separation between the degradation and innovation phases is the death of the former tessera. The principal factors able to produce a zero event are:

- Fire, where raring fires are also fiercer
- Water, as waterlogging of soils or underground runoff
- Wind, in many variable manners
- Air humidity, in extreme situations
- Earthquakes and landslides
- Human actions

If the zero event is accidental, the preparation phase is relatively short and the innovation phase is too. The size of a tessera and the reaction of the propagule bank are the most important factors in preparing the innovation phase : the peripheral belt gains influence in smaller patches.

As Oldeman underlined, the degrading phase of an eco-unit and the character of the zero event which determines it lead to the replacement of an old eco-unit by one or more new ones. This is the only process that is indicated as succession: it needs a recreative phase, not a regenerative one as during fluctuation if a tessera is disturbed (Falinski 1998).

If the degrading tessera was large sized, the zero event may leave intact fragments, while other fragments are already dead. If the old tessera was very small sized, some strong zero event may destroy it together with the neighbouring tesserae, leaving a large empty area on which a larger eco-unit may develop.

4.7.2
Recolonisation

Recolonisation also plays an important role in self reproduction. Pioneer phases tend to be stochastic, but pioneer dynamics are oriented by strain conditions which depend on the landscape unit structure and dynamics (Fig. 4.15). Time has a fundamental functional importance: so, the recreative phases of the elements of a specific landscape have to avoid too rapid transformations.

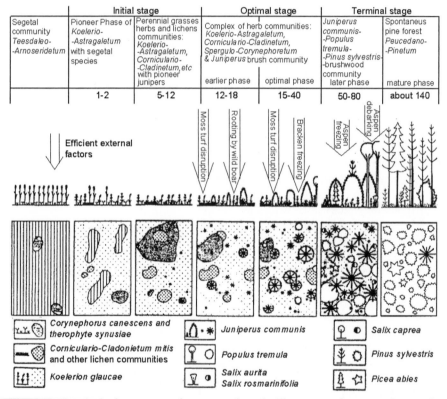

Fig. 4.15 Recolonisation process of a vegetated patch. The course of a succession on the abandoned fields in the Jelonka study area (Poland). Scheme elaborated on the basis of comparative vegetation analysis and dendrochronological research, verified by long-term studies on the permanent plots. (From Falinski 1986, modified)

The most famous studies on colonisation and extinction are contained in the Island Biogeographic Theory (McArthur and Wilson 1967). Even islands are not closed systems, having a permanent relation with a continent. Note that, in some cases, an island may be also a patch surrounded by a different landscape matrix. Because of the constancy of the species number in proportion to the area, invasive species can not establish themselves without the extinction of the local species. Therefore, on islands we find a dynamic equilibrium between colonisation and extinction. The colonisation rate is a decreasing function of the number of local species, while the extinction rate is the opposite. Island area, isolation and age are, in order, the major controls on colonisation and extinction, and hence species number. The limit of this model is found in the hypothesis of an equal probability of extinction for all species. As Blondel (1986) pointed out, this may be true on a geologic time scale, but on shorter time scales this is contradicted by the evolution of endemisms.

Human colonisation is able to reproduce typical ecotopes and landscapes almost all over the world. The *limitatio* (a typical web of fields) utilised by the Romans for their colonisation has been re-utilised in American colonisation and it is still used in Brazil. Human populations acquired early in the Holocene the capacity to change their environment, putting together diverse ecosystem types in order to form their optimal habitat. In doing this, cultural tradition plays an important role, even in an ecological sense.

The abandonment of ecotopes and landscapes due to migration or economic changes in human populations is today frequent, e.g. in mountain agricultural and suburban landscapes. This process generally leads to a re-naturalisation of the landscape, but in case of heavy human influence, alien species may invade these ecotopes.

4.8
Metastability Processes

4.8.1
The Biological Territorial Capacity

As already explained (see Sect. 2.4), we have to relate the landscape equilibrium to the concept of metastability, that is the state of a system oscillating around a central position (steady or stationary state), but susceptible to being diverted to another equilibrium state. Therefore different types of landscapes (or their parts) may be correlated with diverse levels of metastability. This statement has a very important dynamic significance, because it allows knowledge of the transformation modalities of a landscape and consequently (as we will see further) allows its diagnosis.

Trying to evaluate the metastability of a landscape (see Sect. 11.2), one has to refer to the concept of biodiversity (i.e, landscape diversity) and to the concept of latent capacity of homeostasis of an ecocoenotope (i.e. landscape element). We need to start from this second concept.

Referring to a vegetational ecocoenotope, it is possible to define a synthetic function, named *biological territorial capacity* or BTC (Ingegnoli 1980, 1991, 1993, 1999a; Ingegnoli and Giglio 1999), on the basis of: (1) the concept of resistance stability ; (2) the principal types of ecosystems of the ecosphere ; (3) their metabolic data (biomass, gross primary production, respiration, R/PG, R/B). We can elaborate two coefficients:

$$a_i = (R/GP)_i / (R/GP)_{max} \qquad\qquad b_i = (dS/S)_{min}/(dS/S)_i$$

where: R is the respiration, GP is the gross production, dS/S is equal to R/B and is the maintenance/structure ratio (or a thermodynamic order function; Odum 1971, 1983) and I are the principal ecosystems of the ecosphere.

The factor a_i measures the degree of the relative metabolic capacity of principal ecosystems; b_i measures the degree of the relative antithermic (i.e. order) maintenance of the principal ecosystems. We know that the degree of homeostatic capacity of an ecosystem is proportional to its respiration (Odum 1971, 1983). So through the a_i and b_i coefficients, even related in the simplest way, we can have a measure which is a function of this capacity:

$$BTC_i = (a_i + b_i) R_i \ w$$

where w is a variable necessary to consider the emergent property principle and to compensate the environmental difficulties, $w = 0.89 - 0.0054 \ \Omega$
Since $\Omega = (a_i + b_i) R_i$, we can write

$$BTC_i = 0.89 \ \Omega - 0.0054 \ \Omega^2 \ (Mcal/m^2/year)$$

It is possible to calculate an ecological index based on the BTC function (see Sect. 7.5 and 8.4). The BTC indexes, associated with statistical data on the landscape, allow the recognition of regional thresholds of landscape replacement (i.e. metastability thresholds) during time, and especially the transformation modalities controlling landscape changes, even under human influence.

4.8.2
Landscape Metastability

The evaluation of landscape vegetated elements through the BTC shows in any landscape unit a very different structural spectrum, which can be measured with the help of standard classes of BTC (see Sect. 7.5.2). The *landscape diversity* (τ), calculated mainly on this distribution of standard BTC classes (see Sect. 11.2.2) allows the measurement of a *landscape metastability index*, LM, in which τ is related to the mean BTC of the landscape itself.

The change from one level of metastability to another is a very complex process, and we have to underline that the maximum metastability of a landscape generally does not correspond to the sum of the maximum metastabilities of its elements. This is an important statement, because it signifies that an ecological succession (see Sect. 5.1) must not be viewed as simply orientated towards a climax: indeed, in a landscape each component of which is fully at its climax, the maximum of metastability is not usually reached (Ingegnoli 1991, 1993).

A simple model may be used to explain (Fig. 4.16) what has just been affirmed. Compare two landscape units of similar diversity and composed of two repeating types of tesserae: $d=d1+d2$, $c=c1+c2$, where $d1$ and $d2$ are high metastability ecocoenotopes like $c1$, while $c2$ has lower metastability (e.g. oak forest patch, ash patch, beech patch, vs pine forest patches). A strong perturbation destroys part of the d and c basic mosaics with the same proportion.

At time T_2, shortly after the destruction, the two landscape units are still comparable. At time T_3 a difference starts to be evident, because in d the re-development of the destroyed area follows a succession conducive to a new ecocoenotope, different from both the highly resistant-poorly resilient $d1$ and $d2$, while in c the much more resilient $c2$ colonises the destroyed area. At time T_4, the

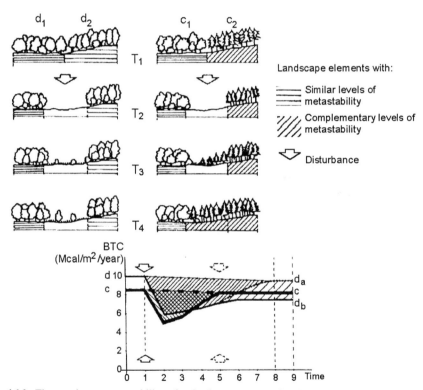

Fig. 4.16 The maximum metastability of a landscape can not be the sum of the maximum metastabilities of its elements (from Ingegnoli 1993). Note the "transformation deficit" (TrD), which is the area of the complex *triangle* below the *line* representing the metastability level during time

difference d-c is augmented, because the new succession in d is slow or stuck, while in c the re-colonisation is complete.

If we quantify with the BTC what has been briefly described here, we may plot the changes in the two landscape units, highlighting the "transformation deficit" (TrD), which represents the quantity of the biological territorial capacity that is lost by each landscape unit during all the processes of transformation and depends not only on the final level of metastability but also on the path of transformation.

The metastability level at time T_1 is higher for d, but at time T_5 it is not. In fact, if the newly formed tessera (d_a) remains at a stage of non-forested ecocoenotope, the BTC of the landscape d should be lower than before the perturbation. Moreover, even if the new tessera (d_b) would be able to recreate a forested ecosystem, that will be at time T_8 at least. Therefore, the amount of the transformation deficit in any case will be greater in landscape d than in c, as we can see, for example, in Fig. 4.16, where the measure of the TrD is:

$$c = 12.56\%, \ d_a = 26.51\%, \ d_b = 21.80\%.$$

We have to underline that even in the case of Ab, of a slow succession, but able to re-create a forest patch, the landscape unit A is less metastable than B, because its capacity to incorporate disturbances is inferior. If a perturbation is repeated at time T5 it should find the landscape unit B already restored, while A is inclined to further degradation. In conclusion, it is important to observe that if the landscape elements have different and complementary levels of metastability, the homeostatic threshold of the entire landscape will be greater.

4.8.3
Landscape Disturbances

The concept of disturbance (see Sect. 2.4.2) becomes in these cases very important, in the sense of the structuring process (O'Neill et al. 1986) incorporated into an ecological system. For example, on a community-ecosystem scale, precipitation must certainly be considered as an external, uncontrollable perturbation. However, considered at the scale of an extensive vegetated landscape, even this aspect can be incorporated.

Naveh and Lieberman (1984, 1994) explained a model in which a Mediterranean grassland landscape is maintained by cybernetic feedback control of grazing pressure. A set point of high spatial diversity and high long-term global stability induced by climatic and rotational fluctuations is maintained by a homeorhetic flow equilibrium, ensuring "order through fluctuation". If we have (see Sect. 2.2.4) a rate of entropy and disorder production $ds/dt < 0$ and a rate of negentropy $d\,info/dt > 0$ we are in presence of an optimun perturbation cycle. If $ds/dt > 0$ and $d\,info/dt < 0$, then the perturbations are too frequent or absent.

Forman and Godron (1986) presented a clear model of the variation of the landscape metastability during time in relation to disturbances. The development of the landscape is expressed as a quantity of biomass in relation to the effects of some perturbations: if the biomass is partially destroyed, landscape metastability diminishes, but development may continue in the same direction as before, if this disturbance can be incorporated. Otherwise, big and sudden perturbations may change the characters of the landscape, even if the incorporation becomes possible, thus development continues along another way. The limits of this model are in the consideration of a proportionality between biomass and landscape evolution: this fact is not always true, as we will see (see Sect. 5.1).

A well known study on disturbances derives from fire in forested landscapes (Turner 1989; Turner et al. 1992). It demonstrates that an increase of biotic diversity is produced exponentially after the burning of diverse ecotopes in a landscape. In the absence of fire perturbations, the biodiversity becomes low. This disturbance may be even necessary, in order to recycle the amount of organic material on poor soils, thus enhancing landscape heterogeneity and ecosystem biodiversity. The attempt of human controllers to avoid all kind of fires in a landscape resulted in exceptionally big fires, of catastrophic force, as in the famous example of the burning of Yellowstone Park (Fig. 2.9).

5 Landscape Transformation and Pathology

5.1
Landscape Transformation

5.1.1
Limits of Traditional Ecological Succession

Calling to mind what was discussed in the previous chapter, let us focus now on ecological succession, which, in general ecology, is the most important process related to transformation. Through serial stages, an ecosystem changes in a predictable way toward a final stage, called climax. After a perturbation, succession returns the ecosystem to the climax. For instance, an abandoned field near a forested patch is re-colonised from the forest edge (i.e. the mantle) and, in a given time, after the re-growth of shrubs and then of trees, the succession restores the climax.

The importance of succession in ecology is so great that it has become the basis of many ecological dynamics, such as in phytosociological sygmeta. But, in reading a text like this, the incompatibility of this kind of succession with the scientific principles of landscape ecology becomes evident.

It should be sufficient to remember the non-equilibrium thermodynamic (see Sect. 2.3.6) with branching points after the instability threshold, or the concepts on landscape metastability just reviewed in the precedent section (see Sects. 2.4.1. and 4.8.2). In the first case, the history becomes the leading criterion of transformation, enhancing the interactions of the ecotissue much more than the reductionist stages of the succession, and determining an unpredictable change. In the second, it is evident that even when a succession to a climax may be considered valid at a single ecocoenotope scale, certainly it is not valid at a landscape scale.

In many "case studies" succession does not work as linear and mechanistic; an example is reported in Fig. 5.1. According to Pignatti (1996b), in the vegetational phytocoenosis of *Cytisus villosus* which follows after a fire of a *Viburno-Quercetum ilicis* patch, for instance in central Italy, or the re-colonisation of *Picea abies* on abandoned alpine pastures in Central Europe, two cases in which normally succession is present, if more than one key factor becomes dominant, the ecological system and its transformation become unpredictable.

Remember that self-organising processes have to be considered at least on three scales (see Sects. 2.2.2 and 2.3.3): if some components of an autocatalytic set have been excluded, the system will appear as linear. It is what happens to the classical theory of succession, because the landscape is never considered as a basic parameter. Therefore, in landscape ecology the importance of ecological succession as linear and divided into primary and secondary phases is drastically reduced. We shall return to this argument after a synthesis of the basic transformation processes.

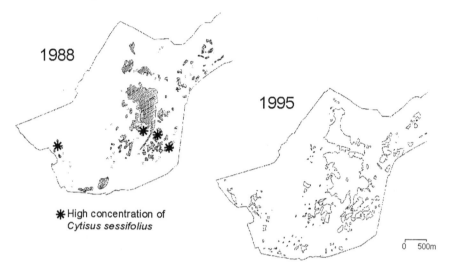

Fig. 5.1 Example of ecological transformation not following the traditional succession theory. Comparison between the distribution of *Cytisus sessilifolius* L. in the natural reserve of Torricchio (Sibillini Mountains National Park, Italy) in 1988 and 1995. (From Canullo and Spada 1996)

5.1.2
Basic Transformation Processes

As sustained in Chap. 2, the principal landscape transformation processes depend on the hierarchical structuring of an ecological system and its non-equilibrium dynamic, metastability, coevolution and evolutionary changes. Let us review the main steps:

- *Hierarchical structuring*. The behaviour of an ecological system is limited by: (a) the potential behaviour of its components on the lower level of scale, (b) the environmental constraints on the upper level of scale. This set of conditions represents the existence field in which the system of ecosystems must reside.

- *Non-equilibrium dynamic.* Thermodynamic bonds may determine an attractor, in its proper existence field, that represents a condition of minimum external energy dissipation. Possible macro-fluctuations produce instabilities, which move the system toward a new organisational state. These new states permit an increase of dissipation and move the system toward new thresholds to reach a new attractor. This could be represented as a cybernetic process of order through fluctuation.
- *Metastability.* An ecological system can remain within a limited set of conditions, but it may show alterations if these conditions change. The system may cross a critical threshold, approaching even radical changes. Different types of landscapes or their parts may be correlated with diverse levels of metastability.
- *Coevolution.* The history of the interactions among the elements of a landscape in a given area shows a particular dominion, that is characterised by the coherence of their reciprocal adaptation. This process leads to a stabilisation of the different homeostatic and homeorhetic capacities of a landscape, which may be expressed with a particular degree of metastability of the entire system.
- *Evolutionary changes.* The structuring of every biological system, such as a landscape, may be pursued, that is the information may be transmitted, only if the final state of the considered system is less unstable (i.e. more metastable) than its initial state. The modalities by which these processes are realised may be different and not limited to a single scale.

5.1.3
Main Transformation Processes in Vegetated Elements

In coherence with the basic transformation processes, we observe that the main dynamics of the vegetated elements of a landscape may be: (1) fluctuation, (2) transitory variations, (3) instability processes, (4) destructive processes, (5) recreative processes.

1. *Fluctuation.* This process is manifested in most ecocoenotopes by a gradual and recurrent exchange of components. The seed bank in the soil contains all the permanent components and also ephemeral species. According to Falinski (1998) the variation in species number seems to be < 12% of their mean.
2. *Transitory variations.* The capacity of incorporation of normal disturbances (external factors) leads to the emergence of *degeneration* and *regeneration* coupled processes. Degeneration is characterised by the advance of invasive species and the penetration of some alien species, sometimes acquiring the status of neophyte. Regeneration implies the reconstruction promoted by internal forces of the degenerated ecocoenotope. This process can be quite fast in fertile and humid habitats, longer in drier and poorer habitats. A simplified vertical structure is generally shown in vegetation ecocoenotopes during regeneration.
3. *Instability processes.* Increased dissipation and instability may occur when

new variations follow during a normally transitory variation. The consequence is an unpredictable transformation. A recreative process is produced when the range and frequency of the disturbance regime changes, often under the influence of the landscape structure, but metastability may increase. By contrast, a destructive process may be the consequence of instability.

4. *Destructive processes.* Out of scale disturbances, the stronger of which produce zero events (see Sect. 4.7.1), can not be incorporated and lead to a *regression* or even to the *death* of the phytocoenosis (e.g. changes in the water table or beaver actions, or heavy human actions). The regression is a process of gradual decline, and may occur also by a too drastic lack of disturbances, leading to a diffuse senescence (Ingegnoli at al. 1995). Soon after the disappearance of the old state, a rapid succession may begin (Falinski 1998).
5. *Recreative processes.* After a destructive process or instability, a forest reconstruction phase begins. In cases like these a direct seed supply from the other landscape elements is very important. The main process remains order through fluctuation explained by Prigogine (see Sect. 2.3.6). A series of changes toward the fluctuation phases may be forecast, but in a very general way, because the variations of ecological parameters are difficult to predict and each site has different landscape conditions and roles. For instance, if the soil characters have been changed in a previous transformation, the recreative process will be often unpredictable. The same unpredictability arises when more than one dominant ecological character is present (Pignatti 1996b).

In any case, we can not distinguish a primary from a secondary succession, because the significance of the primary one is scientifically inconsistent in light of the basic transformation processes.

5.1.4
Landscape Transformation

There are three main ways of landscape transformation (Fig. 5.2). If we plot simple models, in which the position of an ecological system S is shown in relation to its parameters P, its state variable X, its potential V, we note:

1. *Sudden change.* This is generally due to very strong perturbations, which may happen without any previous change in the parameters of the system. For example, after an eruption a new type of vegetated landscape could be formed, or a new type of human landscape after an earthquake.
2. *Gradual change.* This transformation is due to an evolutionary set of processes, in which the values of the system parameters change normally. For example, the passage from a rural landscape to a suburban one, or from abandoned fields to a newly forested landscape.
3. *Temporary unstable change.* Here an ecological system reaches a lower degree of organisation as a necessary step toward a wider and more difficult reorganisation of higher metastability. For example, the change from a wooded-

pastoral landscape to a fire degraded one, toward a forested-agricultural landscape, or the change from an agricultural landscape to a rural-degraded one, toward a suburban park.

Note that the third process is possible mainly in large, complex heterogeneous systems, because in small and more homogeneous systems it would be destructive. These processes are only the most synthetic one and the landscape may follow many diverse ways of transformation, often overlapped. Obviously, the landscape dynamic includes even a form of fluctuation, not dissimilar from what was defined for its vegetated elements: this is for the non-equilibrium thermodynamic principles the so-called order through fluctuation.

Even if it is very difficult to delineate the main principles necessary to control the processes of landscape transformation, it could be useful to remember: (1) the maintenance of congruence between form and ecological characters into a proper range of scales, (2) the presence of a range of disturbances which should be incorporated at the same scales, (3) the maintenance of connectivity, and heterogeneity, in the changing landscape, (4) the limitation of fragmentation and dispersion of natural patches, etc.

As pointed out by Forman (1995), land is transformed in several spatial processes overlapping in order, including perforation, fragmentation, shrinkage and attrition: they increase habitat loss and isolation, but otherwise cause very different effects

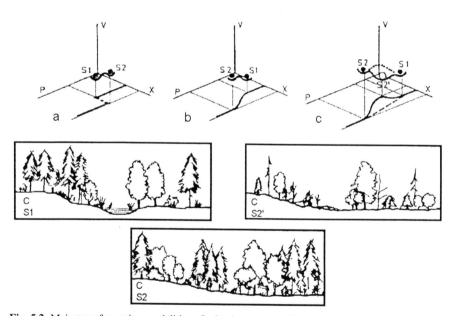

Fig. 5.2 Main transformation modalities of a landscape: *a* sudden change, *b* gradual change, *c* temporary unstable change. *P* parameters, *X* state variables, *V* potential function, *S* system position (Ingegnoli 1993)

on spatial pattern and ecological process. Fragmentation is known; perforation is the process of making holes in an object such as a habitat or land type; shrinkage is the decrease in size of objects, and attrition is their disappearance. These spatial processes overlap through the period of land transformation and increase habitat loss and isolation. The total boundary length between original and new land types increases in the first two processes, and decreases with shrinkage and attrition, while the fractal dimension of the wooded patches generally decreases.

Land is transformed from more to less suitable habitat in a small number of basic mosaic sequences, the ecologically best being in progressive parallel strips from an edge, though modifications of this pattern lead to an "ecologically optimum" sequence. From logging and urbanisation to wildfire, floods or desertification, the landscape is transformed through mosaic sequences. Forman (1995) underlined five sequences.

a. Edge: A new land type spreads unidirectionally in more or less parallel strips from an edge.
b. Corridor: A new corridor bisects the initial land type at the outset, and expands outward on opposite sides.
c. Nucleus: The spread from a single nucleus within the landscape proceeds radially and leaves a shrinking ring of the initial land type.
d. Nuclei: The growth from a few nuclei produces new land type areas expanding radially toward one another.
e. Dispersed: Widely dispersed new patches rapidly eliminate large patches of the initial land type, produce a temporary network of the initial land type, and prevent the emergence of large patches of the new land type until near the end.

Note that, based on the ecological characteristics correlated with the spatial attributes, the edge mosaic sequence is considered ecologically the best of the five. It has no perforation or fragmentation, it is the best for the large patch attributes, and good for the connectivity.

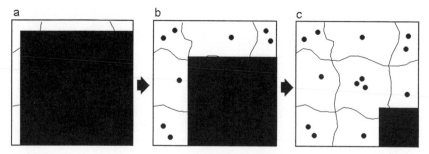

Fig. 5.3A, B, C Model of the "jaws" transformation of a landscape showing three stages of the land transformed from *black to white* land types. *Dots* are small patches and *curved lines* are corridors (from Forman and Collinge 1995). **A** 10% of transformation. **B** 50% of transformation. **C** 90% of transformation.

But the ecologically optimum land transformation could also be the theoretic mosaic sequence proposed by Forman and Collinge (1995). Draw an isolated square landscape of initial land type, where L-shaped or wide-open "jaws" (Fig. 5.3) of new land type will progressively move from upper to left to lower right.

Early in land transformation the jaws appear to grip a huge "chunk" of initial land type, and "bits" of initial land type (small patches and corridors) are scattered over the jaws themselves. In the mid-transformation phase, the thickening jaws covered with scattered bits hold a few large separated chunks. In the late transformation phase the huge jaws covered with bits disappear.

5.1.5
Control of Transformation Processes

It is possible to visualise the regional thresholds of landscape replacement as two co-ordinated fields, monitoring changes in anthropic and natural sets of landscape types, measured on the same scale of BTC (Fig. 5.4).

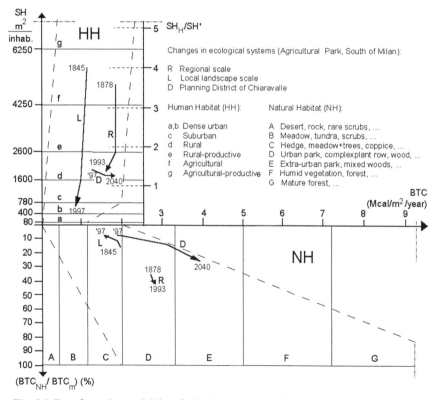

Fig. 5.4 Transformation modalities of a landscape: model of control based on the BTC function. We can see three spatial levels of change: *R* regional scale (Lombardy), *L* local landscape unit scale (suburban landscape of South Milan), *D* district of Chiaravalle, Dotted lines define the more frequent movement field of the system. See also Figs. 13.9 and 13.15

The possibility of being able to analyse large-scale changes allows us to study the past of a landscape, to control the present state and to guide future management. Note that the BTC can be measured also by estimation procedures (see Sect. 7.5). As we may observe in Fig. 5.4

- In the HH the transformation modalities follow generally a cluster of parabolic functions crossing a series of thresholds, from semi-natural protective type of agricultural landscape to the most urbanised one. The mosaic sequence remains that expressed by Richard Forman. Even if the opposite is theoretically possible, in human landscapes these transformations tend to be unidirectional.
- In the NH the transformation modalities are more complex. An ecological succession from a near desert type of landscape to a high BTC mature forest type is theoretically possible, but certainly it is not the main modality and does not follow a straight line. The role and the range of disturbances generally lead the way to change.

The correlation between species and ecocoenotopes biodiversity vs the territorial capacity, or the vegetation biomass vs BTC tends to be almost logarithmic (see Fig. 5.7). But some recurrent modalities of transformation can be represented by closed curves, like in the example of Naveh of homeorhetic Mediterranean landscapes in Israel (Fig.5.5).

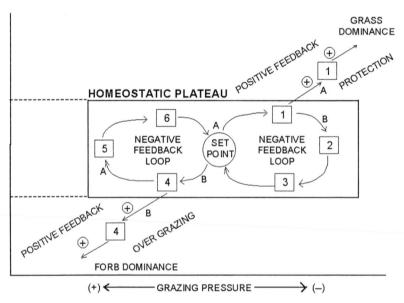

Fig. 5.5 Model of Mediterranean grassland maintained by cybernetic feedback control of grazing pressure. A set point of high spatial diversity and high long-term global stability, induced by climatic and rotational fluctuations, is maintained by a homeorhetic flow equilibrium, ensuring "order through fluctuation" . (from Naveh and Lieberman 1994)

5.2
Landscape Alteration

5.2.1
Principal Conditions of Alteration

The first signs of ecological alteration in a landscape are expressed at the ecocoenotope level, and are generally correlated with the most sensitive populations. If there is a sufficient redundancy, other ecosystems may occupy the landscape degraded niches, but when the alteration influences the entire ecotissue level, then a degradation process is full in action.

Note that the worst degradation seems to be verified in a landscape when the greater part of its dynamic is frozen, that is, when its evolutionary processes have been stopped. This happens when: (1) the metastability level of the upper scale systems is no more able to incorporate the disturbances of their sub-systems, and/or (2) when the biological potentiality of the lower scale components is destroyed, and/or when permanent alterations are caused to the main structures and functions of the landscape. Many researches demonstrated that a landscape crosses critical thresholds (Turner and Gardner 1991, Ingegnoli 1991) in correspondence of which its ecological processes present dramatic qualitative changes: for example, rapid changes in the number and the length of ecotonal edges near a critical threshold and the influence of these modifications on the behaviour of many species, or a drastic change of the biological territorial capacity indicating a change in the landscape from a natural, semi-natural or agricultural to a suburban or urban one.

Let us consider four levels of scale and two series of hierarchically correlated processes for each scale (Table 5.1):

Table 5.1. Example of hierarchically correlated processes in a landscape

Correlated processes (biological and geomorphic)		Scales
B1	Transformation of forested ecotopes	Landscape unit
G1	Soil formation within the ecotopes	(1st level)
B2	Growth of arboreal vegetation complex	Ecotope
G2	Soil pedon processes	(2nd level)
B3	Single vegetation association processes	Tessera
G3	Humus formation	(3rd level)
B4	Growth of the micro-biota within the litter	Dot
G4	Soil micro-grain transportation	(4th level)

If the geomorphic equilibrium (erosion-sedimentation) is altered, for instance G4 is exalted and linked with B2, we will note a lack of incorporation in the third level, thus the production of an "out of scale" disturbance, which may grow as a dangerous perturbation. In cases like these, even more complex and involving many structures in a landscape, an out of scale disturbance may disrupt the organisation of the system of ecocoenotopes; processes of slow frequency, linked with a high degree of landscape organisation, will tend to disappear. Therefore, the components of the system will begin to operate in the fastest way. When the upper scale constraints are lost, the hierarchic organisation of the landscape falls to pieces. If landscape disturbances are incorporated at a regional scale, the biological territorial capacity of the region remains almost constant during a very long period of time, even under strong landscape changes. A good example is the landscape transformations verified in Lombardy from 1878 to 1993, where urban landscapes increased from 1.7% to 9.0%, the agricultural landscapes increased from 39.8% to 44.6% (1928) then decreased to 32.4%, but the main BTC remained almost constant, about 1.9 Mcal/m^2/year (Fig. 5.6).

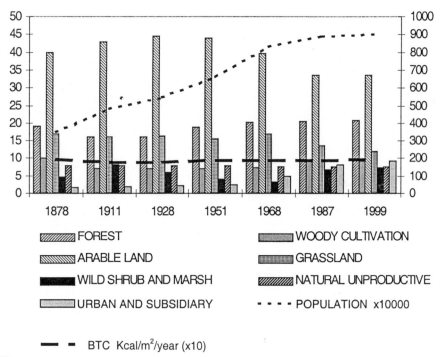

Fig. 5.6 The incorporation of disturbances on large (regional) scale: an example of the extreme transformation of Lombardy from 1878 to 1999. Notwithstanding the decrease of arable land and the increase of urban areas and of population, the regional BTC remained nearly constant

The process of disturbance incorporation on a large scale (e.g. landscape systems scale) is principally due to:

- The amount of transformability of its components; the transformability of a landscape or an ecotope is due to a set of regulating factors and to the structure of the landscape: an ecological system with low metastability (low resistance, high resilience) is more naturally transformable.

- The complementarity of the ecological transformations; the complementarity is due to the dislocation of the disturbances from one type of component to the other. If the entire range of perturbations does not increase too much, or if it does not present an out of scale series of perturbations, then changing the distribution of the disturbances changes the transformations of the ecological system without destroying capacity of incorporation.

5.2.2
Examples of Evaluation of the Levels of Alteration

The incorporation of disturbances on a large scale is a sign of a good metastability capacity, but it can hide significant problems on lower scales, especially when some ecological perturbation presents a discontinuity of incorporation such that it can be incorporated only on a large scale.

Remember all that has already been underlined about metastability (see Sect. 4.8): the change from one level of metastability to another is a very complex process; the maximum metastability of a landscape generally does not correspond to the sum of the maximum metastabilities of its elements; an ecological succession must not be viewed as simply orientated towards a climax; it is possible to measure the "transformation deficit" (TrD) of an ecological system.

All these observations become very important in the evaluation of the level of alteration of a landscape, because a lack of heterogeneity in the metastabilities of the elements or a heavy TrD may be good indicators. Note that the TrD may be important even if the change of BTC is apparently small.

For example, let us examine the results of the analyses of a landscape unit of the southern part of the Serio River Park (Lombardy) made in 1842, 1930 and 1999.

The BTC variations were not so strong, compared with those of many similar landscape units near Milan: 1.99 Mcal/m^2/year in 1842, 1.88 in 1930, 1.62 in 1999, that is a drop of 18.6% in 157 years, a rate of 0.118%/year; only in the last period did the rate reach 0.2%/year. The difference between the two periods seems to be contained: 1 vs 1.69. But the TrD is alarming: it increased from 4.84 to 12.76, that is 1 vs 2.64.

The structural parameters were analysed up too, beginning from the human habitat and the natural habitat. The human habitat components of the same park landscape unit in these periods appeared almost unchanged: 62.3% in 1842, 60.8% in 1930, 63.1% in 1999; but the ratio BTC_{NH}/BTC_{HH} was different: 1.10 in 1842, 1.67 in 1930, 2.221 in 1999. The suspicion of great changes in the landscape structure is proved by comparing the distribution of the land use tesserae divided in seven diagnostic classes of BTC (see Sect. 7.5.2). As shown in Fig. 5.7, these changes are astonishing.

The difficult compromise between heterogeneity and information in the landscape element configuration is disrupted in this case study. Structural alterations are generally perceivable, but more difficult to quantify. This is due to the dichotomy between natural and human sciences, as we will see.

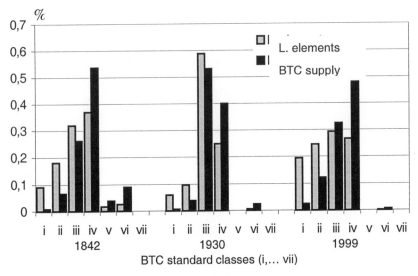

Fig. 5.7 The structure transformation in the landscape unit of the Serio River Park (Lombardy, Italy), going toward an evident alteration

5.2.3
Changes in the Development Direction

Studying the alterations of landscapes and their components allows us to better understand the basic process of transformation. Let us consider an ecotope, and plot some important ecological parameters during its growth: B (biomass), H (biodiversity), BTC (biological territorial capacity) (Fig. 5.8). H and BTC are functions linked to the biomass and are represented by similar asymptotic curves, but they are out of phase. General biodiversity is prevailing in the phases between young and adult (times t_1-t_2), where an excess (redundancy) of diversity is necessary to sustain a not yet completed high level of order, while BTC is prevailing in the phases of maturity (t_2-t_3) which present a higher level of metastability. Note that if we consider only specific biodiversity, going toward maturity (times t_2-t_3) it decreases.

An example can be shown from a Mediterranean type ecotope of *Viburno-Quercetum ilicis* (Br.-Bl. 1936) Rivas-Martinez 1975 presenting three tesserae: very young, near-adult and mature. The biomass volumes (m^3/ha) are generally about 100, 500, 800; the biodiversities (species number) are about 80, 160, 50; BTC (Mcal/m^2/year) about 2.5, 7.5, 12.0.

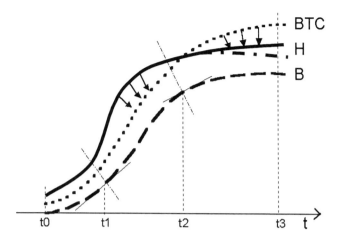

Fig. 5.8 A theoretical example of ecotope growth, in which are plotted biomass *B*, biodiversity *H* and the biologic territorial capacity *BTC*. If H is intended only as specific biodiversity, going toward maturity (times t_2-t_3) it decreases

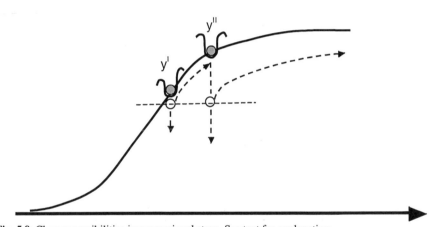

Fig. 5.9 Change possibilities in successional steps. See text for explanation

We have underlined that the proportionality between biomass and biodiversity is a good thing only if the level of order is not properly reached; thus when a strong disturbance occurs the re-growth is said not to go in the same direction as development. In fact (Fig.5.9) given a fixed line of degradation, the ecological system may reach it from a still young-adult phase (Y1) or from a near-mature one (Y2). In the first case, the re-growth can return toward the previous direction of development; in the second case not. Continuing with the *Viburno-Quercetum ilicis* example, its degrading transformation into Macchia (or *maquis*, many types) alters some basic ecological parameters, e.g. the soil, and tends to be persistent.

5.3
Toward a Clinical Pathology of Landscape

5.3.1
Clinical Diagnostic Method

Some considerations on landscape pathology. The definition of landscape as a specific level of life organisation becomes a challenge for environmental evaluation, first of all because man has to pass from a discipline related to technology, economy, sociology, urban design, visual perception, *and* ecology, to another related to biology, natural sciences, medicine, *and* traditional disciplines. Consequently, as we underlined in Chap. 1, analysis, evaluation (and intervention or planning) of the environment require changed methodologies: from engineering, economical and aesthetical rearrangement, to biological diagnosis and therapy. Let us remember that the study of the pathology of any biological system, independently from levels of scale and organisation, needs a basic diagnostic methodology, which can not be avoided. This is true also for landscape dysfunction, and it may be articulated in six phases:

1. Survey of the symptoms
2. Identification of the principal causes
3. Analysis of the reactions to pathogen stimuli
4. Risks of ulterior worsening
5. Choice of therapeutic directions
6. Control of the interventions

Like in medicine, environmental evaluation needs comparisons with "normal" patterns of behaviour of a system of ecosystems.

Therefore the main problem becomes how to know the normal state of an ecological system, and/or, at the same time, how to know the levels of alteration of that system. While in medicine it is the physiology/pathology ratio which permits a clinical diagnosis of an individual, here is the ecology/pathology ratio which permits a clinical diagnosis of an ecocoenotope or a landscape.

For example, dysfunctional landscapes will have less patchiness than normal ones, and any remaining patches will have lower concentrations of soil nutrients, lower water infiltration rates, lower levels of biological activity and lower production cycles (Tongway and Ludwig 1997). We may add: higher transformation deficit, decreasing BTC, natural habitat loss, incorrect ratio between HH and NH, decreasing correlation between heterogeneity and information, higher fragmentation, loss of connectivity, incongruent landscape apparatuses, higher landscape resistance, etc.

Nevertheless, it is sometimes very difficult to perform a correct diagnosis, because some pathologies have "low" symptoms, or apparently not alarming ones. A forested landscape, for instance, may present a high biomass volume and quite high biological territorial capacity, but it may have a too low reproductive rate, too few patches far from the dominant state and a too homogeneous structure of the landscape main mosaic thus presenting an hazardous senescence (Ingegnoli et al. 1995). On the other hand, we have just seen the huge capacity of disturbance incorporation on a regional scale, which seems to be reassuring.

First of all, a more complete framework of clinical diagnosis of the landscapes is needed.

5.3.2
Syndromes Classification

Even if a true classification of the main syndromes has not been elaborated yet, we think that the principal types of landscape dysfunctions can be articulated in six categories, each one of these divided into sub-categories and all related to a particular type of landscape.

The landscape types can be summarised at least in six main classes: (1) high BTC natural (hN), (2) natural (N), (3) semi-natural (sN), (4) agricultural (Ag), (5) sub-urban (sU), (6) urban. A schematic framework may be given, remembering structural and dynamic components (Table 5.2).

Table 5.2. Main landscape syndromes categories and sub-categories

Main landscape syndrome categories	Sub-categories of syndromes
A- Structural alterations	A1- Landscape element anomalies A2- Spatial configuration problems A3- Functional configuration problems A4- Multiple structural degradation
B- Functional alterations	B1- Geobiological alterations B2- Structurally dependent dysfunctions B3- Delimitation problems B4- Movement and flux dysfunctions B5- Information anomalies B6- Reproduction problems B7- Multiple dysfunctions
C- Transformation syndromes	C1- Stability problems C2- Changing process dysfunctions C3- Anomalies in transformation modalities C4- Complex transformation syndrome
D- Catastrophic perturbations	D1- Natural disasters D2- Human-made destruction
E- Pollution degradations	E1- Direct pollution E2- Indirect pollution
F- Complex multiple syndromes	F1- Acute F2- Chronic

A sort of synthetic check-up list for each main landscape syndrome category should be useful in diagnostic evaluation. The present list does not pretend to be an exhaustive work, but it could be good to have an idea of the variety of problems in studying landscape pathologies, and to underline possible critical phenomena.

5.3.3
Main Landscape Syndromes

A. Structural Alterations

A1-Landscape Element Anomalies
- Lack of ecocoenotope structure formation within a tessera, e.g. vertical strata of the vegetation.
- Too geometrical form of landscape elements.
- Abnormal grain of the tesserae or the ecotopes, e.g. too large, in comparison with the same of nearby landscape units.
- Lack of correspondence of form-function in the structure of the elements.

A2- Spatial Configuration Problems
- Low heterogeneity of tesserae and ecotopes, type and form.
- Excess of road density.
- Lack of patchiness among landscape elements.
- Lack or excess of grain contrast in the main ecological mosaic.
- Sharp difference in matrix porosity of vegetation patches.

A3- Functional Configuration Problems
- Excess or insufficiency of network density (Fig. 5.10).
- Presence of some anomalous type of landscape apparatus.
- Lack of natural habitat.
- Lack or excess of ecotissue subsystems (e.g. some landscape apparatuses).

A4- Multiple Structural Degradation
- Loss of congruence between geomorphologic traits and ecotissue structure (even with different levels of scale)(Fig. 5.11).
- Difficulty in locating landscape units or patches of ecotopes.
- Structural subsystems not compatible with the functional role of the land unit within its landscape.

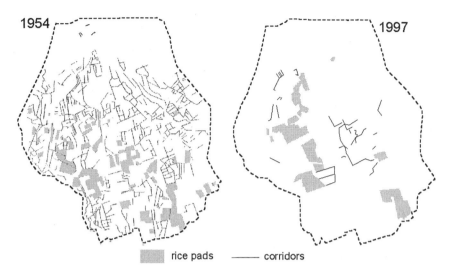

Fig. 5.10 An example of structure alteration in the landscape. Evanescing of the network of corridors and rice pads from 1954 to 1997 in the landscape unit of the South Milan Agricultural Park, Chiaravalle district (Italy)

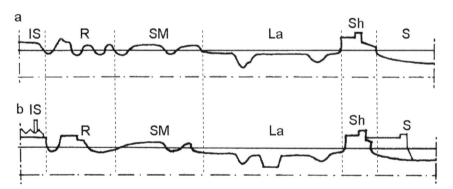

Fig. 5.11A, B A multiple structural alteration in the Venice Lagoon highlighted by the comparison between two transects with the main landscape element types. **A** The situation at the beginning of twenties century. **B** The present situation. The excavation of navigable canals, the formation of a large industrial zone, the building of long breakwaters in the inlets led to the destruction of saltmarshes, the increase of soil subsidence and pollution. This landscape is becoming deeper and flatter, with a loss of biodiversity and BTC. *IS* industrial, suburban and rural landscape elements; *R* near-freshwater and reeds landscape elements; *Sm* saltmarshes and mud flats landscape elements; *La* lagoon typical landscape elements; *Sh* shoreline landscape elements; *S* see (continental shelf) landscape elements

B. Functional Alterations

B1- Geobiological Alterations
- Predominance of morphogenesis, particularly excess instability.
- Excess of soil degradation processes.
- Insufficient or excessive drainage network.
- Erosions and alterations by extractive areas.
- Over-bridling of rivers.

B2- Structurally Dependent Dysfunctions
- Excess fragmentation of landscape components.
- Excess landscape resistance to key species.
- Irregular presence/absence of corridors and connections.
- Negative variation of influence fields among ecotopes or landscape units.
- Problems in strategic points of a landscape matrix.
- Critical thresholds of landscape habitat per capita (in HH or NH).

B3- Delimitation Problems
- Alteration in the boundary formation.
- Alteration of the ecotonal web and/or ecotonal margins.
- Loss of functional characters of landscape boundaries.
- Too high boundary crossing frequency.

B4- Movement and Flux Dysfunctions
- Irregular operations in the main inter-ecosystemic fluxes.
- Presence of perturbation in the advection process.
- Flood anomalies and obstructions (Fig. 5.12).
- Cleaning capacity loss in the excretory apparatus
- Not enough input or output of species and matter in ecotopes or landscape units.
- Too much input or output of species and matter in ecotopes or landscape units.
- Invasion of exotic species.
- Limitations to the inter-regional seasonal movement of species.
- Deviation of migratory flux of species.
- Too high interference by human transport systems.

B5- Information Anomalies
- Loss of genetic variability.
- Lack of speciation processes.
- Loss of natural signs for orientation of moving populations.
- Divergence between human culture and ecological laws.
- Lack of any network for landscape monitoring.
- Degradation of landscape capacity of order accumulation.

B6- Reproduction Problems
- Anomalies in reproduction processes of the landscape components, e.g. propagule banks reduction.
- Obstacles to zero event cyclical occurrence.

- Absence of any senescence phases in the ecotopes and tesserae.
- Interferences and obstacles to recolonisation processes.

B7- Multiple Dysfunctions
- Spreading out of biotic pathologies into the landscape.
- Lack of congruence among inter-scale processes.
- Presence of not-incorporable range of disturbances (out-of-scale).
- Lack of ecotissue structuring disturbances.
- Metastability level not compatible with structure and functions.
- Complex structural-functional generalised alteration.
- Alterations in complex human ecocoenotope/landscape functions.

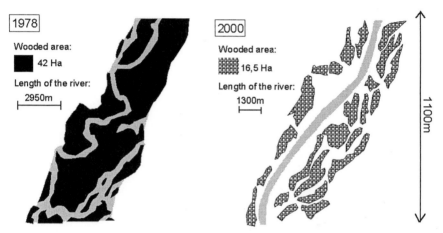

Fig. 5.12 An example of landscape dysfunction. Changes in the fluvial landscape functioning after the destruction of its braided canals. Sangro river, Abruzzo (Italy) (Schipani, unpublished)

C. Transformation Syndromes

C1- Stability Problems
- Break in homeorhetic cycles of landscape maintenance.
- Lack of potentially changeable ecotopes.
- Too homogeneous strategy of equilibrium maintenance of each landscape element.
- Lack of high metastability components.

C2- Changing Process Dysfunctions
- Sudden loss of heterogeneity in the main ecological mosaic.
- Break of connectivity networks in the landscape.
- Too accelerated perforation, fragmentation, shrinkage processes.
- Impossibility to follow the optimum ecological sequence of transformation.
- Loss of complementary successional stages in the patches.

C3- Anomalies in Transformation Modalities
-   Remnants of transformation deficit in the ecotopes.
-   Non-parabolic modalities in human habitat modification.
-   Break in coevolution processes of landscape elements.
-   No gradual change through landscape replacement thresholds.
-   Excess or lack of transformation frequency of the main elements.
-   Sudden excess of human technology plants or buildings (Fig. 5.13).
-   Anomalies in correlation between BTC and plant biomass change.
-   Land abandonment with unbalanced distribution of human population.

C4- Complex Transformation Syndrome
-   Drastic change of typical levels of metastability.
-   Improper role of the ecotissue vs the upper scale landscape/region.
-   Drastic or unbalanced changes in the landscape apparatuses.
-   Loss of historical references and trends (natural or human).
-   Human and natural contrasts in changing structure and functions.
-   Drastic change in the surrounding landscape.

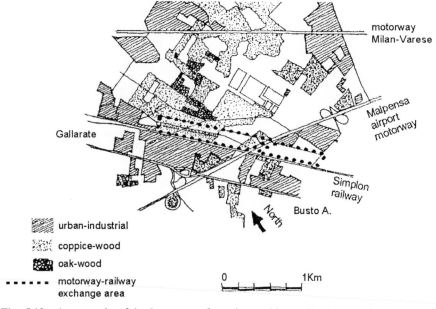

Fig. 5.13 An example of landscape transformation problems. The changes in a wooded landscape unit near Gallarate (Milan, Italy) following the building of a railway-motorway exchange station

D. Catastrophic Perturbation

D1- Natural Disasters
- Strong flood destruction in a landscape.
- Earthquake perturbations on landscape scale.
- Landslide or serious geomorphic instabilities.
- Lava strain and destruction of landscape units.
- Exceptional locust migration.
- Geo-volcanic gas emissions.
- Exceptional drought period.

D2- Human-made Destruction
- Out-of-scale fire degradation.
- Huge technical plant construction and perturbations (Fig. 5.14).
- War destruction and high risk of defence plants.
- Radiation disaster from atomic technical centre.
- Large scale and accelerated deforestation in critical landscapes.
- Fluvial deviation and /or water elimination from entire landscape units.
- Accelerate desertification by wrong rural practices in critical landscapes.

Fig. 5.14 An example of landscape catastrophic destruction. The municipality of Tavernola Bergamasca (Lombardy, Italy) was obliged to install filters in an industrial cement plant, and to restore of the destroyed mountain.

E. Pollution Degradation

E1- Direct Pollution
- Biologic infection of landscape elements.
- Alteration of the web of bio-geo-chemical fluxes.
- Excess of salt in the rising water table.
- Temperature alteration in water bodies.
- Direct chemical or biological pollution of the excretory apparatus.
- Chemical or biological pollution of some patch.
- Excess of noise from the road network.
- Electrical power plants and transportation.
- Waste deposition in critical patches.

E2- Indirect Pollution
- Indirect chemical or biological pollution of the excretory apparatus.
- Microclimatic alteration of the landscape, e.g. acid rains.
- Out-of-scale risk of technical plants.
- Concentration of sand deposition in some patches.
- Gas/oil pipe plant outbreaks.
- Plastic garbage pollution of critical patches by wind and /or water.
Smog excess and fog creation on large landscape units.

F. Complex Multiple Syndromes

F1- Acute
- Catastrophic perturbation in a structural/functional altered landscape.
- Acute pollution in a dysfunctional landscape.
- Heavy,sudden human transformation in a critical landscape unit.
- Serious transformation syndrome in a dysfunctional landscape.
- Generalised structural-functional acute degradation of a landscape unit.

F2- Chronic
- Multiple types of human degradation in a worsening landscape.
- Generalised structural-functional chronic degradation of a landscape unit.
- Suburban-industrial-rural chronic degradation of a landscape unit.
- Continuous and latent de-structuring of a fluvial basin.
- Improper landscape management in a structural/functional critical landscape.

5.4
Perception of Landscape Pathologies

5.4.1
Comparison with Past Situations

According to environmental history, the first observations on landscape destruction, in the nineteenth century, aroused only an aesthetically based reaction. It was probably a consequence derived from the German *Naturphilosophie,* well expressed by von Humboldt in his famous books, such as *Kosmos* (1846), in which an esthetical considerations were mixed with scientific knowledge. The vision of the aspect of wild natural systems and the knowledge of the laws of nature were said to produce joy. Even if the American ambassador in Italy, George Perkins Marsh, noted, in 1864, that man can be considered in many cases as a disturbing agent regarding natural equilibria, especially the landscape, a scientific notion of landscape degradation is quite recent.

Note the E.I.S. (environmental impact statement) that emerged in 1969, in the U.S.A. It tried to value and prevent the destruction of the environment in a more scientific way, but the evaluation of the landscape was again mainly based on visual analysis; or it was limited to just a few components.

For example, deforestation is considered negative, causing landslides or the loss of a rare bird species: but we know that a deforested ecotope is a basic metastable landscape element. On the other hand, there is no doubt that visual perception remains very important, as we can see in Fig. 5.15.

Fig. 5.15 Comparison of the visual aspect of changes in a landscape unit between two periods, 1950-1970 and 1980-2000, in a hilly territory of Europe

It represent a typical European village within its territory (for instance, a pre-Alpine locality in northern Italy, France, Switzerland, southern Germany, Austria): the first view is from the years 1950-1970, the second 30 years after (1980-2000). Since the landscape was not destroyed by war, the first drawing represents a particular landscape condition, the stability of which, excluding small technological changes, has increased over many centuries. The coevolution with nature created a well managed landscape, whose equilibrium is visually immediately perceivable. In the second drawing the same view gives a completely different perception. In a single generation or less, the same territory dramatically changed. The landscape was altered in many components, from the forest and the river, to the fields and the village. The presence of an highway gives the sensation of an out of scale disturbance, as does the new tower built in the centre of the old village.

The visual disfigurement of the landscape moves, in every civilised country, the Association for Monument and Heritage Protection (like the famous British National Trust), as well as the Botanical or Zoological Conservation Associations if rare species are threatened to protest. But nowadays this may be not sufficient. If we remember the landscape ecological principles, we need something more!

5.4.2
Aesthetic vs Semiotic Perception

Visual perception is very important, and it has to do with aesthetics, as architects know. The etymology of aesthetic, following Aristotle, is "what is endowed with sense". Therefore, the visual perception should be very useful also to any ecologist. But this affirmation needs some precise statements.

An ecologist can not be an aesthete: even if beautiful *per se* exists in nature (Lorenz 1978), the ecologist always has to find the sense connected with the aesthetic information. In medicine this fact is well known and the formation of a "clinical eye", has to pass through the medical semiotic, that is the capacity to find the sense of what observed, analysing the information derived from the signs expressed by a biological system. This system is the human body, but it can be also the landscape. Semiotics are necessary to find out all the symptoms and to address the research: analysis and diagnosis.

It is useful to remember that the character of a complex system is given by a specific set of dominant and rare signs, both important in finding the sense of the structure. Moreover, an altered state of a system is not always perceived through its clear disfigurement. This kind of degradation may not be visible, but the syndrome remains serious. The alteration may be perceived by noticing a peculiar lack of information in comparison with what was expected, or noticing the beginning of disorder where not expected (Fig.5.16).

The alteration of an ecological system could even be due to the observation of its uncommonly beautiful aspect, not compatible with the structure and the functions of the examined ecotissue: for instance, the emergence of luxuriant patches of herbs in a prairie landscape, in reality due to the degradation   by overgrazing. The transformation of a forest landscape unit into a golf course may be another case (difficult to limit because people living in the neighbourhood look only to aesthete aesthetics).

Remember, as sustained by Popper (1962) and Lorenz (1973), that the *gestaltic* perception of man is an abstracting process, because it proceeds using so-called "pattern matching", that is a recognition of real elements through the comparison with inside and outside patterns. Therefore this perception is objective, because it recognises the constant characters of things, excluding random variations.

That is why semiotics is a scientifically correct methodology of visual analysis. Note also that it is similar to the visual analysis of painters, as noted regarding Leonardo da Vinci (see Sect. 1.1.2). Besides, this fact explains the use of drawing in scientific representation even today, despite photography.

Fig. 5.16 Alteration of a landscape, not expressed by clear disfigurement, but by a deeper ecological degradation. Photo of an interior agricultural landscape in Campania region (Italy).

5.5
Landscape Pathology and Human Health

5.5.1
Landscape Pollution

In any ecological system, as mentioned before, the relative isolation of hierarchical levels may be altered by "out of scale" disturbances. In these cases anomalous interactions may appear, linking detailed scales with large ones. Many of these interactions are deceitful and dangerous. We are interested here in alterations induced into the organism level by pathologic disturbances on the landscape scale.

Landscape pollution is known to produce toxins, chemical substances that cause any of a wide number of adverse effects in living organisms. The effects may be acute and/or chronic, including changes in living tissues physiology, mutations and consequently cancer.

Many oncologists affirm that cancer is an environmental syndrome: in fact, between urban and rural landscape populations the difference in cancer mortality rate is 1.9 to 1.1, that is about 50% higher in towns, world-wide (IARC 1987).

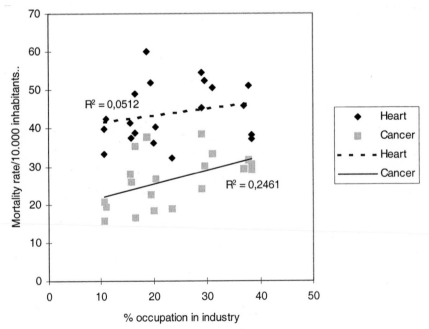

Fig. 5.17 The main causes of death in Italian regions. Note the proportionality between cancer mortality and industrialisation compared with the case for heart diseases for which the result lacks a real significance.

In Fig. 5.17 the Italian situation is reported, plotting the regional parameters regarding the level of industrialisation and the mortality rate for cancer and heart diseases. It is easy to see the direct proportionality between cancer mortality and industrialisation of these territories, while for heart diseases it is hard to say.

Nevertheless, we have to underline that the control of air, water and soil pollution is today the completely dominating criterion of study and application regarding the linkage between human health and landscape.

Alterations due to structural and spatial degradation of ecological systems in a landscape are hardly taken into consideration, or not at all. This is a serious problem, because restoring the landscape will take longer and greater dedication than controlling pollution. It is a much more difficult aim.

5.5.2
Landscape Structure Degradation and Dysfunction

While the effects on human health due to the pollution of the landscape are today quite well known, or at least extensively studied, little has been done to study the effect on health from non-polluting alterations. Note that in landscape pathologies the latter are more numerous (see Sect. 5.3.3). Thus we will try to underline the possible consequences on human health due to landscape degradation with structural and ecological disorder. These involve physiological and ethological behaviours of man, through hormonal and nervous systems disorders.

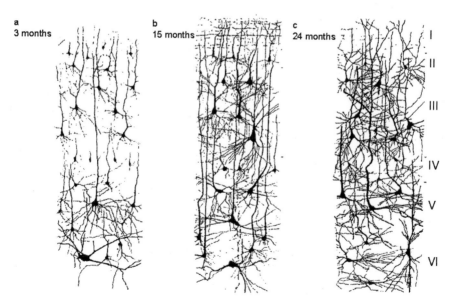

Fig. 5.18 Cerebral cortex maturation in a baby: this is the period of an exceptional plasticity of the brain structure due to indispensable fitting with the environment (From Ramon y Cajal 1909)

Remember that important behavioural functions of man with a genetic and physiologic component may be influenced by the environment, mainly in the first years of life: this process follows diverse steps, from birth to adolescence. We know that from the fourth month to the first year of life the brain increases its volume, the nervous fibres are covered with myelin. The development of the synapses and neuronal webs into the telencephalic cortex is completed, in humans (Fig.5.18), only about at the age of two and a half years.

But an increasing number of nervous fibres cross the *corpus callosum* of the brain linking the hemispheres at age 2 to 5 years. The human brain is considered mature at about 16 years, and some neurones of the grey substance develop till the twentieth year. It is credible to suppose that this long process is done just in order to reach a proper fitness with the environment. In fact the environment is necessary to the developmental program of brain capacities in every mammal, thus even in man. The property to have at birth some non-determined cerebral structures, which have to be specified in relation with the environment, has been emphasised in ethology (Lorenz 1969).

On the other hand, according to Bovet (1969), the learning aptitude is known to be influenced by the environment. It is also well known that social maladjustment, psychic disorders, aggressiveness and cultural desensitisation are evident in many degraded urban peripheries. The same applies to the monotonous grey and noisy landscape of large towns, contrasting with the hyper-exciting super landscape of city centers, the traffic of the every-present roads, the lack of district green, all of which create and exaggerate stress.

The rural landscape outside the towns is not much better: monoculture, destruction of trees and hedgerows, highways, industrial buildings, i.e. another source of stress. The environmental degradation, especially landscape pathologies, certainly is a strong source of stress and when continuous or too strong, becomes harmful to health (Fig. 5.19). According to many physicians (Berne and Levy 1990), the stress disrupts the circadian rhythm of the cortisol secretion and the inhibitor effects of feedback mechanisms.

The sympathetic nervous system and the hypothalamus-pituitary-adrenal axis mediate the integrated responses of the human organism to stress. The responses potentiate each other, both at the central and the peripheral level. The negative feedback exerted by 17-hydroxicorticosterone (cortisol) has the function to limit an excessive reaction, which is dangerous for the organism.

In the case of landscape disorders due to structure degradation and the alteration of normal ecological processes (even in the absence of pollution), the stressors tend to be chronic and the physiologic mediation of cortisol may become insufficient. This could be the reason why behavioural alterations as well as nervous and hormonal diseases are known to be present in every population living in suburban degraded landscapes.

In fact, high levels of cortisol give an effective anti-inflammatory action and activate the immune responses. Therefore, disorders of cortisol mediation processes may also influence the resistance and the immune response to other syndromes, first of all cancer. But even minor diseases such as allergies are dramatically increasing in urbanised landscapes and degraded rural-industrial ones. They do not depend only on air pollution, but it is credible that stress due to landscape pathologies plays an important role.

What has been expressed above is only a synthetic framework of these new health problems. That is why it is necessary to reach a higher control over the state of landscapes, in order to protect our health.

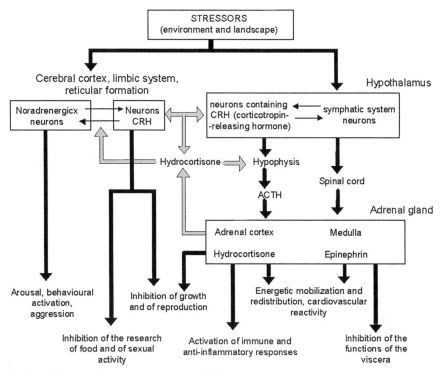

Fig. 5.19 Scheme of the stressor pathways that induce health disorders in human bodies. (From Berne and Levy 1990)

6 Theoretical Influence of Landscape Ecology

6.1
Toward a Unified Ecology

6.1.1
Disciplinary Fields

Landscape ecology is based on the synthesis of polymorphic information treated with a complex approach; it struggles with a unitary vision of ecological problems: therefore, landscape ecology deals with the most advanced fields of ecology, first of all for its theoretical content, while the technical and social importance of its applications is indubitable. Indeed, the theoretical approach of this discipline largely differs from the experimental approach on which most of the present ecological research is based, and consequently it often is not well accepted by the official Academy.

Let us remember some applications such as biological conservation (e.g. minimum preservation areas, ecological networks, etc.), territorial planning (e.g. diagnosis of rural landscapes, urban ecology, etc.) or road ecology: their importance derives from the strength of the practical problems and the urgent necessity of their solutions, which has remained without response from official ecology. This has been one of the major propelling forces for the emergence of landscape ecology theory.

As expressed in Chap. 1, life and its environment represent a unique system organised in hierarchical levels. But general ecology seems not to correspond to this large concept of life, because the majority of ecological texts are limited to traditional chapters regarding organism, population, community- ecosystem and sometimes biomes. When mentioned, landscape ecology is viewed only as an applied field linked with human management.

This is an old question: the history of science forgets too often the enormous importance of technical and practical contributions to the formation of a new theory, of a new disciplinary branch. In addition, theoretical academics do not appreciate being out-dated by a more advanced sector of their discipline that has emerged from lowly origins. The dispute is ancient, and Marcus Vitruvius Pollio, the most famous Roman engineer and architect, warned scientists: practice and theory have to coexist and reinforce each other, otherwise *"umbram non rem persecuti videntur"*, they will pursue the appearance, not the substance of things.

At the end of Chap. 1, we noted that a unified discipline of landscape ecology will change many principles of traditional ecology, going towards a unified ecology. The reasons are manifold:

- Landscape ecology (LE) allows the completion of the biological spectrum, sensu Odum (1971)
- LE revalues the concept of scale, showing that the main ecological processes are scale-dependent (Forman and Godron 1986)
- LE accepts the concept of transdisciplinarity (Naveh and Lieberman 1984) which allows wider application capacities
- LE adopts the new scientific paradigms (see Sect. 1.5) which permit numerous revisions of ecological theory (e.g. from vegetation science to urban ecology)
- LE demonstrates that many classical ecological definitions and principles (population, secondary succession, climax, etc.) are too limited or not applicable to complex systems
- LE demonstrates that space and form may affect many ecological processes, thus the configurations of elements acquire typical characters
- LE pushes ecology to investigate the world of complex adaptive systems
- A true, complete integration among natural ecocoenotopes and human ones is only possible at a landscape level.

On the other hand, it is interesting try to visualise the complementarity of classical general ecology with landscape ecology, therefore the need for their convergence toward a unified ecology.

6.1.2
The Need To Integrate Ecological Fields

Ecology or ecologies ? The question is pertinent because of the many branches of the discipline. To visualise the ecological field of ecology, it is necessary to refer to two main criteria of ordination. One (we will call horizontal) is due to the divisions based on life's organisation levels: organism, population, community-ecosystem, landscape, region, ecosphere.

The other (vertical) is the set of thematic or applied branches: marine, coastal, fresh water, vegetation, forestry, evolutive, animal, ethological, human, environmental design, territorial planning, restoration, conservation, pollution. Crossing these fields it is possible to form a figure (Fig. 6.1) in which the most interesting effect is the comparison between general and landscape ecology.

Note that today general ecology is usually identifiable with community-ecosystem ecology and its linkages with other ecological branches. In this shape, this discipline may cover from 55 to 60% of the total field (represented in the figure as a set of 84 cells) depending on the heterogeneity of texts. On the other side, even landscape ecology may cover 55 to 60% of the same field.

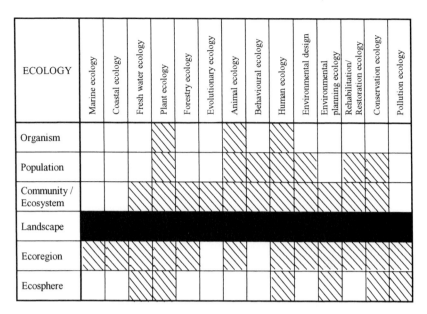

Fig. 6.1 Ecological fields: only together can ecosystem ecology and landscape ecology cover almost all of the disciplinary field

Thus, by only considering a general ecology comprehensive of both, with an overlap of about 25%, it is possible to reach 90 to 95% of the total field. Any other strictly thematic ecology does not exceed 40%.

As it is easy to imagine, the importance of this operation is not only in the wider representativeness of the entire field of ecology. In general ecology it should be obligatory to properly represent the entire field of ecology. The true importance of the enlargement of the field is mainly in the possibility to integrate the component chapters of ecology. Some interesting attempts to proceed this way have recently begun (1998), such as the ecology text written by a group of well known authors (S.I. Dodson, T.F.H. Allen, S.R. Carpenter, A.R. Ives, R.L. Jeanne, J.F. Kitchell, N.E. Langston, M.G. Turner).

Therefore, the large overlap should not be considered a sum, but the occasion to arrive finally at an integration of the ecological processes. For instance, the definition itself of ecosystem has to be enlarged, expressing among the classical components even the new ones due to the interrelation of that system within its landscape.

As we will see, the effect of the new unified ecology will bring unthinkable changes in many collateral fields, like vegetation science, biological conservation, urban ecology, etc.

6.2
Influence on Vegetation Science

6.2.1
The Present Situation and Its Limits

Today, especially in Europe, the vegetation is defined as a set of current vegetable individuals, growing in a determined site and in their natural disposition (Westhoff 1970), and its study is principally founded on phytosociology. The basic unit is the plant association, which indicates a set of plants characterised by a determined floristic composition in correspondence with a determined environment (Braun-Blanquet 1928) and surveyed through sample areas. The combination species-ecological factors is recognised by the means of key species. It is a qualitative-quantitative method. The logic of phytosociology derives from the correspondence between the existence of a given environmental condition of a site and the presence of plant species of a given statistical combination (Pignatti 1980, 1994).

The relation between species and ecological factors, assumed as univocal, permits definition of an n-dimensional ecological space. From a set of auto-ecological spaces the synecological one is defined as the intersection set. For example, projecting on a plane the spaces of five species A,B,C,D,E, the frequencies of which are 0.6, 0.5, 0.4, 0.3, 0.2, the overlapping area may represent an association of these species: the probability of this set to be a casual one is only 0.0072 (Fig. 6.2). This limits the random character of the ecological relation obtained from the presence of species.

Ecological information is often neglected, because of the supposed univocal ecological indication of each species. Thus, the phytosociological model presumes the complete knowledge of an association only through its floristic description. This knowledge is then inserted into syntaxonomy: by the presence of one or few indicator species an association is referred to a syntaxonomic unit, assuming the same ecological characters of this unit.

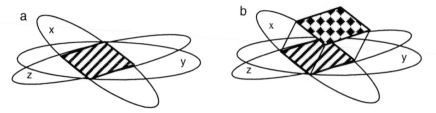

Fig. 6.2A, B Ecological space and the study of vegetation. **A** In the phytosociological model. **B** In the landscape ecological model

The association dynamic is studied based on the concept of ecological succession and climax, assumed to be linear, with a deterministic sense, that is, on the concept of "potential natural vegetation" (Tuexen, 1956). Even the landscape is studied with the sygmetum method, as already discussed (see Sect. 3.5.3).

At the ecocoenotope level, and for a detailed description of the associations of vegetation, this methodology seems to give good results. Qualified supporters of the use of phytosociology even in the study of the landscape are frequent in Europe, so that the Commission of the European Communities (1991) indicated the biotopes with syntaxonomic terms: but not all scientists are in agreement. In fact, the described logic presents certain limitations, especially from the point of view of landscape ecology. The principal criticisms include at least these six.

1. Phytosociology is based on too many deterministic aspects, first of all the importance given to the linear concept of ecological succession, not compatible with landscape ecology (see Sect. 5.1).
2. Until now, even in the representation of the ecological space, it has not been taken into consideration that an association must have an information content that is greater than the sum of the information acquired from the component species (Fig. 6.2). This is what allows an association to become an attractor in its context (i.e. ecotissue), in which it evolves and has to sustain a role.
3. Moreover, the observations of Ellenberg (1960) on relative *Standortkonstanz* of species (relative dependence on site factors) are often not considered. Note that an ecological interpretation of genome redundant size reinforces this concept (Bennett and Smith 1991).
4. It is impossible to show properly the order existing in a vegetational community only with a floristic description (e.g. phytosociologic table). Rather, as expressed in Chap. 2 (see Sect. 2.3.4), if the shorter algorithmic description of a system coincides with the description of the entire system, the system has to be classified as chaotic.
5. The aims of phytosociology are more linked with a description and typing of a supposed natural set of plants than with a study of vegetation in its complete reality. Without an integration, the use of phytosociology in landscape ecology could be in many cases too limited or impossible.
6. Studying landscapes, we must consider as a proper entity also the vegetational new coenosis, created in anthropised landscapes by sets of alien species which have replaced or are replacing autochthonous ones, especially with respect to former natural associations. Information related to natural species are not sufficient from a landscape ecological point of view.
7. The chorological congruence of the species has to be related with the concept of ecoregions (see Sect. 3.6). Thus, e.g. a steno-Mediterranean species should be considered an alien one in Lombardy.

But landscape ecology has to criticise other aspects linked with the phytosociologic method. For instance, the attempt to replace the dogmatic climax concept with the more meaningful "potential natural vegetation" is not yet satisfactory for landscape organisation studies, because the word "potential" is intended to represent undisturbed conditions in a not defined time.

The proposal of Ellenberg (1974), to distinguish among *zonal* vegetation, which expresses the responses of potential vegetation to climatic conditions; *extrazonal* vegetation, responding to local topoclimatic conditions; and *azonal* vegetation, responding to soil moisture conditions, is another good step, but it is again not sufficient for landscape ecological theory. Remember that Ellenberg (1978) already perceived the ecosystem and man's dual part in the structure of a landscape, and Walter (1973) proposed to determine plant formations and types not only in their floristic aspect but also in stability, structure, human influence, diversity, productivity, etc. The reasons for this criticism derive from the self-organisation processes (see Sect. 2.3) especially when the role of disturbances is seen as structuring and when the transgressions in a linear succession are based on the interaction among landscape elements even in the same *zonal* area.

Remember that trying to evaluate the vegetation on the basis of its ecological distance from the potential vegetation is not correct in landscape ecology, because we can not imagine a potential landscape reduced to very few, sometimes only one or two, types of vegetation. This is in contrast with all the main processes and dynamics of the landscape! In facts, it clashes with the non-equilibrium thermodynamic principle and the relative bifurcations of the function of the state of a system in an instability field (see Sect. 2.3.6).

Therefore, the concept of potential vegetation has to be changed. It has to be defined not only for natural cases, but in relation to the main range of landscape disturbances (including man) too, and with defined temporal conditions. It must never be considered as the optimum for a certain landscape (or part of it), but only as a general indication (never to be widely reached) in relation to the climate, the soil and the anthropisation of a certain limited period of time. It could be better named *fittest vegetation*.

This could be a great change of perspective. Note that it signifies also to eliminate, or at least declassify, the concept of primary succession and to change the concept itself of vegetation dynamics.

6.2.2
Evolving Proposals

As written by Pignatti et al. (1998), the floristic approach given by phytosociology seems to show plant communities as a chaotic system. In fact, phytosociological tables are documents which can not be shortened without a loss of information. System theory states that a document which can be reproduced only by its complete description can be considered as a not ordered system. But it is not difficult to demonstrate that vegetation is far from a disordered system. The reason is due to a general concept inherent in all adaptive complex systems in biology (Lorenz 1978; Prigogine 1996; Odum 1971; Giacomini 1982; Pignatti 1988; Ingegnoli 1993).

If vegetation science remains principally based on phytosociology, neglecting the underlying principle of order and allowing an endless analysis, it makes every attempt at synthesis difficult. For these reasons, Pignatti (1998) proposed an alternative approach based on the causal analysis of biodiversity and synecology instead of a mere description of phenomena. This exciting new way is no doubt necessary. But it is not yet sufficient. The main objective of vegetation science must concern description, synthesis and diagnosis of the complex adaptive system of vegetation. For this purpose, it is necessary to integrate:

1. The floristic description
2. The biodiversity and synecology analysis
3. The landscape ecological approach

Vegetation can be considered as a system concentrating order. The flow diagram of a plant community (Pignatti 1996b) very well shows this fact (Fig. 6.3).

The energy collected by the floristic pool and transformed into biomass activates two complementary cycles of self-organisation, which produce specific biodiversity and system synecology. The first cycle leads to the integration of temporal and spatial niches, the second is active at two levels: formation of vegetational layers (microclimate control) and humus formation. The result of these two coupled processes is the distribution of species (A, B, ..., N) which cohabit and form a new reality: the association.

This is a good model for the study of a single ecological community. But if we want to consider the structure and processes regarding the highest level of life organisation (the landscape), we have to refer to Fig. 6.4, which completes and changes the previous model (Ingegnoli 1997).

As evinced from the figure, passing from a community to a system of communities we have to put in evidence at least the following processes:

- The enhancing of complex genetic linkage. In fact, only at landscape level it is possible to interpret the polymorphism and reduce the effects of environmental heterogeneity on metapopulation genetics. Here the same selection process can be completed.
- The necessary addition of the set of communities forming a landscape to the set of species forming a community. For each of these a fluctuation of species and an ecological succession (not linear) are verified. When a critical threshold is reached, a community changes in relatively short time, reaching a diverse level of complexity.
- The influence of the behaviour of permeant animals, among which is generally dominant the human population, on a system of plant communities. It is at landscape level that these influences emerge completely.
- The transformations at ecotissue level which are not limited to the succession of each component community. A process led by geomorphologic conditions of the main mosaic and by the set of local disturbances (natural and human) is correlated with re-colonisation phases of patches and development processes (even phenotypic) of component populations. This process influences the vegetation system, because the optimal conditions of stability and complexity are related to the entire landscape unit, not to the single communities.

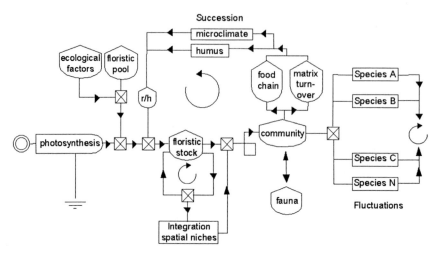

Fig. 6.3 Flow diagram representing a plant community (i.e. association) according to Pignatti (1996). Note the two main complementary cycles: integration of spatial niches and formation of vegetational layers and humus

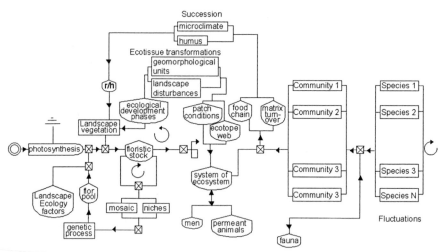

Fig. 6.4 Flow diagram representing a system of plant communities (Ingegnoli 1997). The main difference with the precedent figure is the cycle related to the transformation of the ecotissue

As already discussed, the main mosaic of an ecotissue is generally formed by plant associations, because control of the flux of energy and matter and the capacity to create the proper environment pertain to them. This fact is in accordance with non-equilibrium thermodynamics (see Sect. 2.3.6). Consequently, we need to understand also landscape mosaic characters (e.g. boundary connections, edge/interior ratio, transformations, etc.). In fact, it is not possible to pass from community behaviour to ecotissue behaviour through the same method.

6.2.3
Integrative Approaches

In summary, vegetation science studying complex ordered systems needs an articulate approach. Phytosociology can be utilised only if it is integrated with other methods, especially regarding landscape ecological characters. The vegetation units should be studied by gathering the following information:

- Site data, geographic co-ordinates, altitude, climbing, climate, substrate, etc.
- Descriptors of biological component, height, stratification, biomass, biological forms (sensu Box 1987), L.A.I., interrelation with the landscape, etc.
- Consequences of human actions, stress, forest exploitation, grazing, risks, threatened species, etc.

For this purpose, it is indispensable to use standardised schedules, ecograms and sections. In Fig. 6.5 some ecograms and chorograms proposed by Pignatti (1998) are presented.

The characters relating vegetation to the landscape can be proposed in three forms: (1) a set of eight parameters measuring the relation between a vegetated patch and the landscape through an ecogram, (2) a survey schedule for a vegetated tessera (see Sect. 8.2), (3) the evaluation of the biological territorial capacity of a

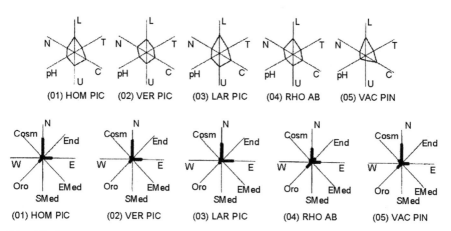

Fig. 6.5 Ecograms and chorograms on forest vegetation. *L* light, *N* nitrogen, *U* humidity, *C* carbon, *T* temperature. (From Pignatti et al.1998)

vegetated tessera (or ecotope) with BTC indexes (see Sect. 7.5). Postponing the points (2) and (3) to the methodological chapters, here it is convenient to propose the set of parameters mentioned in point (1):

1. Frequency of the considered type of tessera in its landscape unit
2. Maximum contrast with the surrounding patches
3. Characters and importance of margins
4. State of the ecological connections
5. Range of landscape disturbances (natural and human)
6. Influence of permeant and/or engineer species
7. Evaluation of the metastability with a BTC index
8. Level of functional roles conferred to the tessera from its landscape

6.3
Influence on Faunistic Studies

6.3.1
Preliminary Considerations

Animal ecology has followed, since the nineteenth century, both the auto-ecological and the syn-ecological approach. In particular great success has been reached in population ecology and community ecology. In all these fields of study, the spatial distribution of animals, the structural pattern of the environments and the importance of scale remained very marginal. An exception was the school of Mac Arthur and Wilson (1967, 1972), which presented island biogeography theory and geographical ecology.

In reality, in the 1970s many scientists began to change direction. Lewins (1976) proposed the term "metapopulation" defining it as a set of different sub-populations separated in patches of non-favourable environments. Odum (1971), in his text *Fundamentals of Ecology*, underlined the concept of permeant animal populations, i.e. species which need to use different ecosystems in a territorial mosaic. Population dynamics may be studied in relation to a new model proposed first by Shapiro (1979) then by Pulliam (1989): the source-sink model. The source patches host a population with a positive reproductive rate, while in sink patches this rate is negative.

In summary, zoologists have begun to consider landscape ecological principles step by step, observing the species specificity of the ecological mosaics, going towards a variegated model of the landscape structure (see Sect. 3.1) in which each animal species can perceive the landscape in a way that can be different from the human perception. Thus, many authors consider the landscape only as the stage, the spatial extent, on which ecological phenomena are played out (King 1997; Farina 1998).

6.3.2
Evolving Theory

The mentioned way of thinking corresponds to the second disciplinary model, as expressed in Chap. 1 (see Sect. 1.3.1): the chorological one. This school affirms that the fundamental themes of landscape ecology are not scale-dependent or limited only to spatial extents of a few kilometres. Not all the researchers are convinced by arguments like these. We may answer that each type of life organisation presents *exportable* characters *and* proper characters (see Sect. 1.6.3). What is more, the species specificity of landscape perception does not imply the existence of many realities. For instance, if we consider an animal perceiving the landscape only in green patches or non-green ones, it does not eliminate the heterogeneity of the different types of vegetation, with their great effects on habitats (temperature, light, etc.), which influence even species perceiving only green *or* non-green.

In a polyvalent logic A *and* Not-A can coexist (see Sect. 9.1.3). Therefore, the human perception of the landscape and any other perception can coexist, without the necessity to reduce the landscape only to a spatial extent. Note that the use of key species and other similar concepts is better understood in a theoretical framework which considers the landscape as an ecotissue (Massa and Ingegnoli 1999).

Strictly linked to the landscape ecological principles is the recent definition of *focal species* instead of umbrella species (Lambeck 1997). In fact, it seems too difficult that a single umbrella species represents the entire heterogeneity of a landscape in a way to protect effectively all the others. It is better to consider a group of different species that is available to identify different spatial and functional habitats. As demonstrated by Massa (1999), with the study of an ecological network in the territory North of Milan, the use of focal species (birds) related to the main ecotope types of that rural-suburban landscape gave very good results.

Landscape ecological principles can also explain the well known equation derived from the studies of Mac Arthur and Wilson (1967):

$$S = CA^z$$

where S is the number of species in a given area A (e.g. an island), and C and z are constants depending on the characters of the territory. Generally C can be considered as 1, and z varies from 0 and 1, thus being logarithmic. The number of species increases with increasing landscape heterogeneity, a process typically scale-dependent, because the landscape *is* a scale-dependent entity.

We can observe that animals are sensitive to the structural pattern of landscapes and they can modify this pattern.

Evident capacities of transformation of large landscapes are demonstrated by mammals populations, such as in African savannas, but behaviour also can be more important than habitat conditions, as shown by beavers in temperate regions.

6.4
Influence on Human Ecology

6.4.1
Culture and Landscapes

Principally, human ecology has been understood as human auto-ecology and as the pollution effects on human health, or even the social-responses to urban and residential problems. It has been mainly studied in Hygiene Departments, Medical Faculties and Social and Urban Departments. The emergence of landscape ecology opened a wide new perspective, linked also to natural fields.

One of the most well known theories was proposed by Naveh and Lieberman (1984), who affirmed that man and his environment are integrated in a "total human ecosystem", considered as the highest level of life organisation on Earth. This vision could be interpreted as too anthropocentric. But Naveh, studying cultural landscapes, underlined that the term culture, derived from the Latin (verb: *colere*) in the sense of cultivation and caring, does not imply the submission of nature to human dominance; by contrast, it indicates an attitude toward conservation and integration. In the spirit of this way of thinking, a particular emphasis has been imposed on cultural landscapes.

If we observe the most typical European landscapes, we may discern immediately the spirit of a particular region, derived from a perfect integration of its human and natural characters and the persistence of its features during many centuries (Fig. 6.6), sometimes millennia, such as in the case of the many remnants of *centuriatio*. This signifies a reached balance in sustainability and cultural cohesive forces. The formation of cultural landscapes needs strong cultural cohesion, because it requires people cooperating and working together, so that competition for space and resources does not destroy the necessity of a sustainable environment.

According to Forman (1995), the term culture is a core subject of anthropology, in which the overall body of extra-somatic and behavioural adaptive measures of human species are represented. In this broad sense the term cultural cohesion appears not to be useful or measurable. An alternative approach for a sustainability analysis uses a narrower operational concept of culture, similar to the one in the dictionary. Therefore, *cultural cohesion* refers to the linking of people by common intellectual, aesthetic and moral traditions. In this manner, culture can be considered as a bonding force in its own right, separate from religion, economics, politics and so forth.

A correct balance between human and natural elements, guided by culture, needs to be referred to the most important components expressing ecological integrity and basic human needs. According again to Richard Forman (1995), ecological integrity and basic human needs can be explained in a useful synthetic way, as we will see. But we think it is important to underline that considerations like these may find their scientific basis in analogy with the concept of focal species and landscape attributes.

Fig. 6.6 An example of a typical cultural landscape: the hilly Tuscany landscape, with its towns and wooded remnants on the top of the hills, and very heterogeneous components

6.4.2
Balancing Culture and Science

A system with ecological integrity has near-natural conditions for four basic characteristics: productivity, biodiversity, soil and water availability. The ecological and human importance of each is extensively described elsewhere. Basic human needs can be restricted to not less than six characters: food, water, health, housing, energy (subsidiary) and culture.

In Sect. II of this book, an effort to win the challenge of the possible overall measures of ecological integrity that are applicable in analysing natural landscapes, but also when expressing basic human needs, is presented: e.g. indexes of biological territorial capacity (BTC, Chaps. 7 and 8), standard habitat per capita (SH, Chap. 8), freshwater functional indexes (FFI, Chap. 7), landscape metastability (LM, Chap. 11), and geomorphologic characters (Chap. 8).

A more general integrative approach to evaluating the human dimension in sustainability is simply to consider the quality of life. This could be estimated in general classes from high to low. Or it could be subdivided and quantified such as infant mortality, life expectancy at birth, literacy rate and per capita income. In doing an evaluation like this we have to understand that technical and scientific parameters can not be sufficient. We have to balance culture and science.

Landscape ecology principles can help us to reduce the dichotomy between artistic and humanistic studies vs technical and scientific fields. This present cultural division represents a great danger, because it breaks the traditional cultural cohesion. As expressed by Ipsen (1998), ecology will not become a driving force in the development of society, and our cities in particular, without aesthetics. However, we believe that ecological aesthetics should not split the world into the beautiful and the necessary if it is not able to see the cities as landscapes. Aesthetics should have the power of bringing things together as a result of their instinctive actions, even when they sometimes only show that here there is no bridge.

Only a transdisciplinary culture can overcome the mentioned dichotomy. An architect and a naturalist can not properly discuss in order to design a good plan if they do not have a common base. Some scientific biological principles have to be studied by architects, and some artistic and social principles have to be studied by naturalists.

6.5
Influence on Conservation Biology

6.5.1
Conservation Biology and Landscape Ecology

The practical effects of preservation began in western civilisation in the second half of the nineteenth century: Yosemite and Yellowstone Parks were founded in the United States in the 1870s. Ecology emerged as a science in the same period in Europe. For about a century both conservationism and ecology were dominated by a sort of near-dogmatic view: to protect nature it is necessary to create and sustain national parks and regional reserves; to study ecology it is necessary to analyse populations, communities and ecosystems which are natural. Two implicit recommendations seemed to be important:

(a)   Act outside human influence
(b)   Act on characterised units, independently from their context.

Parks and reserves were realised mainly and strictly in natural areas, cutting out from preservation boundaries any human village or built patch. Ecological research was carried out natural environments, not considering any possible influence of human factors. Biotopes and reserves, like ecosystem research, did not have to worry about heterogeneity and/or the spatial context, or the scale.

But the reality is different. In their development, human populations were not always so destructive and many species coevolved with man. Thus, conservation biology began to bring new problems to ecology. At the same time, in order to preserve complex zones, it became evident that heterogeneity of ecosystems and spatial scale dependence of processes needed scientific responses different from the classical ones.

The most important response of ecology was the emergence of landscape ecology. Many of the traditional questions of conservation biology found an answer in this discipline, but contemporarily many of the traditional criteria had to change. Think, for instance, of the importance of the ecological corridors and networks. Here let us anticipate some remarks of Chap. 11:

- First of all, let us underline that the fundamental question of biological conservation is not reducible to avoiding the decline of biodiversity, but rather is the conservation of the process, implying fluxes and phases of diversity, that is in the creative and evolutionary capacity of life.
- Note that the basic importance of biodiversity does not require discussion, but only with respect to the diversity of each level of biological organisation and not just to the specific diversity: remember that the diversity of self-organised systems is a complex diversity.
- So, let us agree with Myers (1996), who follows landscape ecological criteria in stating that in conservation planning it is better to preserve the possibility of species radiation than to focus on single rare, even endemic, species.
- Do not neglect the gravity of the destruction of crucial habitats (both natural and human), which constitutes the first reason for the decline of biodiversity and is normally considered less important than pollution problems!
- Consequently, focus on the vision that still considers parks and reserves as islands and confront it with the concept of diffuse naturalness (Giacomini 1965; Erz 1980), founded on the presupposition of the necessity to attend to the entire territory using different degrees of protection. Conservation of nature must be diffused in the human habitat, for example, the conservation of historical or archaeological artefacts has to be exerted even in natural habitats. Furthermore, remember that it is impossible to try to conserve only single details or a single race without conserving their context and the tradition which led to them.
- Also, pay attention to the point that even human culture is not outside of nature. Therefore, nature conservation has to consider also human components of the landscape, when non-altered by the breaking of traditions. In fact, what should be discussed is the modality through which man administrates nature rather than the right to manage nature. The discrimination should not be directed at the presence of man or his changes to wild ecological systems, but the distortion of a correct ethical relationship and all the consequent abuses.

Also, take care of three other considerations strictly related to landscape ecology:

a. The intermediate scale, the landscape or small region, is the optimum scale for managing and planning for conservation and sustainability, as Forman (1995) asserted.
b. Conservation studies and delimitation must be unequivocally referred to the correct landscape unit, refusing administrative limits.
c. According again to Forman (1995), the most important attribute to check, especially for nature conservation purposes, is the measure of *ecological integrity*.

6.6
Influence on Urban Ecology and Territorial Planning

6.6.1
The Strategic Role of Landscape Ecology

Although the foundation of the first graduate course in landscape architecture was in 1900 (Harvard University, USA), we may assert with Goode (1998) that true interest in the relation nature-cities is a quite recent phenomenon, which began about 20-25 years ago. In fact, landscape architecture developed mainly with aesthetic, recreational and cultural aims: the ecological ones were only secondary and/or generic.

After the Stockholm UN Congress on the environment, the UNESCO ecological section promoted the MAB (Man and Biosphere) Project, through which many ecological studies were related also to towns. The capability of nature to survive even in towns was well demonstrated by Sukopp (1984) and Werner and Sukopp (1985) of Berlin University only in the 1980s. Sukopp (1984) was the founder of Urban Ecology. The rise of the IALE, born in about the same year, reinforced and enriched the emergence of urban ecology.

The theoretical principles of landscape ecology influenced deeply urban ecology and territorial planning: first of all through enlargement of the ecological field to include the anthropic components, but also through the importance of space-dependent structure and the dynamics of systems of ecosystems too.

Therefore, conforming adjustments to planning strategies have been led by many municipalities, for instance, the Greater London Plan (1984) is well known. A "Green Paper on the Urban Environment" was published by the EEC in 1990 to underline the importance of nature preservation in towns and the creation of recreational and educational parks. Under UNESCO patronage a World Congress of Urban Ecology was organised in Leipzig by the German Minister of Education, Research and Technology (1997): one of the most salient themes was "Integration of Nature and Landscape into Urban Development".

The impressive importance of landscape ecological theory in relation to urban ecology is based on the concept of landscape as a living system. To have an idea of the potential revolutionary influence of this statement, remember that a territory and a town were considered, until the present, only as geographic, economic and perceptive entities, studied, managed and designed exclusively by architects, engineers, economists and public administrators. Only recently, the contribution of ecology was added just for the necessity of pollution management. The true necessity to change the mentality of planners from a technological to a medical one requires a revolution.

6.6.2
Change in Planning Methodologies

The comparison between a traditional method of territorial planning and a new method influenced by landscape ecology is the best way to understand the above mentioned huge changes (Fig. 6.7).

In traditional planning methods the administrative criteria follow immediately the social demand of territorial management. They are strictly linked with political demagogy and economic exploitation and act as the principal limits to any possibility of a correct relation between man and nature.

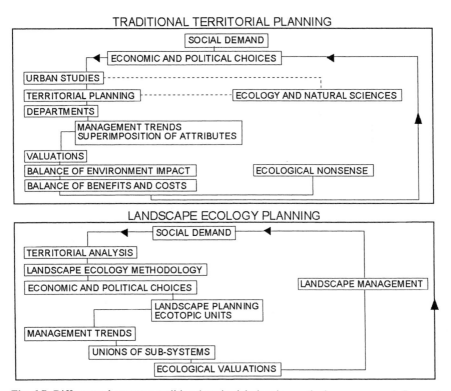

Fig. 6.7 Differences between a traditional territorial planning method vs a new one influenced by landscape ecology

In fact, planners may even ask for the contribution of ecology, but this is intended as a secondary input which may adjust only some evident pollution problem or help to find possible biotopes for preservation. Planning continues to be strictly limited to a municipal area or administrative districts, to be elaborated on by thematically overlapping data, locally using the environmental impact statement criterion. But this may lead to an ecological nonsense. In some cases, e.g. in small towns and rural landscapes, history has taught us that non-planning could be better.

What is suggested by landscape ecology is completely different. A scientific territorial analysis, using landscape ecology criteria, must follow social demand, thus arriving at an evaluation and diagnosis of the landscape unit. Only after a consultation on the ecological state and possible pathologies, can the administrative authorities elaborate proper choices as to the best way to plan. Planning will have to be done following ecotope units and checking an entire range of scales, even outside the municipal borders. The aggregation of ecological subsystems and the evaluation and control of the possible scenarios will be the main method. Landscape and regional ecology offer theory and empirical evidence to study territory.

As we will see in Chap. 12, some models of optimal arrangement of land use in a landscape have been already proposed. The best known is the model following the aggregate-with-outliers principle (Forman 1995) (see Sect. 12.3.2).

Section II
Landscape Methodology and Application

7 Landscape Analysis

7.1
Landscape Survey and Representation

7.1.1
Observation and Control of Landscape Physiognomy

Landscape surveying needs a complex set of operations: observation, measuring, elaboration, registration of data, all supported by proper maps and aerial photographs. But cartography and photo-interpretation are not strictly part of our discipline and we must refer to specific technical books. Moreover today these disciplines are linked with computer science, especially with GIS (geographic information system) programs. While these may be good tools of support, they must remain independent, as today we tend to abuse them.

Instead, the basis of surveying is indeed the capacity to observe and to register information gathered from observation, first of all of landscape physiognomy. As noted in speaking of aesthetical vs semiotic perception (see Sect. 5.4.2), the first analysis is always the semiotic one and is made with photography and frequently with the help of drawing.

Note that following positivism and logical rationalism, the scientific world regarded drawing with diffidence and sometimes with derision. But at present, after changing the scientific paradigm, this kind of analysis is more appreciated. Indeed, remember that many famous naturalists drew well, from Haeckel (a good painter, too) to Lorenz.

In this respect, Konrad Lorenz (1978) underlined that researchers in the natural and biological fields have to understand the methodology of painters. This is important not only for analogies of investigation, but also to develop a sensibility to differences of values in complex organic processes.

Moreover, the use of graphical analysis through drawing permits notable improvement of the capacity of observation and interpretation of the relations between form and function, for instance: effects of erosion, structure of tree patches, ecotonal belts among forest ecotopes, state of the edges, wind effects on vegetation, geomorphologic features, boundary of tesserae, etc. An example is shown in Fig. 7.1.

Fig. 7.1 An example of drawing analysis in a landscape: some characters of the forest landscape of Sila Piccola (South Italy), object of a strategic research project in vegetation and landscape ecology for the Italian Council of Research (CNR), 1993-1995. (Drawn by V. Ingegnoli in 1994)

Also the maintenance of a congruence among the main structure and the local transformations in a landscape can be analysed with drawing. Remember that the relation form-function may change with the scale, but an intrinsic character remains. As expressed even in fractal geometry (see Sect. 7.4.3) some detail remains homothetical to the whole.

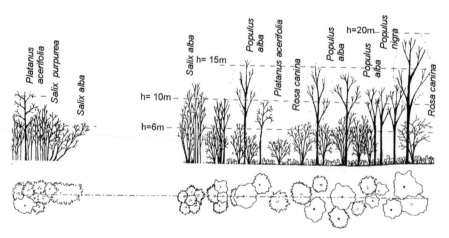

Fig. 7.2 An example of structural differences in a transect of a forested tessera of riparian vegetation in a landscape unit of the Ticino River Park (North Italy). (Donati, unpublished)

Drawing analysis becomes indispensable when there is a problem concerning sections: for instance the section of a valley, or the section of a plant corridor, or the section of a river bank and its vegetation, etc. Particular attention to the structural physiognomy of vegetation is linked to the analysis of the drawings of its transects (Fig. 7.2). Note that the drawings can not be stylised: they must reproduce the exact shape and disposition of the foliage, the number and the position of the trunks and/or the stems and of all the branches, etc.

Information on heterogeneity, biodiversity, vertical structure and growing state of the vegetation derives clearly from these drawings, and helps the evaluation of landscape elements. There are various methods to follow for the survey of a vegetational transect; the most common is to refer to a line perpendicular to the topographic curves, about 150-200 m long, and to assign the disposition of trees and shrubs within a strip of about 10-12 m wide, measuring also the height of canopies and the presence of dead trunks.

7.1.2
Elaborated Drawings

The visual analysis of the landscape and its elements may be articulated in a natural and anthropic semiography. Basic drawings and map are elaborated as regards particular information. For instance, symbols are added to an axonometric drawing to express particular transformation processes (Fig. 7.3)

These symbols can be codified for spatial orientation, geomorphologic signs, margin characters, types of barriers, closed/open spatial characters, main faunal paths, different disturbances, environmental linkages, vegetational structure, visual emergencies, etc. Other similar operations can be done on maps, especially regarding a synthetic morphology and a visual perceptive analysis. Semiography may be very useful for studies on landscape pathology.

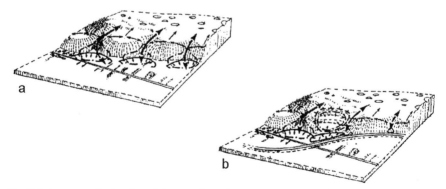

Fig. 7.3 Axonometric drawing of the transformation process on the edge between two types of different landscapes: the karst near Monfalcone and the Isonzo river plain (North Italy). The building of the highway and a wide pit destroyed the natural connections between the landscape

7.2
Ecological Measures

7.2.1
General Ecological Indexes

The following step represents the investigation through ecological indexes, of which many traditional ones may be utilised even in landscape ecology, with some adaptations. In general ecology, the indexes are based essentially on species, while we have to use landscape elements. We describe here only the most important indexes, specifying that the term habitat means landscape element.

$$Relative\ richness \quad R = (S / S_{max}) \times 100$$

where: S is the number of habitat types and S_{max} the maximum possible number of habitat types (Turner 1989).

$$Diversity \quad H = -\sum_{k=1}^{s} p_k \ln p_k$$

where: s is the number of habitat types and p_k the proportion of area in habitat k (Turner 1989). It derives from information theory (see Sect. 2.3.2).

$$Dominance \quad D = \ln s + \sum_{k=1}^{s} p_k \ln p_k$$

where: s is the number of habitat types and p_k the proportion of area in habitat k (Turner 1989). Note that $\ln s$ corresponds to the maximum diversity (H_{max}).

$$Relative\ evenness \quad E = (H_j / H_{max}) \times 100$$

(H_{max}) is reached when each component has the same probability; H_j is the diversity of the studied landscape unit (Forman 1995).

$$Contagion \quad C = 2s \log s + \sum_{i=1}^{s} \sum_{j=1}^{s} q_{ij} \log q_{ij}$$

where: s is the number of habitat types and q_{ij} the probability of habitat i to be adjacent to habitat j (Godron 1968; Turner 1989).

$$Dispersion\ of\ patches \quad R_c = 2d_c (\lambda/\pi)$$

where: d_c is the average distance from a patch (its centre) to its nearest neighbouring patch and λ the average density of patches. If $R_c = 1$, patches are randomly distributed; if $R_c < 1$, patches are aggregated; if $R_c > 1$ (max 2.149), patches are regularly distributed. The equation is also a measure of aggregation (Forman and Godron 1986).

7.2.2
Urban Ecological Measures

Often utilised is the *shifting mosaic*, consisting of a small matrix ($t \times Le$) where columns represent the times of survey expressed in years (t) and rows the landscape element types generally in hectares (Le). It allows the registration of changes in land use and the rate of transformation of one element into others through arrow linkages (Fig. 7.4). It permits control of the real variations in time of those landscape components that apparently do not seem to change. For instance, in the figure the element type "mature forest" measure about 35 ha from 1720 to 1994, but this quantity does not represent the same patches.

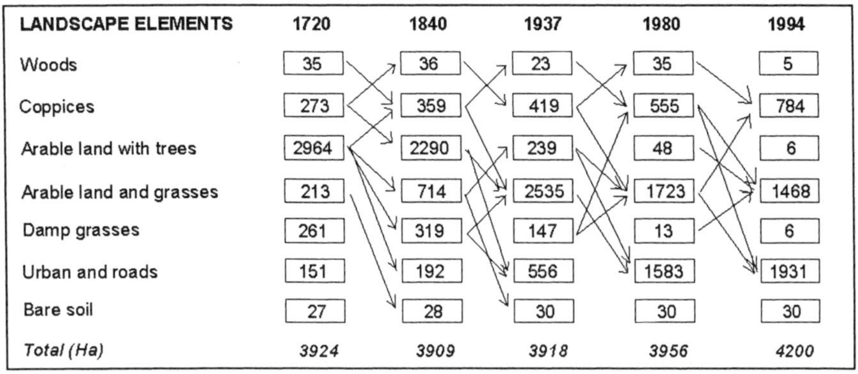

LANDSCAPE ELEMENTS	1720	1840	1937	1980	1994
Woods	35	36	23	35	5
Coppices	273	359	419	555	784
Arable land with trees	2964	2290	239	48	6
Arable land and grasses	213	714	2535	1723	1468
Damp grasses	261	319	147	13	6
Urban and roads	151	192	556	1583	1931
Bare soil	27	28	30	30	30
Total (Ha)	*3924*	*3909*	*3918*	*3956*	*4200*

Fig. 7.4 Shifting mosaic of the landscape unit of suburban Monza Park (Milan, Italy)

A very interesting index is the *Biotopflaechenfaktor* (BFF) which is an index applicable to monitor the ecological state of an urbanised tessera or ecotope (Ermer et al. 1996) through the naturalness degrees of the types of surfaces which form the landscape element. The most important natural function guiding the attribution values for the types of surfaces is the possibility of water percolation into the soil: The measure of BFF is the relation of the effective natural surface vs the entire urbanised area. The attribution of the BFF indexes to the main type of surface is shown in Table 7.1.

Fig. 7.5 presents a practical example: given a lot area of 750 m^2, in a suburban landscape, the built surface measures 280 m^2, the terraced garage 60 m^2, the paved entrance courtyard 195 m^2, the garden 215 m^2. The BFF was calculated as follows:

$$BFF = [(280 \times 0) + (60 \times 0.5) + (195 \times 0.3) + (215 \times 1)] / 750 = 0.405$$

A value of BFF of 40.5% is quite high: the most common examples in Europe may be about 20-30% in open urban landscapes but even less than 5-10% in a closed one. By contrast, in the United States of America BFF can reach 50%. The importance of an index like this is not only in order to control urbanised or semi-urban landscapes, but to permit a more correct planning and design of the lots in relationship to their landscape.

Table 7.1. Attribution of values/square metres for measuring BFF (from Ermer et al. 1996)

Type of surface	Current components	BFF values
Total waterproof	Asphalt, concrete, house foundations	0.0
Almost waterproof	Paved grounds, flat stones	0.3
Semi-opened paved	Flat stones with greenery, or similar	0.5
Vegetated pavements	Garage greenery covers, etc.	0.5
Vegetated terraces	Green terraces, > 0.8 m of soil	0.7
Garden vegetation	Vegetation on normal soil	1.0
Gravel areas	Water percolation areas with gravel/sand	0.2
Climbing vegetation	Walls covered by climbing greenery	0.5
Roof greenery	Roof gardens	0.7

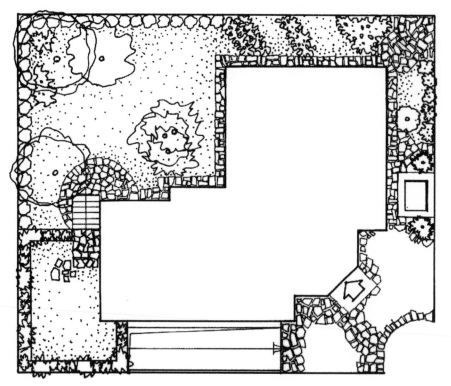

Fig. 7.5 Example of application of the BFF index to a suburban landscape element

7.3
Measures of Vegetation

7.3.1
Leaf Area and Ellenberg Indexes

Many botanical measures related to vegetation are useful in landscape analysis. For instance, the LAI (*leaf area index*) defines the leaf area capable to intercept the light as a relation between leaf and soil area covered by plants (Table 7.2).

If 110 plants, each one with 12 leaves of 0.0035 m^2 of surface, are present on 1 m^2, we will have:

$$LAI = 110 \, x \, 12 \, x \, 0.0035 \, / \, 1 \, m^2 = 4.62$$

Another way to utilise traditional data on vegetation is the information value given by plant communities, on the basis of the studies of Ellenberg (1979) and Landolt (1977). Information increases the more precisely species reflect the environmental characters. Three classes may be distinguished: class 1 contains species not very indicative; class 2 species indicative for 1-2 environmental factors; class 3 species indicative for 3-4 factors.

Table 7.2. An example of LAI measuring in the Alps (from Pignatti 1995, modified)

Elevation (m above sea level)	Vegetation types	LAI (m^2/m^2)	Species diversity
800-1000	Prairie	11.60	35-45
1000-1700	Spruce wood	10.20	21-35
1700-1900	Green alder wood	11.10	25-50
1900-2200	Rhododendron shrub	5.30	20-30
2200-2400	Small snow valleys	1.26	15-25
2400-2800	Alpine vegetation	1.34	25-40
2800-3100	Snow vegetation	0.45	15-30

Indicator values of plant species may be registered with reference to their preferences towards the factors: light, moisture, humus and pH. For example, Bas Pedroli et al. (1988) suggested:

- L = light: 1= full light, 2= half light, 3= indifferent, 4= shade, 5= full shade

- M = moisture: ip= inundation preferring, it= inundation tolerant, oh= O_2 - rich subsurface flow, b= spring biotope, s= stagnant O_2 - poor water, v= weak moisture indicator, x= xerophyte, - = moderately moist

- pH: 1= pH > 7,   2= 5< pH <7,   3= pH < 5

- H = humus: H= mull (H1) and moder (H2) indicators, M= mineral soils indicators, R= mor indicator, G= on H, M, as well as on R

7.3.2
Vegetation Form

Also the relationships among *vegetation forms* can help in the study of a landscape. For instance, the proportion between therophyte (Th) and hemicryptophyte (He) may suggest changes in an examined patch under disturbance, with respect to regional media. In Europe, this quantity changes in a range from Th/He = 0.2-0.3 in Alpine regions, and Th/He = 0.5-0.6 in Central Europe, to Th/He = 1.5-1.8 in Mediterranean regions (Pignatti 1988).

More interesting approaches in eco-physiognomy are based on the idea that plant form, in combination with climatic conditions, dictates plant water and energy budgets, and therefore plant distributions in regions and landscapes. Box (1987) detailed these relationships between form and function of plants.

General form relationships are: LAI increases with increasing plant size; evergreen leaves have longer potential growing seasons than deciduous; evergreens have hard leaves, deciduous soft; conifers have higher LAI; GPP (gross primary production) increases with increasing AT (actual transpiration); R (respiration) increases with increasing temperature (exponentially); GPP, via increased LAI and
AT, increases with plant size; GPP, via AT, is greater for soft leaves; softer leaves usually have shorter growing seasons (i.e. annual AT and GPP reduced); R increases with increasing B (plant biomass). As a result, it appears that:

1. There exist various configurations of form characters which have different functional limits and optima and therefore represent different "basic ecological types".
2. These basic configurations have different advantages and disadvantages in different situations.
3. The existence of such basic "ecophysiognomic" types implies an ecology and plant geography at this level.

In summary, a landscape ecologist must be able to describe plants in terms of the form features necessary to help in the ecological diagnosis. Six characters were proposed by Box that pertain to our analysis (Table 7.3).

1. Structural type (e.g. tree, shrub, graminoid, stem-succulent, epiphyte).
2. Relative plant size (relative to other plant types of the same structural type).
3. Type of photosynthetic organ (e.g. broad or narrow leaf, photosynthetic stem, etc.).
4. Relative size of photosynthetic organ (relative to others of the same general type).
5. Consistency of the photosynthetic organ (expressing resistance to exchange of water and CO^2, e.g. malacophyllous, sclerophyllous, ligneous, succulent).
6. Photosynthetic habit (seasonal/aseasonal variation in deployment and/or activity of photosynthetic organs, e.g. evergreen, summergreen, ephemeral).

Table 7.3. Dichotomic sequence of plant forms (from Box 1987, modified and integrated)

1.	Vascular plants (kormophyte)................................	2
	Non-vascular plant (cryptogam)	..Thallophytes (incl. mosses, algae, liverworts, lichens)
2.	Plant rooted in ground..	3
	Plant not rooted in ground, sitting on other plants	...Epiphytes (incl. parasitic)
3.	Plant self-supporting (at least at maturity)...................	4
	Plant permanently sprawling, climbing, or otherwise not self-supporting	...Vines / Lianas
4.	Plant with stems..	5
	Stems essentially absent above ground (ab.gr.); leaves in terminal rosette at ground level (often succulent)	...Rosette-shrubs
5.	Stems not permanently woody or succulent (ab. gr.)........	6
	Stems permanently woody (ab. gr.) or succulent	...10
6.	Plant totally herbaceous.......................................	7
	Plant not totally herbaceous	...9
7.	Spermatophyte (especially angiosperm).....................	8
	Pteridophyte	...Ferns
8.	Plant grass-like..	....................Graminoids
	Plant not grass-like	...Forbs (incl. geophytes)
9.	Plant perennial with progressively sclerotic culm, >0.8m tall (or similar)	...........................Reeds
	Plant perennial from woody *xylopodium* (or similar)	...Semi-shrubs (incl. trees *plantulae* or seedlings)
10.	Stems succulent, with or without woody skeleton; plant perennial	..............Stem-succulents
	Stems permanently woody (ab. ground)	...11
11.	Plant with multiple main stems, often with none dominant; plant generally < 5m tall	12
	Plant with a single main stem (trunk); plant generally > 5m tall	...13
12.	Plant with 1 or more main stems, < 1m (except for emergent trunks), prostrate, usually highly branched	.....Krummholz (e.g. twisted shrubs, incl. cushion-shrubs, dwarf-shrubs)
	Plant with 1 or more main stems, > 1m; growth form indeterminate (e.g. overgrown bushes, scandents)	...Arborescent shrubs (incl. cushion-shrubs, dwarf-shrubs)
13.	Plant growing from terminal bud, with wood produced secondarily, no bark; rarely branching; leaves in rosette(s), usually terminal	......................Tuft-trees (incl. tuft-treelets, tree-ferns)
	Plant with true wood growing outside; usually with bark, plant usually branching	...Trees (incl. treelets, small trees)

7.3.3
Plant Biomass

Another field of vegetational analysis of great importance for landscape ecology is the measure of *plant biomass*, or phytomass. As we will see in the next chapter, this quantity is needed for the evaluation and diagnosis of tesserae and ecotopes, thus for the understanding of the landscape ecological state. The phytomass quantification is not always simple, even if generally only the above-ground biomass is required. In the case of a *prairie*, a few samples of small squares consisting of cut-down herbs (e.g. 0.25 to 4 m^2) can be easily dried, weighed and expressed as kg/m^2. In the case of a *shrub* formation, the work is harder, because the sample squares are bigger (e.g. 25 to 100m^2) and it could be difficult to take measures by hand: stem height (h), diameter (d), stem number per shrub (ns).

It is also necessary to have or quantify a plant geometrical coefficient (A = allometric, see Table 7.4) to estimate the volume of plant biomass (PB = m^3/ha), with the simple equation:

$$PB = \pi (d/2)^2 \times ns \times h \times A$$

In the case of a *forest* it is generally convenient to use a good instrument, such as a mirror relascope. This is an optical instrument, sufficiently precise, which permits the basal area of trees to be evaluated following the equation:

$$B_a = (k \, \Sigma \, z) / n$$

where: B_a is the basal area of trees (m^2/ha); k the angular factor [1,2,3,4]; z the number of selected trees (the value of which will be 1 if the trunk is > finder band; ½ if equal); n the number of surveyed sites. This number n is suggested to be n = 8 $\sqrt{S}$, S being the forest patch surface (ha), or (better):

$$n = (t_s \times s_x) / \varepsilon \%$$

where: t_s is Student's test; s_x the standard deviation; ε the sampling error (Hellrigl 1990). With the mirror relascope it is easy to measure the height of trees (h) too; thus the PB can be calculated knowing the allometric coefficient: if the measures are very few (e.g. in a small tessera) they need a reductive coefficient (0.85-0.95).
Frequently, the measure of the attributes of vegetation needs spatial statistic methodologies. In this text we will only indicate a common case of *sampling*, through an example of estimation of an attribute of vegetation, the PB, in a landscape main mosaic along a trait of a fluvial corridor (Fig. 7.6).

First we point out the type of patches resulting from the field survey on a map: A = *mixed shrub grove* and B = *willow grove* patches. Then, a grid of a known measure is overlapped for the evaluation of the surfaces in proportion to the number of the median points of the cells covering the patches of interested e.g. A = 26, B = 39, corresponding to A = 13.5 ha, B = 20.5 ha. A random selection of 12 samples (usually a number between 10 and 30) of small areas (e.g. 0,01 ha) is then allocated in proportion to the surfaces of patches A and B: in practice 5 for A and 7 for B.

Table 7.4. Allometric coefficients for plant biomass evaluation, referred to European vegetation (main species) (from Susmel 1980 modified and integrated). These values can be used only as a reference, because each forest may present allometric variations

Plant species	Foliage/trunk (mass ratio)	Allometric coefficient (A)	(A) for an isolated tree	Specific weight
Abies alba Miller	Ø 20 = 0.167	0.47		
	Ø 50 = 0.152	0.46	0.51	0.44
	Ø 95 = 0.127	0.45		
Picea abies (L.) Karsten subsp. *abies*	Ø 20 = 0.500	0.55		
	Ø50 = 0.467	0.54	0.60	0.44
	Ø95 = 0.445	0.53		
Larix decidua Miller	Ø 50 = 0.450	0.57	0.62	0.60
Pinus sylvestris L.	Ø 50 = 0.300	0.46	0.50	0.53
Fagus sylvatica L.	Ø 20 = 0.267	0.71		
	Ø50 = 0.528	0.84	0.95	0.74
	Ø95 = 1.450	1.35		
Prunus avium L.	Ø 30 = 0.186	0.73	0.80	0.66
	Ø50 = 0.410	0.70		
Robinia pseudacacia L.	Ø 30 = 0.170	0.65	0.75	0.78
Populus alba L.	Ø 50 = 0.450	0.77	0.85	0.50
Alnus glutinosa (L.) Gaertner	Ø 30 = 0.180	0.60	0.70	0.56
Quercus robur L.	Ø 50 = 0.560	0.85	1.05	0.75
Quercus pubescens Willd.	Ø 30 = 0.190	0.70	0.80	0.96
Quercus ilex L.	Ø 30 = 0.250	0.75	0.88	0.90

The field survey gave a PB per each sample area. The total results (sample mean PB: x and variance: s^2) for our samples were: x_A = 0.0948 (m³/100m²) with s^2_A =0.00101 (m³/100m²)² and x_B = 0.0789 with s^2_B = 0.0013, being s^2

$$s^2 = \Sigma^n (x_i - x)^2 / n - 1$$

where: n is the sample size; x_i the value of the attribute (here PB) in the i sample unit; $n-1$ the number of degrees of freedom of the estimation. At this point it was necessary to calculate the sample mean variance (s^2_x):

$$s^2_x = (s^2 / n) (N-n) / N$$

where: N is the statistical population; n the sample size; thus $s^2_x(A)$ the 0.00016, $s^2_x(B)$ the 0.00015. Evaluation of the sampling error (e) became possible:

$$e = t_s \sqrt{s^2_x}$$

where: t_s is the Student test. With a statistical confidence of 90%, $t_s(A)$ was 2.132, $t_s(B)$ was 1.943, $e(A)$ was 0.0269, $e(B)$ was 0.0238 and the percent errors were: 28.4 (A) and 30.2 (B). The estimations were consequently: PB_A = 9.48 ± 2.69 (m³/ha), PB_B = 7.89 ± 2.38 (m³/ha).

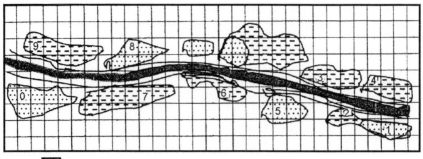

$\boxed{\vdots}$ A= Crataegus monogyna, Acer campestre, Pyrus pyraster, Salix sp.

$\boxed{\vdots}$ B= Salix purpurea, Salix elaeagnos

Fig. 7.6 An example of estimation of plant biomass with a sampling method. Shrub patches along a river bed

7.3.4
Vertical Structure of Vegetation

The evaluation of vegetated tesserae for landscape ecological aims needs to elaborate some survey schedules necessary to integrate the phytosociological data and especially to measure the BTC of the vegetation (see Sect. 8.3). Among the parameters of the schedule, the vertical structure of the vegetation is to be reported; on the other hand, it is a significant measure per se, as evinced by Falinski (1986).

Here we show two examples: Fig. 7.7 for the forest structure and Fig. 7.8 for prairie (also with shrubs, if the case), in order to have a reference.

In Fig. 7.7 there are six layers (numbers are only a reference, h is height):

a1. Dominant trees, first order height: h >30 m
a2. Intermediate trees, second order height: 20 m < h <30 m
a3. Small trees, third order height: 5 m < h <20 m
b. Shrubs and /or shrubby treelets: 0.8-0.9 m < h <5 m
c. Herbs and /or small shrubs: 0.1 m < h <0.8-0.9 m
d. Moss, lichens and very small herbs: h <0.1-0.15 m

In Fig. 7.8 there are four layers:

(D) Shrubs and /or small isolated trees: h > 1.0 m
A   Tall herbs and small scattered shrubs: 0.4 < h < 0.9-1.0 m
B   Small herbs and very small scattered shrubs: 0.1 < h < 0.35-0.40 m
C   Moss, lichens and very small herbs: h < 0.1 m

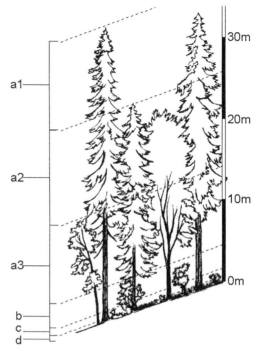

Fig. 7.7 Example of the vertical structure of a mixed forest (boreal and temperate). The six main layers are indicated with *dotted lines*

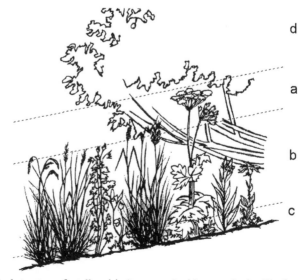

Fig. 7.8 Example of the vertical structure of a tall prairie (temperate) with some shrubs. The four main layers are indicated with *dotted lines*

7.4
Neutral Models for Measuring Landscape

7.4.1
Gravity Model and Transition Matrix

A configuration of patches in a landscape sometimes needs to be evaluated in prevision of a possible transformation. In cases like these, it could be useful to understand the effect of the totality of interactions (unknown process) among the patches or between them in relation to another element which remains unchanged. It is possible to utilise a neutral model, the *gravity model*, which is derived from classical physics and based on gravity between two masses:

$$I_{ij} = K\,(P_i \times P_j)\,/\,d^2$$

where: I_{ij} is the interaction measure; P_i the attributes of the patch i ; d the distance between two patches and K is a constant.

It is necessary to appropriately choose the attributes of the patches and a correct setting of the K index. For instance, K can be calibrated on the interference of barriers and/or possible disturbances in the area.

The *transition matrix* may be used to control whether the transformation of a landscape unit is independent or not. At time T_1, n elements present a certain distribution within their landscape. At time T_2 each element may remain intact or be replaced (at least partially) by another element. It is possible to form a matrix ($n \times n$) in which the rate of change is registered (each row will sum 1). Through a matrix like this we can verify some properties, for instance the Marcov one, i.e., if consecutive transitions are independent. The mentioned rates can be considered as relative frequencies of transition between two periods (T_1-T_2) in a matrix F.

Assuming that F corresponds to transition probabilities, and extending the analysis to a second period (T_2-T_3) the Marcov hypothesis states that new probabilities of transition can be generated calculating the square of the matrix F (row per column): F^2. Comparing F^2 with the real frequencies of the second period registered in the matrix F' it is possible to see if there is accordance or not.

7.4.2
Measure of Connection

The analysis of a landscape network may be quite complex and may range from application of the autocatalytic system criteria of Ulanovicz (see Sect. 2.3.3) to more simple empirical considerations. Useful suggestions come from the application of the topological *theory of graphs* (Forman and Godron 1986; Forman 1995). A graph is a geometrical entity applicable to make explicit the relations among objects in a general way, reducing the elements to points (vertices or nodes, *V*) and their relationships to arcs (linkages, *L*). A graph is important for finding paths among its nodes: a path may have a determined length or be closed, therefore forming a cycle. A graph is connected if its nodes are all linked by one or more paths. A complete graph with *K* nodes has by definition a number of arcs equivalent to:

$$L = [K(K-1)]/2$$

L depends on a binomial coefficient (*K* choices 2): it describes the situation in which all the couples are in relation; the results is that the number of linkages is < K^2. The degree of a node is the number of arcs connected to it. Graphs derivable from maps are called planar graphs. In these graph the arcs do not cross over.

Thus a complete graph is no longer planar for *K* = 5. If a graph accepts the case of arcs departing and returning in the same node (loop) it is called multigraph. A tree is a connected graph without cycles. A simple graph is called Hamiltonian if it has a cycle of Hamilton, that is a closed path crossing all the nodes only once each.

From this theoretical framework (it is much more complex) we can derive two useful indexes: for a network connectivity (γ) and for a network circuitry (α).

$$\gamma = L/3(V-2); \qquad \alpha = (L-V+1)/2V-5$$

The conversion of a landscape unit basic mosaic into a network is needed to identify key gaps and to apply the connectivity and circuitry indexes. In general, it is necessary to put in evidence the vegetation patches and the main species sources as nodes and the existing corridors and sporadic routes as linkages (Fig.7.9).

The connection of the analysed landscape structure is always monitored by comparison with past configurations. Frequently, these data can be used to design a restoration plan for natural conservation purpose or for planning a new suburb. In cases like these, we may observe that the mentioned γ and α indexes do not always have a clear significance, because they are not related to the types of studied landscape.

The proper connection of a landscape represents another classical expression of what Zonneveld (1995) called the main law of ecology: not too much, not too little, just enough. Therefore we need to plot a field in which it is easier to compare the dynamic of the analysed connections with the γ and α indexes. This can be done by observing that in a network the relations between *L* and *V* can be plotted to form a family of curves in the field formed by the γ and α indexes (Fig.7.9). The family of curves becomes interesting when considering the relation *L/V*, because this could be coupled with the main types of landscape, at least for the control of the vegetation network. In fact, both theoretically and practically it is possible to ascertain that a network of an urban landscape may generally have *L/V* = 0.25-0.75 (max 1.0) while an open forest landscape can reach even *L/V* = 2-2.5. In this range it is possible to differentiate suburban, agricultural and recreational park landscapes. Thus, monitoring connection transformations and/or design the restoration plan may acquire a useful tool.

An example is provided by monitoring Monza Park (Lombardy, Italy). At present we can measure a broken network of vegetation: *V* = 50, *L* = 35, γ = 0.24,

$\alpha = -0.13$; ($L/V= 0.7$). Comparing with Fig. 7.9, we can see the absolute insufficiency of the situation, representing a level of urban landscape, not of a recreational park. Thus, after considering possible new patches of trees and new linkages in order to restore the historical park, some scenarios could be designed giving a solution of at least $V = 60$, $L = 69$, ($L/V= 1.15$) as the first step. The second step should overcome $L/V= 1.5$.

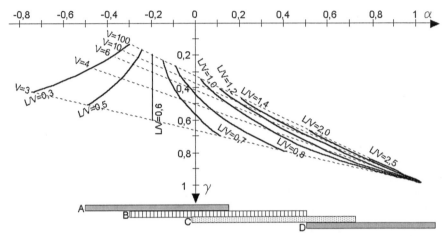

Fig. 7.9 Control of the connections in a landscape: relationship between the γ and α indexes. The *dotted lines* represent the different numbers of nodes, the *curved lines* the values of the ratio L/V. A is the urban landscape, B the suburban-rural landscape, C the cultivated landscape, D the open forested landscape

7.4.3
Fractal Geometry Measures

Many characters of a system may change with scale, but the intrinsic part of them does not. This is in accordance with the description of the landscape given in Chap. 1, as a specific level of life organisation, and in Chap. 2, as a self-organising system. A similar process which links form and function is verified also in geographical aspects, such as the coast line: changing the scale, the line changes, but something similar remains and may be measured.

In mathematics is possible to describe a process like this, applying the concept of homology, in particular the invariant of homothety. This is what Mandelbrot (1980) did in the theory of fractals. A fractal is a geometric Euclidean dimension which is not an integer but fractionary (from Latin *fractus*). A fractal dimension expresses a rule between form and functions.

The concept of homology is part of projective geometry and it is defined as a bi-univocal correspondence among Euclidean elements maintaining its bi-

relations; homothety is an homology with an improper axis. Thus an homology is characterised by a centre, an axis and a pair of corresponding points. In an homothety the relation r between two corresponding segments is constant (homothetic invariant); looking for instance at Fig. 7.10:

$$r = (OR_\infty AA') = (AA'O \, R_\infty) = (AA'O) = AO/A'O$$

In the case of a parallelepiped (Euclidean dimension = 2) the homothetic relation r referred to the N parts in which a figure is related to the others is:

$$r(N) = 1/N^{1/2} \quad \text{from which} \quad log \, r(N) = log \, (1/N^{1/D}) = -(log \, N)/D \quad \text{therefore:}$$

$$D = - \, log \, N/log \, r(N) = log \, N/log \, (1/r)$$

This is the most well known Mandelbrot equation, in which D is the fractal dimension.

Thus D is calculated as a relationship between two logarithms quite easy to be identified. Note that the rule between form and functions may be constant or variable. The fractals can be exact when exhibiting a very regular structure because they are formed on the basis of a simple and constant rule, like snow crystals; or they may be irregular when formed on the basis of variable rules, like a coast line or vegetated patches. For instance, the rule may be a landslide process on a mountain chain, thus the fractals may have also a statistical nature.

The meaning of D may be represented in comparison of Euclidean dimensions. For example: $D = 0$ is the point, therefore $D = 0.55$ is a set of points along a line; $D = 1$ is a line, $D = 1.29$ is a curve with fluctuations at all scales; $D = 2$ is a plane, $D = 2.4$ is a rough surface, like a folded sheet.

Observing self-similar figures, it is possible to generalise the fractal dimension. A segment is a mono-dimensional self-similar figure, composed of two parts measuring one half. A bi-dimensional figure is a square, composed of

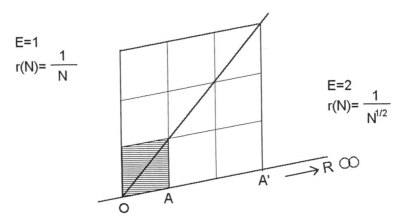

Fig. 7.10 Homothety on a plane, with an improper axis. See text for explanation. (From Ingegnoli 1993)

four parts of one half. A three-dimensional figure is a cube, composed of eight parts of one half. Thus, a self-similar figure of dimension D is formed by n^D parts of magnitude $1/n$. So, the fractal dimension D may be interpreted as a relation between a quantity Q and the scale of magnitude L at which Q is measured: $Q(L) = L^{Dq}$. This is applicable also to irregular fractals. Even the relation area/perimeter of patches (A/P) can be measured with fractal dimensions (Milne 1991), with β, η as constants:

$$A = \beta L^{Da} \quad P = \eta L^{Dp};\ \text{since}\ D = Da/Dp,\ \text{it is possible to write: } log\ A \cong D\ log\ P$$

Voss (1988) presented a fractal as a set of S points in a space of dimension d, for instance $(d=2)$ in a map. The spatial dependence among the points was studied by measuring the probability $P(m,L)$ of m points observed with a window of size L^2 centred on individual points of the set S (e.g. a landscape cover type). The measure of the mass $M^q(L)$ of the S points at a given L is:

$$M^q(L) = \Sigma^N_{m=1}\ m^q\ P(m,L)$$

where q is an index of the moment of the fractal distribution on the map and $N(L)$ is the number of different values of m observed for a given L. This formula has the significance of statistical moments. When $Q = 1$, $Q(L)$ indicates the means of the points found in a window of size L; when $Q = 2$, $Q(L)$ indicates the variance of points.

One of the most important utilisations of fractals in landscape ecology is the evaluation of the irregular dispersion of patches in a territory, measured by counting the number of grid cells occupied by the mosaic, from low to high resolution. Given a certain patch distribution on a map, e.g. forested tesserae in a landscape unit, we have to overlap a grid of an appropriate size able to contain the mosaic. This grid has low resolution and is proportional to the scale.

Each cell measures L per side, and L will be divided hierarchically; for instance $r(N) = 1/2, 1/4, 1/8, 1/16, 1/32, 1/64$. Thus, each grid cell is divided in N parts: 4, 16, 64, 256, 1024, 4096. Stating a value of significance (e.g. > 25%) of the attribute (e.g. vegetation) per cell, we calculate per each passage of scale the number of N parts occupied by the patches.

Then, we have to apply the Mandelbrot equation: $D = log\ N\ /\ log\ (1/r)$. A simple example (Fig. 7.11; Table 7.5) is presented.

Table 7.5 Example of calculation of the fractal dimension D related to Fig. 7.12.

$1/r$	$log\ 1/r$	N'	$log\ N'$	D'
2	0.301	3	0.4771	1.585
4	0.602	11	1.0414	1.729
8	0.903	35	1.544	1.709
16	1.204	115	2.06	1.711
32	1.505	396	2.598	1.726
64	1.806	1604	3.205	1.7774

N' Number of the N parts occupied by the vegetation; D' Fractal dimension per each N'

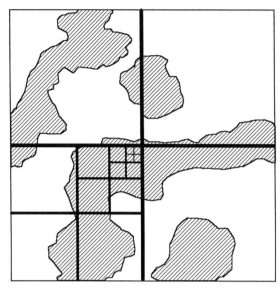

Fig. 7.11 Example of the distribution of vegetated patches and the grid for fractal analysis. The value of significance per cell was 30%

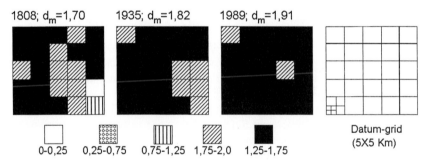

Fig. 7.12 Measures of the fractal dimensions of a grid to survey the character of a landscape unit near the Po river, Italy. The figure shows the tendency toward the homogeneity of this landscape unit, expressed by the change of the fractal value of its forested and wooded tesserae. (From Ingegnoli 1993)

In many cases it could be useful to calculate not the D of the entire distribution of patches, but the fractal dimension of each of the first division cells. This is useful to compare the transformations of the distribution of patches, as shown in Fig. 7.12. The increase of homogeneity of the analysed landscape unit is expressed by the uniformity of the fractal dimensions, from 1808 to 1989.

The meaning of the use of D in ecological systems investigation does not refer to the simple measure per se, but to its variation in time. For instance, in this example (Fig.7.12) we did not have a strong decrease in the wooded patches surface, but a clear variation in their distribution, which led to a decrease of heterogeneity. The linear regression of *D'* leads to the equation: $y = 0.089x + 1.6119$, from which D = 1.701.

7.5
Biological Territorial Capacity Indexes

7.5.1
BTC Evaluation

As discussed in Chap. 4 (Sect. 4.8.1), important dynamic processes linked to the metastability of the landscape can be expressed by the biological territorial capacity or BTC. Remember that this synthetic function, referred to the main vegetational ecosystems, is able to compare the states of a/more landscape/s, because it can measure the relative relationships between *R/GP* and *R/B* (metabolic capacity and anti-thermal maintenance).

As shown in Table 7.6, from the BTC function it is possible to obtain very useful ecological indexes, remembering the equation:

$$BTC_i = 0.89 \; \Omega - 0.0054 \; \Omega^2 \; (Mcal/m^2/year)$$

where $\Omega = (a_i + b_i) \, R_i$ with $a_i = (R/GP)_i / (R/GP)_{max}$ and $b_i = (dS/S)_{min}/(dS/S)_i$; with *R* the respiration, *GP* the gross productivity, $dS/S = R/B$ which is maintenance to structure ratio and *i* the principal plant ecosystems of the ecosphere.

Even if the BTC index have been calculated near their maximum values, obviously referred in the table to B_e (not to B_b), those values may be used only at a scale of synthesis, the calculation being based on synthetic data. Moreover, we need to know the ranges per each type of ecosystem: at a general scale, the main BTC ranges of evaluation for temperate regions are shown in Fig. 7.13.

On the other hand, note that in general we do not need an high precision in the measure of BTC, because many of the important data necessary for landscape analysis are intrinsically not precise (e.g. the historical reconstruction of a landscape) and because of the seasonal variation of vegetation metabolism. The most crucial character of an ecological function like BTC remains a good proportionate evaluation among the landscape components.

Anyway, as we will see in the next chapter, an useful method of estimation of the state of vegetation in a landscape permits also a valid estimation of its biological territorial capacity. Obviously, the wide range of values is concerned with forested vegetation, because forests present clear differences in growing phases: young, adult, mature, senescent.

Depending mainly on these phases, BTC may range from 2 to 12 as in temperate deciduous forests. The other types of vegetation have shorter BTC ranges but are more numerous and may represent local types or human alterations.

Table 7.6. Plant biomass (at ecosphere scale: PB_b, at ecocoenotope scale: PB_e), net primary production (NP), respiration vs gross primary production rate (R/GP), respiration vs biomass rate (R/B, where $B = PB_e$), and calculated BTC values (Mcal/m²/year).

Ecosystem types of vegetation	Plant biomass PB_b	PB_e	NP	R/GP	R/B	BTC
Tropical rain forest	6-80	105	3.5	0.75	0.078	20.5
Tropical seasonal forest	6-60	80	2.5	0.70	0.072	12.8
Dry forest savannah	5-30	38	1.7	0.42	0.032	3.3
Shrub savannah	2-15	18	2.8	0.42	0.111	3.0
Graminoids savannah	1-4	4.5	2.6	0.40	0.377	1.9
Mangrove forest	5-45	55	1.2	0.65	0.040	6.2
Tropical marsh and swamp	2-30	36	4.5	0.39	0.083	4.7
Shrub semi-desert	0.1-4	4.2	0.25	0.45	0.059	0.5
Sand warm desert	0.01-0.2	0.25	0.02	0.45	0.140	0.05
Tropical yearly culture	0.5-2	2.8	1.8	0.41	0.464	1.4
Tropical perennial culture	2-12	15	1.9	0.55	0.190	4.4
Chaparral bush-land	3-15	20	1.4	0.65	0.130	5.0
Mediterranean forest	6-60	85	2.2	0.71	0.063	12.6
Dry temperate grass	0.5-2.2	2.8	0.8	0.40	0.179	0.6
Moist temperate grass	1-3	3.5	1.4	0.40	0.266	1.0
Temperate deciduous forest	6-60	80	2.2	0.69	0.061	11.7
Temperate woodland	6-40	50	1.9	0.69	0.054	8.7
Temperate marsh and swamp	2-15	18	3.3	0.39	0.117	3.0
Temperate yearly culture	0.2-2	3	1.4	0.41	0.333	1.1
Temperate perennial culture	2-10	15	2.3	0.60	0.227	5.5
Suburban green	0.5-5	6	0.8	0.55	0.167	1.6
Urban green	0.1-2	2.5	0.4	0.50	0.160	0.6
Boreal and alpine forest	6-55	73	1.8	0.71	0.060	10.7
Boreal open forest	5-35	40	1.0	0.71	0.062	6.2
Tundra and alpine vegetation	0.5-2.8	3	1.0	0.35	0.167	0.6
Cold desert	0.01-1	1.3	0.1	0.38	0.061	0.15
Peat bog	1-10	12	1.2	0.40	0.067	0.9
Lake and stream	0.01-0.4	0.5	0.1	0.40	0.160	0.1
Estuaries	1-15	18	2.0	0.39	0.072	2.2
Continental shelf	0.1-1.5	2	0.4	0.42	0.140	0.4

NP Net primary production (kg/m²/year) and PB_b (kg/m²/year) normal ranges at ecosphere scale (Golley and Lieth 1972; Golley and Vyas 1975; Whittaker and Likens in Lieth and Whittaker 1975; Kimmins modified in Piussi 1994). PB_e (kg/m²/year) near mature stages at ecocoenotope scale

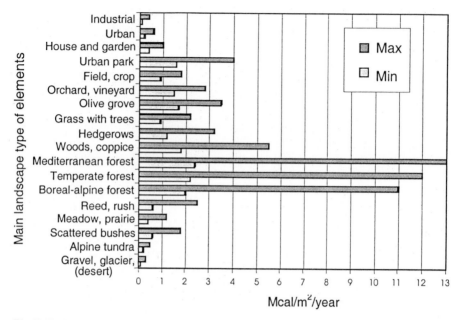

Fig. 7.13 Indicative range of values of the biological territorial capacity (BTC) for temperate regions. (From Ingegnoli 1999a)

7.5.2
Standard BTC Classes and Landscape Types

Therefore, the analysis (and diagnosis) of BTC needs to have a standard subdivision of the BTC values into classes, through which it is possible to compare the studied cases of landscape evaluation. The proposed standard is composed of eight BTC classes (Mcal/m^2/year), because of their good significance:

- Class I: 0 - 0.4 (*mean value 0.2*) Desert, semi-desert, lake and stream, continental shelf, but also degraded grass or tundra, and suburban scrub.
- Class II: 0.4 - 1.2 (*mean value 0.8*) Prairie, tundra, cultivated field, and also urban greenery, degraded shrub, etc.
- Class III: 1.2 – 2.4 (*mean value 1.8*) Bushed grass, reed thicket, low shrub, graminoid savanna, tree-cultivated plantation, orchard and garden, urban greenery.
- Class IV: 2.4 – 4.0 (*mean value 3.2*) Young forest, dry forest savanna, shrub savannah, temperate marsh and swamp, temperate coppice, semi-natural orchard, semi-natural suburban park.

- Class V: 4.0 – 6.4 (*mean value 5.2*) Near-young natural forest, adult forest partially degraded, mangrove forest, tropical marsh and swamp, tropical perennial culture, chaparral bush-land (macchia mediterranea), temperate perennial culture, semi-natural olive-culture, boreal open forest.
- Class VI: 6.4 – 9.6 (*mean value 8.0*) Adult natural forest, mature forest partially degraded, temperate woodland.
- Class VII: 9.6 – 13.2 (*mean value 11.4*) Tropical seasonal forest, tropical rain forest partially degraded, mediterranean mature forest, temperate mature deciduous forest, boreal-alpine mature forest.
- Class VIII: 13.2 – 20.4 (*mean value 16.8*) Tropical mature rain forest.

These standard BTC classes are useful in many landscape analyses and applications: one of the first utilisations concerns the possibility to evaluate standard types of landscape. A theoretical reference of the range of BTC values (and of mean plant biomass value) of the main types of landscape in the temperate regions is shown in Table 7.7. These values can be used in a preliminary synthetic comparison with the surveyed landscape, representing a simple theoretical model also applicable to give a standard sequence of diversity.

The biological territorial capacity is a crucial ecological function deeply concerned with landscape characters, thus its index allow a wide range of analyses, directly or not. We have already mentioned: (a) the transformation deficit, the measure of which is allowed by BTC (Sect. 4.8.2.), (b) a control of landscape structure alteration using the standard BTC classes and land-use data (Sect. 5.2.3).

Chapter 6 described the need for some new perspectives and referred to BTC. These new approaches include the evaluation parameters of vegetation (see Sect. 8.3), a landscape diversity and metastability index (see Sect. 11.2.2), new method of planning dimensioning and control (see Sect. 12.3). Even at the ecosphere scale (see Sect. 10.2.3) the variation of the BTC of the main landscape types can give good results. Other analyses in which the BTC plays an evident role will follow.

Table 7.7. Standard temperate landscape types and their BTC and plant biomass values

Main temperate landscapes	Normal mosaic of standard BTC classes (d = dominant, r = rare)	BTC values mean and variance Mcal/m^2/year		Plant biomass mean and variance kg/m^2	
Semi-desert	I(d) +II	0.30	0.06	0.47	0.10
Grassland	I +II(d) +III(r)	0.79	0.03	1.39	0.07
Reed thicket	I +II +III(d) +IV(r)	1.63	0.09	3.74	0.14
Bush-land	I +II +III(d) +IV(d)	2.17	0.27	5.95	0.98
Closed forest	I +II +III +IV (d) +V +VI(d) +VII	4.80	0.49	25.50	3.53
Open forest	I + II(d) +III +IV +V(d) +VI	2.93	0.20	12.08	1.02
Agricultural	I + II(d) +III +IV +V	1.16	0.27	2.74	1.11
Rural-orchard	I +II(d) +III +IV(d) +V +VI(r)	2.41	0.35	8.68	1.83
Suburban	I(d) +II(d) +III +IV +V(r)	1.00	0.32	2.68	1.20
Open urban	I(d) +II(d) +III +IV	0.90	0.20	2.07	0.61
Closed urban	I(d) +II(d) +III	0.63	0.14	1.13	0.32

7.6
Other Analyses

7.6.1
Influence Fields of Network Elements

Prominent patches or corridors generally acquire the role of attractors and direct many landscape patterns (see Sect. 4.3.1). These functions may be put in evidence by the emergence of *influence fields*: the intensity of this influence varies with the distance from the corridor and it is proportional to the degree of its role. As we know, a vegetated corridor crossing cultivated fields acquires a prominent natural role. The measure of its influence field requires a complex analysis regarding:

- Microclimate influence, where a distance is given over which a variable differs significantly from conditions in the open (Forman 1995)
- Soil erosion and control, depending on prevalent winds, type of soil, porosity of the hedgerow and aridity
- Structure, form, porosity, width and height of the corridor
- Home range of key species using the corridor as their own habitat
- Heavy seeds dispersal of ecologically indicative species
- Number of species in the corridor vs number of field species
- Landscape contrast among the surrounding elements
- The medium biological territorial capacity of the landscape
- The presence of a canal within the corridor
- The presence of a range of local disturbances

At landscape scale it is often impossible to measure this influence field, therefore it is useful to propose a synthetic index. This can be done referring to the potential capacity of ecological re-balancing of the corridor vs the landscape matrix and to its main microclimate influence. The BTC of the corridor (see Sect. 8.3.3) is compared with the medium BTC of the landscape (or even the region) and a balance is made: the BTC deficit of the adjacent fields and roads, such as in Fig. 7.14, results in a measure of distance from the boundary of the corridor. Another distance is given by the microclimate influence, referred to the height of the corridor and its porosity. Thus we measure two factors:

$$Lb = [BTC(+) \times Wc/2] / BTC(-); \qquad Lm = 6 H$$

where Lb is the length of BTC balance, Wc the width of the corridor, Lm the length of the microclimate's normal influence, H the height of the corridor trees, $BTC(+)$ the BTC exceeding the regional or landscape media, $BTC(-)$ the BTC deficit of the adjacent landscape elements. Then we consider the mean of the two factors multiplied by the eventual presence of singular elements of disturbances crossing the matrix near the corridor (field roads, power-lines, small buildings):

$$CISF = \frac{1}{2} \, (Lb + Lm) \times 0.8 \, 1/d \quad (m)$$

CISF is the corridor synthetic influence field, *d* the number of singular disturbances (see Fig. 7.14). This ecological index is particularly useful in the control of complex networks, in which are present both natural (or semi-natural) and human (technical) corridors.

Note that the concept of influence field of the corridors can be used also for the technical elements. In these cases it is necessary to refer to other specific measures. For instance, from 2 to 4 μT (micro Tesla = 0.796 A/m) for electrical power lines of high tension (e.g. 300 kV), corresponding to a distance of about 25-100 m from the line. For the roads, the influence field it is more variable, from 40 to 320 m (see also Chap. 12).

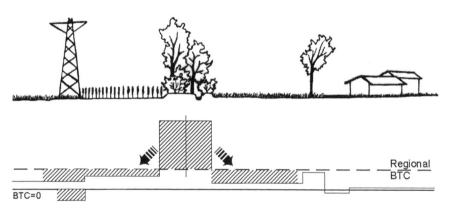

Fig. 7.14 Example of the dimensioning of the influence field of a corridor

7.6.2
River Functionality Index

The importance of the fresh water network in landscape ecology was already emphasised in relation to the excretory apparatus (see Sect. 3.4.1; 4.2.2). The analyses of the components, the rivers, have long been considered to be a competence of limnology, which is only interested in the water body and the riverbed: however, in limnology none of the principle of landscape ecology has been taken into consideration. But about ten years ago a few scientists noted the necessity to study the river functionality in a more complete way, considering the entire river corridor: the vegetation conditions of the riparian belt and the landscape characters in which the river flows, the morphological structure of the riverbed and the biological condition of the macro-benthos.

From the studies of Robert C. Petersen of the Limnologic Institute of Lund, Sweden (1991), continued by Siligardi and Maiolini (1993) of the Istituto Superiore di Ricerche Agrarie of San Michele all'Adige (Trento), Italy, an ecological index of river functionality, the *Indice di funzionalità fluviale* (IFF), was recently presented (Table 7.8).

Table 7.8. Survey schedule for the estimation of the River functionality index (IFF) (from ANPA 2000, modified).

Characters of the river corridor	LB	RB
River:		
Site:		
Tract:		
Altitude:		
1. Main composition of the surrounding landscape		
a. Forests and woods ecotopes	25	25
b. Meadows and pastures	20	20
c. Cultivated ecotopes and/or scattered urbanisation	5	5
d. Urbanised ecotopes and industrial suburban areas	1	1
2a. Vegetation of the natural peri-fluvial belt		
a. Arboreal riparian plant formations (dominant)	30	30
b. Shrubby riparian plant formations and/or reed thickets	25	25
c. Non-riparian arboreal formations	10	10
d. Non-riparian shrubby or herbaceous plant formations (or absent)	1	1
2 b. Vegetation within the artificial river-banks		
a. Arboreal riparian plant formations (dominant)	20	20
b. Shrubby riparian plant formations and/or reeds thickets	15	15
c. Non riparian arboreal formations	5	5
d. Non riparian shrubby or herbaceous plant formations (or absent)	1	1
3. Breadth of the vegetation peri-fluvial belt (trees and/or shrubs)		
a. Vegetation belt > 30 m	20	20
b. Vegetation belt 5 – 30 m	15	15
c. Vegetation belt 1 – 5 m	5	5
d. Absence of vegetation belt	1	1
4. Continuity of the vegetation peri-fluvial belt (trees and/or shrubs)		
a. Without interruptions	20	20
b. With interruptions	10	10
c. Frequent interruptions or only herbaceous	5	5
d. Bare soil or rare herbaceous	1	1
5. Water conditions of the riverbed		
a. Moderate flow width < triple of the wetted riverbed	20	
b. Moderate flow width > triple of the wetted riverbed (seasonal)	15	
c. Moderate flow width > triple of the wetted riverbed (frequent)	5	
d. Wetted riverbed very reduced or absent (or impermeablised)	1	
6. Riparian morphology		
a. Presence of arboreal vegetation and/or rocks	25	25
b. Presence of shrubs and herbs	15	15
c. With thin herbaceous strate	5	5
d. Bare shores	1	1

7. *Trophic input retention structures*		
a. Riverbed with rocks and/or embanked trunks or reeds and hydrophytes	25	
b. Rocks or branches with sediment or reeds and hydrophytes	15	
c. Retention structures dependent only on floods (or reeds absent)	5	
d. Sand sediments without algae	1	
8. *Erosions*		
a. Non-evident and non-relevant	20	20
b. Only in curves and narrows	15	15
c. Frequent, with riparian excavation	5	5
d. Very evident, with excavations and landslides	1	1
9. *Transversal section*		
a. Natural	15	
b. Semi-natural (few artefacts)	10	
c. Semi-artificial (few natural remnants)	5	
d. Artificial	1	
10.*Structure of the bottom of the riverbed*		
a. Stable and diversified	25	
b. Partially movable	15	
c. Easily movable	5	
d. Artificial or with concrete	1	
11.*Scrapings, puddles or meanders*		
a. Clearly distinguished and recurrent	25	
b. With irregular succession	20	
c. Few meanders, or long puddles and short scrapings	5	
d. Absence of meanders, scrapings and puddles (canalised)	1	
12a. *Lotic waters vegetation*		
a. Very scarce periphyton and macrophytes	15	
b. Some presence of periphyton and macrophytes	10	
c. Discrete periphyton and high presence of macrophytes	5	
d. High presence of periphyton and macrophytes	1	
12b. *Lentic waters vegetation*		
a. Very scarce periphyton and tolerant macrophytes	15	
b. Some presence of periphyton and tolerant macrophytes	10	
c. Discrete periphyton and high presence of tolerant macrophytes	5	
d. High presence of periphyton and tolerant macrophytes	1	
13. *Debris*		
a. Vegetal fragments (recognisable and fibrous)	15	
b. Vegetal fragments (fibrous and pulpous)	10	
c. Pulpous fragments	5	
d. Anaerobic debris	1	
14. *Macrobenthonic community*		
a. Well structured and diversified, in accordance with the river type	20	
b. Quite diversified but with altered structure	10	
c. Badly balanced community with pollution tolerant taxa	5	
d. Absence of community, few pollution tolerant taxa	1	

LB left bank of the river, *RB* right bank

Note that two categories (2 and 12) present two alternative versions: only one of these should be considered. Moreover, the distinction between lotic and lentic water flow is done looking at the water surface: if it is rippled or flat. The score values of the river functionality index (IFF) can be ordered in five levels of functionality (Table 7.9).

Table 7.9. Evaluation and levels of functionality of the IFF

RFI values	Functionality levels	Sense of functionality	Colour
261 – 300	I	High	Blue
251 – 260	I – II	High – good	Blue – Green
201 – 250	II	Good	Green
181 – 200	II – III	Good – mediocre	Green – Yellow
121 – 180	III	Mediocre	Yellow
101 – 120	III – IV	Mediocre – low	Yellow – Orange
61 – 100	IV	Low	Orange
51 – 60	IV – V	Low – bad	Orange – Red
14 - 50	V	Bad	Red

7.7
Summary of the Most Important Indexes

Table 7.10 reports the most important landscape ecological indexes presented in this text and the relative chapter of references.

Table 7.10. The most important ecological indexes related to landscape ecology.

Index	Description	Author/s	Location in the book
R	Relative richness	Turner 1989	Sect. 7.2.1
H	Diversity	Turner 1989	Sect. 7.2.1
η	Specific landscape diversity	Brandmayr 1990	Sect. 11.2.1
τ	Landscape diversity	Ingegnoli this text	Sect. 11.2.2
C	Contagion	Godron 1968; Turner 1989	Sect. 7.2.1
D	Dominance	Turner 1989	Sect. 7.2.1
E	Relative evennes	Forman 1995	Sect. 7.2.1
BTC	Biological territorial capacity	Ingegnoli 1980; 1991; 1993	Sect. 7.5.1
LM; LM*	Landscape metastability	Ingegnoli this text	Sect. 11.2.2
γ	Connectivity	Forman and Godron 1986	Sect. 7.4.2
α	Circuitry	Forman and Godron 1986	Sect. 7.4.2
CISF	Corridor synthetic influence field	Ingegnoli this text	Sect. 7.6.1
IFF	Indice di funzionalità fluviale	Siligardi and Maiolini 1993	Sect. 7.5.2
BFF	Biotopflaechenfaktor	Ermer et al. 1996	Sect. 7.2.2
SH; SH*	Standard habitat	Ingegnoli 1980; 1993	Sect. 8.5

8 Landscape Components Evaluation

8.1
Climatic and Geomorphologic Aspects

8.1.1
Climate Evaluation

The analysis of the landscape and its components represents the fundamental phase of the study of its ecological state, but a subsequent elaboration of data is always necessary for evaluation and diagnostic purposes. A complete treatment of the evaluation of all the components would be too onerous for this book: so, only some aspects will be studied more in-depth, noting for the others only the prevailing indications.

The present and past climatic conditions hold great importance in helping us to understand the history and the evolution of a certain landscape, in relation both to geomorphic processes and to the presence of man (man is one of the factors influencing the climate, the climate is one of the factors influencing the presence and the history of man).

The scientific concept of climate was born when man began to acquire experience of the conditions in which he was living and of the differences compared to other conditions: in ancient Greek the word κλιμα (-τοσ, το coming from κλινω= to tilt, to incline) was used to indicate the inclination of the Earth toward the pole too, so shifting from the original meaning of region, geographic zone to latitude and then to climate.

There are many classifications of the climates of the Earth, each one related to a specific purpose. The words macro-, meso-, topo- and microclimate themselves present different spatial limits. Landscape ecology is more interested in bio-climatic studies, linking the climate with living systems. Among many authors, we mention Rivas-Martinez (1995), whose five bio-climatic macro-regions of the Earth are: tropical, mediterranean, temperate, boreal, polar. Each of these regions is articulated in various bio-climates, characterised by vegetational formations and biocoenosis. For instance, his bio-climatic map of Europe shows a very high variability of climates, the highest in the world. Four of the macro-regional climates are present on a relatively small continent. The influence of the North-

Atlantic current only in Europe reaches up to the polar circle, and the mountain chains are disposed in a way that does not impede the Atlantic warm and humid winds that flow. These effects produce a difference between western and eastern Europe, that results in a yearly thermal excursion of 10-12° C vs 20-25° C. Many depressions arrive during the year, thus the climate is variable, especially in western Europe and in winter.

No doubt that regional climates provide only a general indication for landscape ecological analysis. Many sub-climates have been classified, even by Rivas-Martinez, but the true interest of our discipline is linked especially with local climate.

The evaluation of local climates is possible by combining a simple analysis of monthly and seasonal temperature and/or rainfall data (the last one being particularly important in hilly and mountainous landscapes) with specific climatic indexes, mainly based on temperature and precipitation. Note that each index has a specific range and its own ecological meaning, given by its inventor, away from which it is no more valid. We will limit this synthetic section only to some indexes, one group related to aridity and humidity, the other only to temperature.

Defining: P as the annual rainfall (mm), T as the annual mean temperature (°C), M as the mean maximum temperature (°C) of the warmest month, m as the mean minimum temperature (°C) of the coldest month, we will find:

a. *Rain-thermal quotient (Q)* $= [P / (M^2-m^2)] \times 100$ (Emberger). The values of Q are low toward aridity, high toward humidity: e.g. 15 = Baghdad, 48 = Madrid, 85 = Rome. The value of m can be useful to classify the climate: < (-3) °C very cold;   (-3)–(0) °C cold;   (0)–(+3) °C cool;   (+3)–(+7) °C temperate; > 7 °C warm.

b. *Evapotranspiration index (ETP)* $= 0.1645 [P / (T+12.2)]^{10/9}$ (Thornthwaite). This allows the evaluation of a *HI* (humidity index)= $100s/n$ and of an *AI* (aridity index)= $100d/n$, where: s is water excess; d is water deficit; n is ETP.

c. *Aridity index (i)*$= P / (T+ 10)$ (De Martonne). The simple relation of Lang (P/T) was modified. The ecological values of i are: 5 = desert vegetation; 5-10 = steppe; 10-20 = prairie; >20 = forest.

d. *Ombrothermic index Io $= Pp / Tp$* (Rivas-Martinez), where: Pp is the sum of the mean precipitation (mm) of the months with mean temperature > 0 °C; Tp the sum of the mean monthly temperature >0 °C. Proposed to determine the climate region, the Io establishes that: 0.1 < Io < 1.5 = Mediterranean climate; 2< Io <3.8 = temperate climate. For the transition zones between temperate and Mediterranean situations (1.5< Io <2) a compensating *Iovc* was added.

e. *Monthly drought stress (MDS)*$= 2 (50-p)$ (Mitrakos), with p definied as monthly rainfall (mm). MDS defines intensity and duration of monthly aridity, 0 being the minimum value corresponding to aridity absence. Rivas-Martinez added SDS (summer drought stress) as the sum: June MDS + July MDS + August MDS.

A. *Warmth index (WI)* = $\Sigma\,(t - 5)$ (Kira). It expresses the total quantity of warmth available for plant growth, *t* being the mean monthly temperature.

B. *Thermic index (It)* = $(T + m + M) \times 10$ (Rivas-Martinez). This puts the Emberger index in relation to the annual temperature. It is used to determine the bio-climatic belts and horizons and underlines the importance of the minimum temperature of the coldest month as a constraint for vegetation.

C. *Continentality index (Ic)* = *Tmax* − *Tmin* (Rivas-Martinez), where: *Tmax* is the mean temperature of the warmest month; *Tmin* is the mean temperature of the coldest month (°C); 0<Ic<21= oceanic climate, 21 < Ic < 65 = continental.

D. *Monthly cold stress (MCS)* = *8 (10-t)* (Mitrakos) with *t* being the mean minimum monthly temperature (°C). It expresses the intensity and duration of monthly cold stress, 0 being the minimum value corresponding to cold absence. Rivas-Martinez added WCS (winter cold stress) as the sum of December MCS + January MCS + February MCS.

Other important indexes, especially as limiting factor detectors, are: the snow cover durability (Fig. 8.1), the distribution of frost days, the length of vegetation period, the duration of sunshine, the number of summer days.

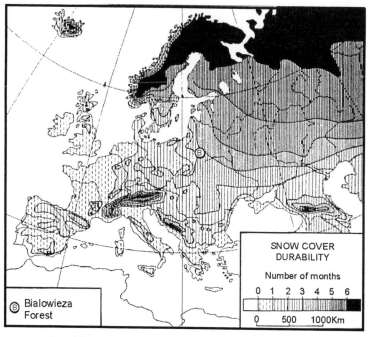

Fig. 8.1 The snow cover durability according to Falinski (1994a)

All the characters of wind, the frequency, length and seasonal nature of storms and of high intensity precipitation can be considered too.

For landscape studies it can be useful to plot graphs of the local climatic control, applying the described indexes. The concept of topoclimate is also very important: it is what we need to monitor.

8.1.2
Geomorphologic Evaluation

The situation is similar to the preceding one. Here, only some synthetic notes are reported, just to underline the importance of geomorphologic evaluation of the landscape. There are two main arguments, which often need an expert consultant: assessment of geomorphic danger and soil evaluation.

As regards the geomorphologic aspects useful in the study of a landscape, a first step could be a synthetic morphology map (Fig. 8.2) as proposed by Romani (1988). It consists of a codified survey of the main emergencies, like slope inclination classes, ridges, terraces and characteristic ambits or large visual references too.

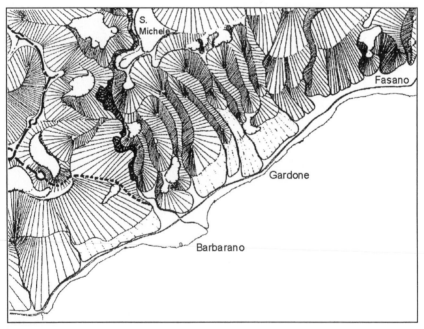

Fig. 8.2 Morphologic map of synthesis of a landscape (from Romani 1988). This area is from Lake Garda, near Brescia (North Italy)

But, at the base of all the information, the map of geomorphologic processes on the landscape remains; the data to be represented on this map are:

- *Topographic data*: essentially the contour lines and altitudes
- *Lithologic data*: divided in surface formations (autochthonous and allocthonous) and geological formations (intrusive, effusive, metamorphic, calcareous, arenaceous, etc.)
- *Structural data*: horizon attitude, faults, fold axis, etc.
- *Morphogenetic data*: types of processes
- *Endogenous processes*: volcanic forms
- *Littoral processes*: the actions of the sea and of the lakes, such as on the seashore, etc.
- *Fluvial processes*: erosions, meanders, etc.
- *Karstic processes*: dolines, etc.
- *Aeolian processes*: dunes, basins, etc.
- *Glacial processes*: moraines, cirques, etc.
- *Cryo-nival processes*: loess accumulation, ice-flow, etc.
- *Slope processes*: ravines, landslides, etc.
- *Anthropic processes*: artificial basins, relief modifications, roads, etc.
- *Polygenetic forms*: depositions and forms generated through the concourse of many processes
- *Structural forms*: fault scarps, selective erosion steps, etc.

The evaluation of the geomorphologic danger is obviously a very important component of the landscape diagnosis. To elaborate a complex map of this argument, we have to apply a method available to analyse both causes and effects.

The analysis of the causes of geomorphologic dangers considers all the causes of instability of natural and anthropic sets. For instance geologic, morphologic, hydro-geologic, climatologic, etc. These data are ordered in classes and elaborated as a function of their relative importance to produce an "integrated analysis map" through two phases (Panizza 1988):

1. The assemblage of the parameters of each thematic map in four main geomorphologic processes (e.g. bank erosion, washing away, plastic deformations, ravines and drains); the surfaces of maximum (M) and minimum (m) potential instability have to be underlined.
2. For each geomorphologic process, the location of the potential instability areas only after having considered the parameters predisposing the instability (geo-lythologic, hydro-geologic, climatologic, vegetational, by slope inclination).

The areas of M potential instability, considering the four geomorphologic processes, form the final map.

The analysis of the effects is based mainly on the geomorphologic map, from which a current geomorphologic dynamics map will be derived (e.g. the erosion and accumulation forms and the recurrence of their instabilities). To elaborate the map of the geomorphologic dynamics, after having compiled the geomorphologic map, it is necessary to consider five genetic groups of forms connected to action: (a) surface current waters, (b) gravity, (c) ice and snow, (d) wave movement and coastal currents, (e) wind.

Table 8.1. Soil fertility classes

Class	Description	Depth (cm)	Skeleton (%)
I	Soils with almost no limitations, good for many types of cultivation. Deep, well drained soil, few slopes. To maintain their fertility they need traditional culture practices: manure, mineral fertiliser, etc.	> 100	< 15
II	Soils with a few limitations, which reduce the choice of cultivation and/or need light conservation practices. The types of limitations may be: slope, moderate potential erosion, non-optimal structure, traces of salinity, some problems of draining, topoclimatic adversity.	80-100	Superior 15-35 Deep 35-70
III	Soils with serious limitations reducing cultivation types or agricultural practices and/or that need specific conservation practices. Possible limitations: more slope, receptivity to erosion, flooding, moderate salinity, more topo-climatic adversity.	50-80	35-70
IV	Soils with strong limitations, available only for a very few types of cultivation. Conservation could be very difficult. Possible limitations: too much slope, erosion, salinity, frequent flooding, bad drainage, topo-climatic adversity.	25-50	Superior 35-70 Deep > 70

Soil evaluation may be a very difficult and specialised subject. Remember what was synthesised in Chap. 4: soil dynamics are quite complex. Anyway, we suggest that all landscape ecologists acquire at least the capacity to discern the main soil horizons and the four principal soil fertility classes (Table 8.1). As we will see, the main classes of land capability will be utilised in the vegetation schedules regarding field evaluation in agricultural landscapes (see Sect. 8.2.3).

8.2
Evaluation of the Vegetation Components

8.2.1
Methodological Criteria

In Chap. 6 we explained the necessity for integrative approaches (see Sect. 6.2.3) in studying vegetation. In fact, for landscape ecological principles it is not acceptable to reduce all the information regarding vegetation to the phytosociological criteria. It must be clear that we appreciate the discipline of phytosociology but, as shown in Sect. 6.2, this is not enough.

One of the useful forms in which vegetational characters can be related to landscape ecology is through a survey schedule for the evaluation of a vegetated tessera (Ingegnoli 1999a).

The schedule has been designed to check the organisation level and to estimate the metastability of a tessera considering both general ecological and landscape ecological characters:

- A = Landscape element characters (e.g. tessera, corridor)
- B = Plant biomass above ground
- C = Ecocoenotope parameters
- D = Relation among the elements and their landscape parameters

The parameters for each A,B,C,D group range from 2 to 12, therefore reaching a value of about 26-32 (see Table 8.3). There are four evaluation classes, the weights per class depending on an evaluation model designed as shown later on.

Remembering the well known relationships among gross productivity, net productivity and respiration in vegetation ecosystems (Odum 1971; Duvigneaud 1977), the development of a vegetation community may be synthesised in:

1. The growing phases from young-adult to maturity, expressed by an exponential process
2. The growing phase from maturity toward old age, expressed by a logarithmic process

As expressed in Sect. 7.5.1, the reference values of BTC generally have to be considered near the threshold of maturity of the main vegetation ecosystems. Following this consideration, it is possible to design a sufficiently credible model and to calculate, for each one of the main types of vegetation ecosystems, an exponential-logarithmic curve of development having an adapted temporal dimension.

Each curve presents in the transition phase (1-2) its own BTC values, defined after monitoring through the field study, of critical points referred, for instance, to plant biomass relations and structural and ecological parameters.

The behaviours of these curves have been subdivided into four intervals of the same breadth, corresponding to four evaluation classes. Thus the derived values are the weights (scores) to be coupled to the A, B, C, D ranks of parameters, which represent the self-organisation level and the metabolic potentiality related to a system of ecosystem.

This research is at present extended to temperate and boreal forests, shrublands, green-lands, vegetated corridors, agricultural fields and gardens and urban arboreous green. The study of schlerophyll forests and reed thickets is on going.

The evaluation schedule for forested tesserae is presented in Table 8.2.

Table 8.2. Schedule for the evaluation of boreal and temperate forested tesserae, applicable also for the estimation of their biological territorial capacity (BTC)

	1	5	12	22	score
Temperate forest	1	5	12	22	score
Boreal forest	1	5	14	25	score
A. TESSERA (*Ts*) CHARACTERS					
A1- Vegetation height (m)	< 9	9.1-18	18.1-29	> 29.1	canopy trees
A2- Canopy cover (%)	< 30	> 90	31-60	61-90	*Ts* surface
A3- Management	simple coppice	complex coppice	wood	natural forest	or similar
A4- Permanence (years)	< 80	81-160	161-320	> 321	*Ts* real age
A5- Structural differentiation	low	medium	good	high	age, groups
B. VEGETATIONAL BIOMASS (ABOVE GROUND)					
B1- Dead plant biomass in *Ts*	near 0	> 10	2-6	6-10	% of pB
B2- Litter depth of the *Ts*	near 0	< 1.5	1.6-3.5	>3.5	cm
B3b - pB volume (m^3/ha)	< 200	201-500	501-1000	> 1000	boreal forest
B3t - pB volume (m^3/ha)	< 150	150-400	400-650	> 650	temperate f.
C. ECOCOENOTOPE PARAMETERS					
C1- Dominant species	> 3	3	2	1	pB of trees
C2- Key species presence (%)	< 5	6-40	41-80	> 80	botanical
C3- Diversity	< 15	16-30	31-40	> 40	n° sp./*Ts*
C4- Plant forms (n°)	< 3	4-5	6-7	> 7	Table 7.3
C5- Dynamic state	degrading	recreation	regener.	fluctuat.	Sect. 5.1.3
C6- Vertical stratification	2	3	4	>4	Fig. 7.7
C7- Renewal capacity	none	intense	sporadic	normal	dominant sp.
C8- Allochthonous sp. (%)	> 10	10-4	< 4	0	not regional
C9- Threatened plants	evident	suspect	risk	0	even acid r.
C10- Infesting plants	near all	> 25	< 25	none	% of cover
D. LANDSCAPE UNIT (*LU*) PARAMETERS					
D1- Boundary connections	0	< 25	26-75	> 76	% perimeter
D2- Interior ratio (%)	absent	< 30	31-89	> 90	vs ecotope
D3- Source (vs surroundings)	sink	neutral	partial	effective	
D4- Role in the landscape unit	reduced	minor	evident	important	
D5- Disturbance incorporation	none	scarce	normal	high	local disturb.
D6- Geo-physical instability	evident	partial	risk	none	physiotope
D7- Permeant fauna interest	none	medium	near good	attractive	*Ts*/key sp.
D8- Transformation reason of the *Ts* as landscape element	strong disturban.	gradual change	tempor. instability	fluctua-tion	*LU* trend Sect. 5.1.4
D9- Landscape pathology interference	extremely serious	near chronic	easy to recover	none	coming from surroundings
D10- Permanence (years)	< 100	100-300	300-1200	> 1200	age of *LU*
E. RESULTS OF THE SURVEY					
E1- Total score Y (= a+b+c+d)	a=	b=	c=	d=	Y=
E2- Quality of the Ts			Q= Y/ 616 (700)		

A1, height weighted average; *A5*, e.g. not coeval plants, groups of the same species presence; *B1-B3*, *pB* is plant biomass volume: see Sect. 7.3.3; *C1*, tree/s species the *pB* volume of which clearly exceeds the equitability (eq) of the total canopy *pB*. If none exceeds, sign the first column (a); *C2*, presence related to phytosociological association/s or to phytocoenosis of reference; *C8*, not autochthonous of the ecoregion species presence; *C9*, present situation; *C10*, even key species (not dominant) but with a too massive cover; *D1*, with analogous vegetation *Ts*; *D2*, surface % of *Ts* with interior vs edge characters: edge= 1.5 h; *D3*, source-sink theory; *D4*, functional role of the tessera in the ecotope or in the entire landscape unit; *D6*, physics or hydrogeologic or climatic instabilities; *D10*, years of permanence of the ecotope as forest

8.2.2
Forest Evaluation

The model for normalised temperate and boreal forest development is shown in Fig. 8.3: the reference values of BTC are, respectively, 12 and 11 Mcal/m²/year (see Sect. 7.5.1).

The development time considered in the model is 120–150 years, expandable for another 50-60%. Thus the equations resulting, for temperate (*tp*) forest:

$$BTC\ (tp_1) = t^{0.52} - 0.8 \quad + \quad BTC\ (tp_2) = 2.45\ ln\ t$$

where: $BTC\ (tp1) = 0–12$ (i.e. young-adult phase), $BTC\ (tp2) = 12–13$ (i.e. mature phase); for boreal (br) forest:

$$BTC\ (br_1) = t^{0.50} - 0.7 \quad + \quad BTC\ (br_2) = 2.25\ ln\ t$$

where: $BTC\ (br_1) = 0–11$ (i.e. young-adult phase), $BTC\ (br_2) = 11–12$ (i.e. mature phase).

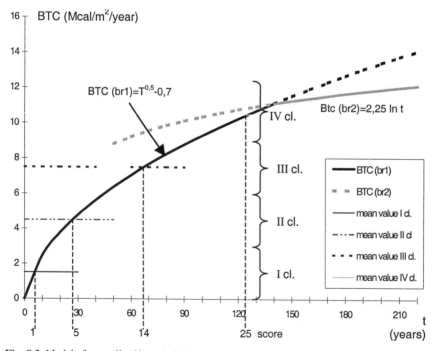

Fig. 8.3 Model of normalised boreal-alpine forest development

Dividing the relative maximum of BTC into the four evaluation classes we obtain respectively (mean values):

Temperate forest: 1.6, 4.9, 8.1, 11.4, with an interval of 9.8 Mcal/m^2/year
Boreal forest: 1.5, 4.5, 7.5, 10.5, with an interval of 9.0 Mcal/m^2/year

From the precedent equations the four evaluation classes lead to the scores pointed out at the top of the schedule (see Fig. 8.3). The possible value of the 28 parameters of the schedule varies from 28 to 616 and from 28 to 700. Therefore, the valuation equations and their parameters result in:

Temperate forest $BTC\ (tp) = 0.01667\ (y - 28) + 0.15\ (pB / 65)$ Mcal/m^2/year
Boreal forest $BTC\ (br) = 0.01339\ (y - 28) + 0.10\ (pB / 75)$ Mcal/m^2/year

where the angular coefficient is the ratio BTC interval/score interval (e.g. temperate: $9.8/(616-28) = 0.01667$), and the second part of the equations depends on plant biomass volume (pB) related to a minimum standard value (65) in m^3/ha.

For example, consider a tessera of boreal forest (*Homogino-Piceetum* Zukrigl 1973 or subalpine spruce) of about 1.5 ha and a mean height of 33.4 m in the Paneveggio Natural Park (Trentino, 1,450 m above sea level). Given a schedule score of 493 (quality = 70.04%) and a plant biomass volume of 1,361.3 m^3/ha , the application of the equation gives: $BTC = 6.23 + 1.81 = 8.04$ Mcal/ m^2/year. We should have expected more, because of the high pB of the tessera and the protection of the park: but the forest management was not completely natural (quality of the tessera not very good).

Note that following this model, it is possible to establish a comparison between the surveyed data and an optimal theoretical condition per tessera. For this purpose the pB has to be measured in kg/m^2; this passage can be synthesised:

$$pB\ (kg/m^2) = pB(m^3/ha) \times 0.095 \times Sp\ W$$

where SpW is the specific weight and 0.095 the reduction coefficient.
The mentioned condition can be calculated as a diagnostic relationship:

$$DR\ (tp) = 1 - (BTC_S / pB^{0.56} - 0.2)$$
$$DR\ (br) = 1 - (BTC_S / pB^{0.55} - 0.1)$$

where BTC_S is the surveyed value: thus the theoretical BTC is calculated on the surveyed plant biomass pB. An available tolerance interval has been estimated on the basis of field observations and the mean and variance on 20 samples and may be ± 12% for temperate forests and ± 10% for boreal forests.

Continuing with our example, the theoretical BTC of a forest tessera of good natural quality presenting a $pB = 1,361$ m^3/ha = 56.9 kg/m^2 should have been 9.13 Mcal/m^2/year, not 8.04. Consequently, its $DR = 11.9\%$ (> 10%): this forest tessera is out of the possible tolerance range, confirming our supposition but giving it an actual measure.

Figure 8.4 represents a more complete example of the application of this method of evaluation of the vegetation following landscape ecology principles.

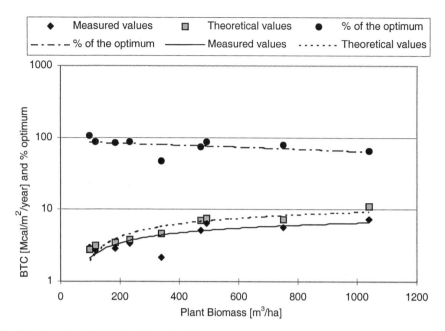

Fig. 8.4 Samples of the biological territorial capacity of boreal forest tesserae vs plant biomass above ground, in Trentino-South Tyrol. Logarithmic regression curves have been plotted both for the real samples and for theoretical ones. The straight line of regression (*above*) represents the differences with theoretical values

8.2.3
Other Vegetation Types

Following models similar to the one set out in the precedent section, a series of schedules have been designed, one for each vegetation type: shrub-lands, tree corridors, agricultural fields, gardens and urban gardens, green-lands.

Therefore, considering the data of Table 8.3, the BTC evaluation equations result, respectively, in:

- *Shrubs,* e.g. willow groves, juniper bush, alder groves, thorn thickets, mountain pine shrubs, etc. (Table 8.4): angular coefficient $3.2 / (928\text{-}29) = 0.00356$;

$$BTC\,(sh) = 0.00356\,(y - 29) + 0.1\,(pB\,/\,17)\ (\text{Mcal/m}^2/\text{year})$$

- *Vegetated corridors,* e.g. hedgerows, remnant wooded corridors, fences, boulevards, etc. (Table 8.5): angular coefficient $7.6 / (1023\text{-}31) = 0.00766$;

$$BTC\,(vc) = 0.00766\,(y - 31) + 0.1\,(pB\,/\,75)\ (\text{Mcal/m}^2/\text{year})$$

- *Agricultural fields,* e.g. cornfields and fields of vegetables, sunflowers, cotton, etc. (Table 8.6): angular coefficient $1.5 / (806\text{-}26) = 0.00192$;

$$BTC\ (af) = 0.00192\ (y - 26) + 0.09\ B3 \quad (\text{Mcal/m}^2/\text{year})$$

- *Urban gardens,* e.g. urban parks, home gardens, sportfield greenery, etc. (Table 8.7): angular coefficient $6.0 / (1170-30) = 0.00526$;

$$BTC\ (ug) = 0.00526\ (y - 30) + 0.1\ (pB / 45) \quad (\text{Mcal/m}^2/\text{year})$$

- *Prairies* as natural and semi-natural grasses (Table 8.8); pB measured in kg/m^2; angular coefficient $1.2 / (928-29) = 0.001335$;

$$BTC\ (pr) = 0.001335\ (y - 29) + 0.02\ (pB / 0.14) \quad (\text{Mcal/m}^2/\text{year})$$

- *Reeds* as natural and semi-natural reed thickets and tall sedges (Table 8.8); angular coefficient $2.4 / (1073-29) = 0.0023$; pB measured in kg/m^2;

$$BTC\ (rd) = 0.0023\ (y-29) + 0.0\ 4\ (pB / 0.3)\ (\text{Mcal/m}^2/\text{year})$$

Table 8.3 summarises the characterisations of the mentioned models.

Table 8.3. Main characterisation of the models linked to the schedules of evaluation of vegetated tesserae

Vegetation types	Reference value of BTC	Time of development (years)	BTC interval	Evaluation classes				Parameters of the schedule
Temperate forest	12.0	120-150	9.8	1	5	12	22	28
Boreal forest	11.0	120-150	9.0	1	5	14	25	28
Shrubs	4.0	30-40	3.2	1	6	16	32	29
Prairie and pasture	1.4	20-24	1.2	1	4	13	32	29
Reed thicket	2.8	36-48	2.4	1	6	17	37	29
Vegetated corridors	9.5	90-130	7.6	1	7	17	33	31
Agricultural fields	2.0	10-20	1.5	1	5	15	31	26
Urban gardens	8.0	70-110	6.0	1	7	20	39	30
Hydrophyte prairies	1.8	15-20	1.7	-	-	-	-	-
Chaparral	5.0	35-45	4.0	-	-	-	-	-
Schlerophyll forest	13.0	120-150	10.5	-	-	-	-	-
Bushed savanna	3.5	15-30	3.0	-	-	-	-	-
Tropical rain forest	17.5	130-180	15.0	-	-	-	-	-

Note: Research on the five last types of vegetation have not been started yet

The last four types of vegetation have been presented only to show the possibility of extending research like this to all the main vegetation types usually considered in the ecosphere (e.g. the 30 types in Table 7.6).

The following pages will report the main types of schedules designed for shrublands, tree corridors, agricultural fields, gardens and urban gardens, green-lands (Tables 8.4–8.8). Obviously, each one of the surveyed tesserae has to be identified firstly through:

- Name, localisation, date; altitude; slope angle and exposure
- Surface of the tessera; type of substratum and soil
- Synthetic description of the vegetation type and/or phytosociological relevés

Table 8.4. Schedule for the evaluation of shrub tesserae, applicable also for the estimation of their biological territorial capacity (BTC)

Shrubs	1	6	16	32	score
A. TESSERA (Ts) CHARACTERS					
A1- Vegetation height (m)	< 0.7	0.71-1.5	1.51-3.0	> 3.01	weighted av.
A2- Shrubs cover (%)	< 30	31-60	61-90	> 90	Ts surface
A3- Management	planted	managed	semi-nat.	natural	or similar
A4- Presence of grass patch %	30-45	21-30	11-20	< 10	max 45%
A5- Permanence (years)	< 20	21-80	81-120	> 120	Ts real age
A6- Structural differentiation	low	medium	good	high	age, groups
B. VEGETATIONAL BIOMASS (ABOVE GROUND)					
B1- Dead plant biomass in Ts	near 0	>10	2-6	6-10	% of pB
B2- Litter depth of the Ts	near 0	< 1.5	1.6-3.5	>3.5	cm
B3- Plant biomass volume	< 25	26-75	76-125	> 125	m^3/ha
C. ECOCOENOTOPE PARAMETERS					
C1- Dominant species	not clear	1	2	> 2	Pb
C2- Key species presence (%)	< 5	6-20	21-80	> 80	botanical
C3- Diversity	< 15	16-30	31-44	> 45	n° sp./Ts
C4- Plant forms (n°)	< 3	4-5	6-7	> 7	Table 7.3
C5- Dynamic state	degrading	recreation	regener.	fluctuat.	Sect. 5.1.3
C6- Vertical stratification	1	2	3	4	Fig. 7.7
C7- Renewal capacity	none	sporadic	normal	intense	dominant sp.
C8- Allochthonous sp. (%)	> 10	10- 4	< 4	0	Not regional
C9- Threatened plants	evident	suspect	risk	0	even acid r.
C10- Infesting plants	near all	>25	< 25	none	% of cover
D. LANDSCAPE UNIT (LU) PARAMETERS					
D1- Boundary connections	0	< 25	26-75	> 76	% perimeter
D2- Interior ratio (%)	absent	<30	31-89	>90	vs ecotope
D3- Source (vs surroundings)	sink	neutral	partial	effective	
D4- Role in the landscape unit	reduced	minor	evident	important	
D5- Disturbance incorporation	none	scarce	normal	high	local disturb.
D6- Geo-physical instability	evident	partial	risk	none	physiotope
D7- Permeant fauna interest	none	medium	near good	attraction	Ts/key sp.
D8- Transformation reason of the Ts as landscape element	strong disturban.	gradual change	tempor. instability	fluctua-tion	LU trend Sect. 5.1.4
D9- Landscape pathology interference	extremely serious	near chronic	easy to recover	none	surrounding ecotopes
D10- Permanence (years)	<50	51-150	151-300	>300	Age of LU
E. RESULTS OF THE SURVEY					
E1- Total score Y (= a+b+c+d)	a=	b=	c=	d=	Y=
E2- Quality of the Ts			Q= Y / 928		

$A1$, weighted average of shrubs (h); $A4$, % of covering: if the grass patch cover exceeds 45% of the total Ts, use the schedule for prairie and pastures (Table 8.9); $A6$, presence of not coeval plants and of groups of the same species; $B3$, pB is plant biomass volume: see Sect. 7.3.3; $C1$, shrub species the biomass volume of which clearly exceeds the equitability (eq) of the total pB. If no species exceed, sign the first column (a); $C2$, presence compared with the phytosociological association/s or the phytocoenosis of reference; $C8$, not autochthonous of the region; $C9$, present situation; $C10$, even character species (not dominant) but with a too massive covering; $D1$, with analogous vegetation Ts; $D2$, surface % of the Ts with interior vs edge characters: edge = 4 h; $D3$, following the source-sink theory; $D4$, functional role of the tessera in the ecotope or in the entire landscape unit; $D6$, hydro-geologic and/or climatic and/or physical instabilities; $D10$, years of permanence of the ecotope as shrub land

Table 8.5. Schedule for the evaluation of vegetated corridors, applicable also for the estimation of their biological territorial capacity (BTC)

Vegetated corridor	1	7	17	33	score
A. CORRIDOR (*Cd*) CHARACTERS					
A1- Corridor width (W)	< 2.5	2.6 -10	10.1 -20	> 20.1	metres
A2- Vegetation height (h) (m)	< 3	3.1 -12	12.1 -24	> 24.1	weighted av.
A3- Presence of water (w)	none	1 canal	2 canals	river	w < W
A4- Interruptions (> h)	> 4	3-4	1-2	0	n° in 10 h
A5- Management works	cutting	pruning	marginal	none	
A6- Tree cover (%)	< 30	31-60	61-90	> 90	medium
A7- Permanence (years)	< 60	61-120	121-180	> 180	*Cd* real age
A8- Linearity	rectilinear	semi-rectil	mixed	irregular	entire tract
A9- Presence of road (rd)	traffic rd	paved rd	rural road	no or path	rd < W
B. VEGETATIONAL BIOMASS (ABOVE GROUND)					
B1- Dead plant biomass in *Cd*	near 0	exceeding	near norm	normal	wood
B2- Litter depth of the *Cd*	near 0	< 1.5	1.6-3.5	>3.5	cm
B3- Plant biomass volume	< 100	101-300	301-600	> 600	m^3/ha
C. ECOCOENOTOPE PARAMETERS					
C1- Dominant species	1	1-2	2-3	> 3	studied tract
C2- Key species presence (%)	< 5	6-20	21-80	> 80	botanical
C3- Diversity	< 15	16-30	31-44	> 45	n° sp/stud. tr
C4- Plant forms (n°)	< 3	4-5	6-8	> 8	Table 7.3
C5- Dynamic state	degrading	recreation	regener.	fluctuat.	Sect. 5.1.3
C6- Vertical stratification	1	2	3	≥4	Fig. 7.7
C7- Renewal capacity	none	sporadic	normal	intense	domin. sp.
C8- Allochthonous sp. (%)	> 10	10-4	< 4	0	not regional
C9- Infesting species	many	few	sporadic	none	even local
C10- Threatened plants	evident	suspect	risk	none	even acid r.
D. LANDSCAPE UNIT (LU) PARAMETERS					
D1- Source (vs surroundings)	sink	neutral	partial	effective	
D2- Connections of the *Cd*	0	1	2	>2	nodes
D3- Interior species in the *Cd*	0	sporadic	few	many	forest sp.
D4- Network participation	not	neutral	potential	effective	
D5- Disturbance incorporation	0	scarce	normal	high	local disturb.
D6- Technologic interferences	≥2 crosses	1 cross	very near	far/or no	
D7- Faunal exchange	poor	low	normal	high	vs matrix
D8- Lichens presence on trees	0	1-15	16-35	>36	n° species
D9- Type of matrix	urban	suburban	rural	Semi-nat.	matrix
E. RESULTS OF THE SURVEY					
E1- Total score Y (= a+b+c+d)	a=	b=	c=	d=	Y=
E2- Quality of the corridor			Q= Y / 1023		

A1, Projection of the width of leafage on the ground; *A2*, weighted average of canopy trees (h); *A3*, canal and riverbed width smaller than W; *A4*, no. of interruptions in a tract 10 times (h) the lenght; *A9*, road width smaller than W; *B3*, *pB* is plant biomass volume: see Sect. 7.3.3; *C1*, tree/s species the biomass volume of which clearly exceeds the equitability of the total canopy *pB*. If no species exceed, sign the fourth column (d); *C2*, presence compared with phytosociological association/s or phytocoenosis of reference; *C8*, presence of species not autochthonous of the ecoregion; *C9*, even character species (not dominant) but with a too massive covering; *C10*, present situation; *D6*, by technological network; *D8*, following Nimis (1990); *D9*, *Semi-nat.* is Semi-natural

Table 8.6. Schedule for the evaluation of agricultural fields, applicable also for the estimation of their biological territorial capacity (BTC)

Agricultural fields	1	5	15	31	score
A. TESSERA (*Ts*) CHARACTERS					
A1- Vegetation height (m)	< 0.5	0.51-1	1.01-2	> 2.01	weighted av.
A2- Field form	geometric	polygonal	near-irreg.	near-natur	of *Ts*
A3- Tree or shrub presence	None	one	few	scattered	within *Ts*
A4- Management	industrial	paraindus.	traditional	biologic	cultivations
A5- Permanence (years)	< 10	11-50	51-100	> 100	age of *Ts*
B. VEGETATIONAL BIOMASS (ABOVE GROUND)					
B1- Dead plant biomass	None	low	medium	high	on ground
B2- Litter depth of the *Ts*	near 0	< 1.5	1.6-3.5	>3.5	cm
B3- Plant biomass (kg/m^2)	< 1.0	1.01-2	2.1-3	> 3.1	above gr.
C. ECOCOENOTOPE PARAMETERS					
C1- Diversity	< 10	11-20	21-30	> 30	n°sp./*Ts*
C2- Sp. of natural phytocoen.	None	sporadic	marginal	patchy	
C3- Genetic characters	transgenic	allochth.	current	traditional	of cultivars
C4- Chemicals	> 2	2	1	0	types
C5- Allochthonous sp. (%)	> 10	10-2	< 2	0	not regional
C6- Threatened plants	evident	suspect	risk	0	even acid r.
C7- Soil limiting factors	big patch	small pat.	marginal	none	
C8- Land capability classes	IV	III	II	I	Table 8.1
D. LANDSCAPE UNIT (*LU*) PARAMETERS					
D1- Contagion (semi-nat. *Ts*)	0	< 10	11-50	> 50	% perimeter
D2- Margin around cultivation	0	< 50	> 50	complete	% perimeter
D3- Irrigation type	technical	near tech.	canals	near nat.	
D4- Role in the landscape unit	reduced	minor	evident	important	
D5- Type of tillage	technical	mixed	marginal	none	soil
D6- Geo-physical instability	evident	partial	risk	none	physiotope
D7- Hedgerows network	no	marginal	partial	complete	presence
D8- Faunal micro-habitat	none	medium	near good	attraction	*Ts*/key sp.
D9- Landscape pathology interference	extremely serious	near chronic	easy to recover	none	surrounding ecotopes
D10- Permanence (years)	<25	26-100	101-200	>200	age of *LU*
E. RESULTS OF THE SURVEY					
E1- Total score Y (= a+b+c+d)	a=	b=	c=	d=	Y=
E2- Quality of the Ts			Q=Y/806		

A1, weighted average of the higher layer (h); *A4*, *industrial*: monoculture, with chemicals and mechanical tillings; *para-indus.*: not completely industrial; *A5*, years of permanence of the *Ts* as agricultural field, independent from the present type of cultivation; *B3*, *pB* is the total plant biomass above ground: see Sect.7.3.3; *C2*, species presence compared with the phytosociological association/s or phytocoenosis of reference; *C3*, *allochth.* is allochthonous; *C4*, e.g. herbicides, fertilizers; *C5*, not autochthonous of the ecoregion; *C7*, evidenced in the field; *small p.* is small patches; *D1*, with natural or semi-natural vegetation *Ts*; *D2*, presence of a margin between the cultivation and the border of the field; *D4*, functional role of the tessera in the ecotope or in the entire landscape unit; *D5*, machining vs handwork; *D10*, years of permanence of the ecotope or landscape unit as agricultural land.

Table 8.7. Schedule for the evaluation of urban arboreous green tesserae, applicable also for the estimation of their biological territorial capacity (BTC)

Urban gardens	1	7	20	39	Score
A. LANDSCAPE ELEMENT (*Le*) CHARACTERS					
A1- Vegetation height (m)	< 6	6.1-12	12.1-24	> 24	domin. trees
A2- Canopy cover (%)	< 25	25.1-50	50.1-75	> 75	*Le* surface
A3- Green substrate	concrete	terrace	mixed	natur. soil	
A4- Management works	cutting	pruning	marginal	near 0	
A5- Permanence (years)	< 50	51-100	101-250	> 250	*Le* real age
A6- Gardening composition	technical	formal	mixed	English	
A7- Botanical species	common	monum. trees	not common	ecotype or rare	
B. VEGETATIONAL BIOMASS (ABOVE GROUND)					
B1- Litter depth of the *Le*	near 0	< 1.5	1.5-3.5	>3.5	cm
B2- pB volume	< 100	101-250	251-500	> 500	m³/ha
C. ECOCOENOTOPE PARAMETERS					
C1- Dominant species	not clear	3	2	1	pB trees
C2- Key species presence (%)	< 5	6-20	21-50	> 50	botanical
C3- Diversity	< 10	11-25	26-40	> 40	n° sp./*Le*
C4- Plant forms (n°)	< 3	4-5	6-7	> 7	Table 7.3
C5-Vertical stratification	1	2	3	4	Fig. 7.7
C6- Renewal capacity	none	helped	low	normal	dominant sp.
C7- Allochthonous sp. (%)	> 20	11-20	5-10	< 4	not regional
C8- Threatened plants	evident	suspect	risk	0	even acid r.
C9- Structure differentiation	low	medium	good	high	age, groups
D. LANDSCAPE UNIT (*LU*) PARAMETERS					
D1- Boundary connections	0	< 10	11-50	> 51	% perimeter
D2- Interior species	0	sporadic	few	many	forest sp.
D3- Network participation	none	possible	minor	evident	
D4- Ratio green/urban	< 0.3	0.31-0.5	0.51-0.70	> 0.71	*Le*/ecotope
D5- Role in the landscape unit	reduced	minor	evident	important	
D6- Disturbance incorporation	none	scarse	normal	high	local disturb.
D7- Lichens presence on trees	0	1-15	16-35	>35	species/tree
D8- Geo-physical instability	evident	partial	risk	none	physiotope
D9- Landscape apparatus	unclear	1	2	> 2	functions
D10- Permeant fauna interest	none	medium	near good	attraction	*Le*/key sp.
D11- Transformation reasons of the *Le*	built equipment	people disturban.	gardening	self-developm.	main processes
D12- Landscape pathology interference	extremely serious	near chronic	easy to recover	none	surrounding ecotopes
E. RESULTS OF THE SURVEY					
E1- Total score Y (= a+b+c+d)	a=	b=	c=	d=	Y
E2- Quality of the Le			Q= Y / 1209		

A6, technical e.g. sportfield (greenness), *formal* e.g. French gardening, *English* or natural shaped; *A3, nat. soil* is natural soil; *A7*, peculiar or important plants; *monum. trees* is monumental trees; *B2, pB* is plant biomass volume: see Sect. 7.3.3; *C1*, tree/s species the biomass volume of which clearly exceeds the equitability of the total canopy *pB*. If no species exceed, sign the first column (a); *C2*, compared with local vegetation; *C7*, species presence not autochthonous of the ecoregion; *C8*, present situation; *C9*, presence of not coeval plants and of groups of the same species; *D1*, with analogous vegetation Ts; *D7*, following Nimis (1990); *D8*, hydrogeologic and/or climatic and/or physical instabilities; *D9*, n° of landscape apparatuses to which the functions of the landscape element belong; *D11*, *self-developm.* is self-development

Table 8.8. Schedule for the evaluation of prairie and pasture tesserae, applicable also for the estimation of their biological territorial capacity (BTC).

Prairie and pasture (p)	1	4	13	32	score
Reed (r)	1	6	17	37	score
A. TESSERA (*Ts*) CHARACTERS					
A1p- Vegetation height (m)	< 0.2	0.21-0.6	0.61-1.0	> 1.01	weighted av.
A1r- Vegetation height (m)	< 0.7	0.71-1.5	1.51-4	> 4	weighted av.
A2- Soil cover (%)	< 30	31-60	61-90	> 90	*Ts* surface
A3- Shrub presence (%)	< 5	5-15	15-25	> 25	max 30%
A4- Management	artificial	semi-natural	natural-grazed	natural	or similar
A5- Permanence (years)	< 20	21-60	61-120	> 120	*Ts* real age
A6- Structural differentiation	low	medium	good	high	groups, etc.
B. VEGETATIONAL BIOMASS (ABOVE GROUND)					
B1- Dead plant biomass %	> 60	60-21	< 20	near 0	dried/green
B2p- Litter depth of the *Ts*	near 0	< 1.5	1.6-3.5	>3.5	cm
B2r- Litter depth of the *Ts*	near 0	< 5	5-10	>10	cm
B3p- Plant biomass	< 0.6	0.61-1.2	1.21-2	> 2.01	kg/m^2
B3r- Plant biomass	< 1	1.0-2.5	2.5-4	> 4	kg/m^2
C. ECOCOENOTOPE PARAMETERS					
C1p- Dominant species	not clear	1	2-3	> 3	cover
C1r- Dominant species	not clear	>2	2	1	cover
C2- Key species presence (%)	< 5	6-20	21-80	> 80	botanical
C3p- Diversity	< 15	16-25	26-40	> 40	n° sp./*Ts*
C3r- Diversity	< 10	10-20	21-30	> 30	n°sp./*Ts*
C4- Plant forms (n°)	≤ 2	3	4	> 4	Table 7.3
C5- Dynamic state	degrading	recreation	regener.	fluctuat.	Sect. 5.1.3
C6- Vertical stratification	1	2	3	4	Fig. 7.8
C7- Renew capacity	none	sporadic	normal	intense	domin. sp.
C8- Allochthonous sp. (%)	> 10	10-2	< 2	0	not regional
C9- Threatened plants	evident	suspect	risk	0	even acid r.
C10- Infesting plants	near all	>25	<25	none	% of cover
D. LANDSCAPE UNIT (*LU*) PARAMETERS					
D1- Boundary connections	0	< 20	21-80	> 80	% perimeter
D2- Margins	not clear	> 30	30-10	< 10	% area
D3- Source (vs surroundings)	sink	neutral	partial	effective	
D4- Role in the landscape unit	reduced	minor	evident	important	
D5- Disturbance incorporation	none	scarce	normal	high	local disturb.
D6p- Geo-physical instability	evident	partial	risk	none	physiotope
D6r- Hydrologic instability	heavy floods	lack of water	irregular frequence	normal seasonal	water level fluctuations
D7- Permeant fauna interest	none	medium	near good	attraction	*Ts*/key sp.
D8- Transformation reasons of the *Ts* as landscape element	strong disturban.	gradual change	temporary instability	fluctua-tion	Sect. 5.1.4
D9- Landscape pathology interference	extremely serious	near chronic	easy to recover	none	surrounding ecotopes
D10- Permanence (years)	<70	70-150	150-300	>300	age of *LU*
E. RESULTS OF THE SURVEY					
E1- Total score Y (= a+b+c+d)	a=	b=	c=	d=	Y=
E2p- Quality of the Ts			Q=Y/928		
E2r Quality of the Ts			Q=Y / 1073		

A1p, A1r, higher layer (h) weighted average; *A3*, % of covering. If the shrub cover exceeds 30%, use the schedule for shrubbery (Table 8.5); *A6*, presence of not coeval plants and of groups of the same species; *B3p, B3r, pB* is plant biomass: see Sect. 7.3.3; *C1*, herb species the biomass of which clearly exceeds the equitability of the total *pB*. If no species exceed, sign the first column (a); *C2*, presence compared with the phytosociological association/s or the phytocoenosis of reference; *C6*, do not consider an isolated tree; *C8*, not autochthonous of the region; *C9*, present situation; *C10*, even character species (not dominant) but with a too massive covering; *D1*, with analogous vegetation *Ts*; *D2*, surface % of the *Ts* with edge characters: edge= from 5 to 10 h; *D3*, following the source-sink theory; *D4*, functional role of the tessera in the ecotope or in the entire landscape unit; *D6p*, hydrogeologic and/or climatic and/or physical instabilities; *D10p*, years of permanence of the ecotope as prairie or pasture; *D10r*, years of permanence of the ecotope as reeds.

8.3

Evaluation of the Faunal Components

8.3.1
Presence of Permeant and Engineer Species

In Chap. 1 we underlined that landscapes have their own characters, for instance they can support species that require forage in a grassland ecocoenotope and cover in a forest ecocoenotope, or can direct ecocoenotope transformation to maintain a steady state of the landscape itself. Thus, one of the most important criteria to evaluate faunal aspects in a landscape is to study the so-called permeant species (Odum 1971) or engineer species (Sanderson and Harris 2000). These species use different ecocoenotope and change the landscape pattern: they can not live in a single ecocoenotope, therefore they are typical species of landscapes. Note that man is one of these species, even if his ability and power is indeed much greater.

At the landscape scale these engineer species modify and create mosaics of habitat and enhance species richness and biodiversity. The transformation created by elephants in African landscapes is sometimes impressive (Fig. 8.5). Large organisms such as elephants, corals, beavers are known to be engineer, but there are many others. For example, wide-ranging predators, such as wolves, and migratory herbivores, such as wildebeest, might better be referred to as landscape engineer, because their activities strongly impact the environment.

Evaluating the presence of engineer species is certainly necessary in each study of landscapes, even in Europe. Some of these species are quite common (e.g. wild boars, red ants: *Formica rufa*); others are quite rare (e.g. ibexes). Many forested and/or wild landscapes present a serious lack of engineer species, sometimes even an absence, made worse by removing domestic herds (to preserve nature!).

Permeant species (species that use many different habitats) or, better, typical landscape species are especially necessary for the evaluation of faunal conditions in a landscape. Many species of birds belonging to important *taxa* are permeants.

We may consider as "umbrella species" the species having the greatest ecological needs, hence the first to decrease when their environment is altered. But generally these species are not truly permeant: they have distinct habitat exigencies, thus the umbrella effect is limited because they can not properly be used to protect all the species living in a landscape. It is better to refer to the concept of "focal species" (Lambeck 1997) which concerns a small set of species representing the landscape heterogeneity of habitats, thus usable to protect the spatial and functional exigencies of a landscape. In the case of various species limited by the amount of resources, the less abundant one will be the focal species on which the landscape rehabilitation must be designed.

The study of the presence or the absence of dragonfly species (*Odonata*) in a landscape unit helps to evaluate its ecological conditions. For example, in the central Chiaravalle district of the South Milan Agricultural Park 13 species of dragonflies were found (Ingegnoli 1998), among which only two that were quite uncommon: but compared with a potential 20-25 species in this humid field landscape (Ott 1997) that value was very meaningful.

The *Odonata* expert Juergen Ott published in 1995 a table of human impacts on dragonflies and a key species table (Table 8.9) for different biotopes, applicable to central European landscapes (data from Rheinland-Palatinate).

Fig. 8.5 An example of the transformation of the environment made by a well known engineer species, like elephants, in the Serengeti Park in Africa: two tracks in a steep river bank

Table 8.9. Biotopes, habitat characters and key species of *Odonata* (from Ott 1997)

Biotope type	Landscape elements (habitat)	Key species of *Odonata*
Lakes	Colonisation zone, riparian reeds, semi-submersed vegetation, open water	*Coenagrion puella, Enallagma cyathigerum, Erythromma najas, Anax imperator, Aeshna grandis, Somatochlora metallica, Orthetrum cancellatum*
Small temporary water basins	Partially shaded by vegetation	*Lestes dryas, Sympetrum sanguineum*
Small perennial water basins	Partially shaded by vegetation	*Laestes sponsa, Platycnemis pennipes, Aeshna canea, Cordulia aenea*
Marshy waters and peats	Acid waters, partially shaded, presence of *Sphagnum*	*Coenagrion hastulatum, Aeshna juncea, Somatochlora arctica, Leucorrhinia dubia*
Secondary brooks and canals	Rapid change of vegetation types, with also hydrophites	*Erythromma viridulum, Libellula depressa, Crocothemis erytraea, Gomphus pulchellus*
Rivers and current waters	With meanders, mainly shaded, few sand banks, hydrophytes	*Calopteryx virgo, Calopterys splendens, Orthetrum coerulescens, Cordulegaster boltoni, Gomphus* sp.

Focal species can be used for the evaluation of the state of a landscape, as in the case of the distribution of 140 species of birds in Lombardy in accordance with the portions of the landscape fittest for them, elaborated on by Renato Massa and his collaborators. Only five species were found in areas with less than 10% of forest cover, while at 20% of forest cover 35 species were present and at 30% only 24. Massa also proposed the indicator species of mature plane forest in Lombardy (Table 8.10), selected within a group of species the abundance centre of which is a forest cover of over 50% and an altitude centre which is under 700 m above sea level.

Table 8.10. Bird species in Lombardy (Italy) living in landscapes units with more that 50% forest cover and indicator species* of mature plane forests (from Massa 1998)

Bird species and indicators*	Forest cover (%)	Frequency (%)	Abundance (%)	Distribution (%)
Chaffinch (*Fringilla coelebs*)	50.02	44.34	5.10	85.00
Blue tit (*Parus caeruleus*)	58.73	7.97	0.58	41.67
Long-tailed tit (*Aegithalos caudatus*)	65.12	5.20	38.33	38.33
Jay (*Garrulus glandarius*)*	62.23	4.01	0.16	32.50
Red woodpecker (*Picoides major*)*	57.38	2.88	0.12	25.00
Green woodpecker (*Picus viridis*)*	54.09	2.62	0.14	17.50
Marsh tit (*Parus palustris*)*	80.39	2.16	0.14	20.83
Nuthatch (*Sitta europea*)*	75.75	1.85	0.11	15.83
Buzzard (*Buteo buteo*)*	55.36	1.65	0.06	19.17
Sardin. warbler (*Sylvia melanocephala*)	50.32	0.72	0.07	3.33
Tree-creeper (*Certhia brachydactyla*)	81.67	0.21	0.09	2.50
Dabchick (*Tachybaptus ruficollis*)	61.25	0.21	0.01	3.33
Blue rock thrush (*Monticola solitarius*)	55.00	0.10	0.01	1.67
Grey warbler (*Sylvia nisoria*)	75.00	0.10	0.01	1.67

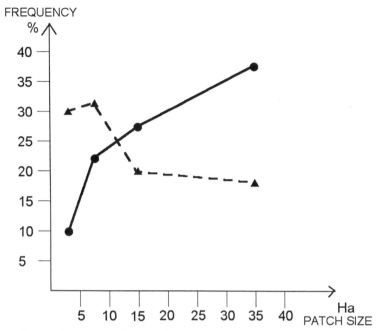

Figure 8.6. Distribution of capercaillie locations in a forested landscape in the Bavarian Alps in relation to availability of old forest stand. (From data of Storch 1997).

Another example of focal species is the capercaillie (*Tetrao urogallus*), a quite rare and large grouse living in Eurasia. This species, once considered as local, shows a landscape-scale home range with a radius of about 4 km from its main lek (or arena, that is its gathering place for courtship). Therefore the study done by Storch (1997) in the Bavarian Alps becomes particularly useful. Different habitat types were considered, such as clear-cut, thicket, pole stage wood, middle aged forest, old forest and mixed: for each type the presence of capercaillie was registered, evidencing variations in winter and summer and between males and females. In relation to the availability of old forest stands of various size (Fig. 8.6), the distribution of the capercaillie locations could be a good indication when evaluating alpine landscapes.

The evaluation of faunal landscape characters depends also on the type of margins of the fluvial corridor. For instance, studies of the Rhone river (France) and Danube (near Vienna, Austria) demonstrate the importance of the length of the river banks per unit of length of the fluvial corridor. In the first case (Pont and Persat 1990), the mean number of fish captured per fishing station double when the length ratio (km/km) increased by about 40%. In the second case (Schiemer 1991), the number of species found in a Danube channel varied from five to ten passing along a length ratio (vs 100 m of fluvial course) of 120 to 320 m.

8.3.2
Faunal Sensitivity in a Landscape Unit

There are many methods based on the evaluation of faunal quality and faunal vulnerability. With our colleagues (R. Massa and L. Fornasari), we measured the faunal sensitivity $S(L)$ in a landscape unit. This sensitivity was evaluated in two steps, the first being synthesised by this equation:

$$S(f) = (Q1 + Q2 + Q3) \, x \, (V1 + V2 + V3)$$

where: $S(f)$ is the faunal sensitivity (generic), $Q1$ the general rarity, $Q2$ the chorology, $Q3$ the critical state, $V1$ the abundance, $V2$ the habitat extension and $V3$ the fragility. The schedule for evaluation of the $S(f)$ significance is shown in Table 8.11.

Table 8.11. Evaluation of faunal sensitivity in a landscape unit

Faunal attributes	0	1	2	3	score
Q. FAUNAL QUALITY (general characters)					
Q1- General rarity	not determin.	rare, risk	vulnerab. (general)	threaten.	species rarity
Q2- Chorology	cosmo-polites	Continent	regional	endemic	areal
Q3- Critical state: local presence vs other areas	other	rare in district	rare in region	doubt in region	presence in other areas
V. FAUNAL VULNERABILITY (local characters)					
V1- Abundance	abundant > 0.5	common < 0.5	scarce < 0.25	rare < 0.125	$i/p°$
V2- Habitat extension	not selective	rural habitat	ecotonal habitat	forest habitat	habitat sensitivity
V3- Fragility: living conditions	any habitats	not degraded	colonial	optimal habitat	species needs vs habitat
R. RESULTS					
R1- Total scores	a	b	c	d	S(f)

FAUNAL SENSITIVITY		TOTAL SCORES
I	Critical S(f)	S(f) > 72
II	High S(f)	S(f) < 72
III	Relevant S(f)	S(f) < 5 4
IV	Mean S(f)	S(f) < 36
LEVEL OF ATTENTION		
V	Low S(f)	S(f) < 18
VI	Negligible S(f)	S(f) < 9

Note $i/p°$: If not birds, related to survey squares

The second step of this methodology can be described through this equation:

$$S(L) = \Sigma \; Si \; x \; Kil$$

where: $S(L)$ is the landscape unit sensitivity, Si the i-species sensitivity, Kil the relative density weight of the i-species within the landscape unit. This last parameter is considered: sporadic if $Kil = 0.5$, reduced if $Kil = 0.75$, normal if $Kil = 1$, high if $Kil = 1.25$.

Note that this ecological index evaluates the importance of faunal species also for conservation, and not only for gathering information about a community, such as the Shannon index H.

8.4
Human Elements and Their Evaluation

8.4.1
Theoretical Standard Habitat

We know that the carrying capacity of a territory is not free from limitations, despite our increasing technology. It is essential that the relationship habitat-individual human is based on ecological principles instead of economical and urban standards. Note that in the history of western civilisation it is possible to find empirical measures regulating the relation habitat-individual.

The most important of these was the codification of the minimum inheritable piece of land by ancient Roman laws: the *heredium* = 5,040 m², divisible in 2 *jugera* or 4 *acta*. The Roman agronomists, having theorised the rotation of cultures and the best procedures of fertilisation, defined this quantity as being able to sustain a new family. In fact the heredium was widely adopted by planners in the colonisation of Europe, North Africa and the Middle East in the form of the *limitatio centuriata* methods.

The quantification of an index like the minimum standard habitat per capita (SH*) can be very useful even for the comparison of the transformations in human landscapes. To calculate the SH* it is usually necessary to consider the following arguments, related to the studied territory: the mean food consumption, the mean caloric need per capita, optimal agricultural production (not forced), the statistical needs in the areas of buildings, industry, services, the mean regional BTC, the ratio HH/NH.

As expressed in Table 8.12, the evaluation of the SH* is referred to the main human landscape apparatuses: residential (RSD), subsidiary (SBS), productive (PRD) and protective (PRT). The choice of the mean regional BTC is due to its constancy: it remained almost unchanged for about two centuries despite the wide-ranging transformations of the industrial era, and is thus a good disturbance incorporation example (at regional scale).

Table 8.12. Evaluation of the theoretical standard habitat per capita (SH*) in Lombardy

Landscape apparatuses	Surface m^2	BTC of the apparatuses	BTC – BTC$_R$	BTC deficit per apparatus	% SH
SH-PRD	1043	1.60	- 0.50	521.5	73.09
SH-RSD	105	0.85	- 1.25	131.25	7.36
SH-SBS	79	0.50	- 1.60	126.4	5.54
Deficit	1227	1.47	- 0.63	779.15	-
Needed SH-PRT	200	6.0	+ 3.9	780	14.01
SH*	1427	2.1	0	0	100

SH(standard habitats): *PRD* productive, *RSD* residential, *SBS* subsidiary, *PRT* protective; *BTC$_R$* optimal regional mean of biological territorial capacity (Mcal/m^2/year).

To get an idea of the dimensions of the SH* related to regions other than Lombardy without a long calculation is possible in general. Referring to four classes of regions (boreal, temperate, sub-tropical and tropical) we may observe the variation of the main parameters related to human landscape apparatuses (Table 8.13): the mean caloric need per capita and the mean BTC of yearly cultures (PRD), the minimum need of residential and subsidiary habitats (RSD, SBS), the mean BTC of protective habitat (0.2 forest + 0.4 wood + 0.4 shrub) and consequently the variation of a mean regional BTC$_R$ as references for the dimensioning of SH*. The BTC$_R$ can be calculated starting from the temperate region data (from European Union countries) multiplied by a small adjusting coefficient. This value BTC$_R$ = 2.20 is not too dissimilar from that of the mean terrestrial ecosystems BTC = 1.80 (1990).

We can profit from the use of the theoretical standard habitat, too, in relation to the standard habitat per capita (see Sect. 4.3.4.): this ratio (σ = SH/ SH*) gives us the value of the carrying capacity of a landscape unit. Two examples will be presented in the following sections (Tables 8.15; 8.16).

Table 8.13. Evaluation of the theoretical minimum standard habitat per capita (SH*) related to the main types of ecological regions

Ecoregion types (f)	Boreal	Temperate	Sub-tropical	Tropical
Edible Kcal/day per capita	3100	2850	2600	2350
Yearly cultures BTC	1.12	1.20	1.27	1.40
SH-PRD m^2	1228	1050	903	735
SH-PRT BTC	5.6	6.0	6.0	7.4
Deduced BTC (r)	2.35	2.2	2.3	2.35
SH-RSD m^2	110	105	100	95
SH-SBS m^2	84	80	76	72
Mean BTC (PRD-RSD-SBS)	1.37	1.46	1.53	1.66
Needed SH-PRT m^2	429	241	193	112
Resulting SH* m^2	1851	1476	1272	1014

8.4.2 Human Habitat Measure

The evaluation of the human habitat (HH) of a landscape unit is one of the first studies to do: the best way to measure it is to refer to the landscape apparatuses (see Sect. 3.4.1). An example will be shown in Table 8.15.

Nonetheless, the HH is generally measured based on the ecological mosaic of land-uses (summing the surfaces pertaining to urban, industries, roads, railways, fields, etc.), because it is more readily available. But in the real situation we have many overlapping conditions (e.g. mountain pastures with *malghe*, that are shepherd's huts in the Alps, or managed forests are partially HH, too): therefore the real HH values are usually different from the ecomosaic ones (HH_m).

These considerations confirm again the importance of the ecotissue concept to study landscapes, where a great number of processes or characteristics can not be mapped. So, a model is needed to estimate this difference and to correct the values, which is useful also in the evaluation of past HH and/or BTC, when only land use data were collected, and in some applications (e.g. territorial planning), when it is impossible to get a correct survey of landscape apparatuses.

The quantification of these difference may be obtained by applying to the standard classes of BTC (representative of the major types of ecosystems) the following values of HH: 98% (I class, if not water); 95% (II class); 90% (III class); 60% (IV class); 20% (V class); 8% (VI class); 3% (VII class), instead of the HH_m percentage (see also Table 8.16). These values have been calculated using both experimental data on different landscape types and their elaboration. A simple expression can model this process:

$$HH_e = 0.85 (HH_m + 10)$$

HH_e being the "ecotissue HH" correct value and HH_m the "ecomosaic HH". As noted in Table 8.14 , a HH_m of 5% corresponds to a HH_e of 12.6% (difference = + 7.6%) but this difference becomes near zero at about 65% and it becomes negative at HH_m of 100% (HH_e is 93.9: the difference = - 6.1).

Table 8.14. Correspondence between the mosaic and the ecotissue percentage of HH

% HH_m	0	5	10	20	30	40	50	60	70	80	90	100
% HH_e	<5	12.6	16.9	25.7	34.5	43.3	52.2	61.0	69.8	79.4	85.2	93.9

8.4.3 Human Habitat Evaluation and Control

Let us show an example in Table 8.15, using the central district of the suburban South Milan Agricultural Park, in which HH has been referred to the Landscape apparatus and was estimated starting from the present situation; its value gives immediately the complementary value of natural habitat NH. These two values could be related to obtain a first measure of the degree of human transformation of the landscape: HH / NH = 8.7 (1845), 13.5 (1950), 17.9 (1997).

The measure of the ratio between the percentage of the BTC of NH and the mean BTC (%BTC$_{NH}$/BTC$_{mean}$) can give additional insight into the process of urbanisation of this agricultural area. In this case, note the extremely low level of BTC$_{NH}$. Obviously, its value depends on the simple relation:

$$BTC_{mean} = BTC_{HH} \times HH\% + BTC_{NH} \times NH\%$$

The σ (see Sect. 8.4.1) gives the value of the carrying capacity of that landscape unit: in 1845 the agricultural landscape was still productive, showing the capacity to maintain four times its population; in 1950 it became suburban-rural and was available to maintain only its inhabitants; in 1997 it became part of the agricultural park belt of Milan, but in fact the area is no longer able to maintain its population.

Table 8.15. Transformations in the suburban landscape unit of the current South Milan Agricultural Park

Landscape apparatuses	1845 (%)	HH	BTC	1950 (%)	HH	BTC	1997 (%)	HH	BTC
Residential	2.1	1	0.75	7.3	1	0.7	15.9	1	0.7
Industrial	0.1	1	0.2	2.5	1	0.2	12.2	1	0.2
Cultivated	72.4	0.93	1.1	80.3	0.95	1.1	63.4	0.95	1.1
Marsh grass	23.8	0.85	1.5	8.2	0.85	1.3	7.6	0.85	1.3
Semi-natural	1.6	0.3	2.1	1.7	0.2	2	0.9	0.2	1.8
Total	100	0.897	1.20	100	0.931	1.08	100	0.947	0.95
Natural habitat		0.103	2.33		0.069	2.23		0.053	2.1
BTC$_{NH}$/BTC$_{mn}$ (%)	17.2			12.4			9.8		
σ =SH / SH*	4.02			1.07			0.47		

SH standard habitat per capita, *SH** its minimum value; BTC (Mcal/m^2/year)

The evaluation of the effects of the increase of HH in a landscape needs a specific model: it must be referred to a territory in which at least a part of its forested ecotopes represent some primeval landscape unit. A unit like this, mainly composed of a complex mosaic of mature forest patches in dynamic fluctuation and regeneration is rare in Europe. Very instructive were the experiences in beech-fir forests (*Asyneumati-Fagetum abietetosum* Gentile 1964) in the Apennines (e.g. Sila Piccola, Italy), in spruce forests (*Homogino-Piceetum* and *Veronico-Piceetum* Ellenb. and Klötzli 1972) in the Alps (e.g. Trentino, Italy) and especially a survey in the meso-eutrophic deciduous *Tilio-Carpinetum* of Bialowieza Forest (Poland).

In particular, the Bialowieza landscape presented two interesting situations: the National Park Forest (4700 Ha) and the contiguous wide clearing (1400 Ha) of the Polanie Bialowieska, about at the centre of the landscape. Many interesting data were collected and synthesised by Falinski (1994a, 1998) on the forest and by Pabjanek (1999) on the rural town and its clearing (see Sect. 13.2). The elaboration of some crucial landscape ecological measures on vegetation (following the methodology presented in this book) and their integration with the mentioned data allowed the formulation of this model (Table 8.16, Fig. 8.7). This research is referred to a landscape of sub-regional scale. The structure of a landscape of mature deciduous meso-eutrophic forest without the presence of HH was normalised through the seven standard BTC classes (see Sect. 7.5.2).

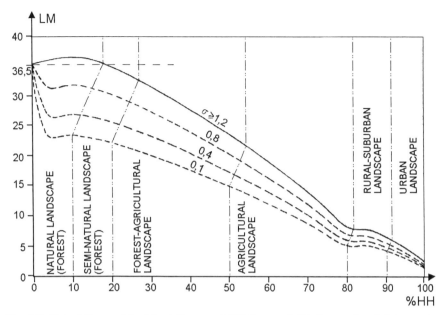

Fig. 8.7. Plot of the model of the influence of HH (human habitat) on a forested temperate landscape. The LM (landscape metastability) reaches its maximum when HH = 10%, and it begins to decrease below its starting point when HH=18.5%, if the σ = SH/SH* is $\geq$ 1.2; if σ = 0.1 the decrease starts from 10%

Table 8.16. A scheme of the model proposed to demonstrate the effects of the variation (step by step) of the human habitat, starting from a forested landscape (Central European vegetation) to an urban landscape

Mosaic HH (HH$_m$) %		0	5	10	20	30	40	50	60	70	80	90	100
Mcal/m^2/year	BTC classes	Relative percentage of presence of the different standard classes of BTC											
10.0	VII	41.0	40.0	37.0	32.0	27.0	22.0	17.0	12.0	7.0	2.0	-	-
8.4	VI	47.0	45.9	44.0	39.0	34.0	29.0	24.0	19.0	14.0	9.0	4.0	-
5.6	V	7.5	7.5	7.5	7.5	7.5	7.5	7.5	7.5	7.5	6.0	8.0	3.0
3.5	IV	-	0.5	1.0	1.5	2.0	2.5	3.0	3.5	4.0	4.0	3.5	1.5
2.2	III	-	-	0.5	1.0	1.5	2.0	2.5	3.0	3.5	3.0	4.5	5.0
0.9	II	3.0	4.1	8.0	15.5	23.5	31.5	39.5	47.5	55.5	70.0	53.0	30.5
0.2	I	1.5	2.0	2.5	3.5	4.5	5.5	6.5	7.5	8.5	6.0	27.0	60.0
Ecotissue HH (HH$_e$) %		<5	12.6	16.9	25.7	34.5	43.3	52.2	61.0	69.8	79.4	85.2	93.9
Mosaic BTC		8.53	8.37	7.97	7.16	6.35	5.54	4.73	3.92	3.10	2.20	1.61	0.76
Ecotissue BTC		8.5	8.22	7.76	6.93	6.09	5.24	4.42	3.62	2.82	1.97	1.41	0.65
LD (τ)		4.18	4.37	4.72	5.05	5.23	5.32	5.32	5.24	5.22	4.05	4.60	3.90
LM' [$\sigma \geq 1.2$]		35.6	36.0	36.6	35.0	31.9	27.9	23.5	19.0	14.2	8.0	6.6	2.5
LM [$\sigma = 0.8$]		-	31.2	31.7	30.3	27.6	24.2	20.4	16.4	12.2	6.9	5.7	2.2
LM [$\sigma = 0.4$]		-	26.4	26.8	25.6	23.4	20.4	17.2	13.9	10.4	5.8	4.8	1.9
LM [$\sigma = 0.1$]		-	22.8	23.2	22.1	20.2	17.6	14.9	12.0	9.0	5.1	4.2	1.6

σ SH/SH* (standard habitat ratio); *LM* Landscape metastability index; *LM'* normal LM

As shown in Table 8.16 the mean BTC of a landscape like this is very high (8.50 Mcal/m^2/year), but 88% of the ecotopes are condensed in the upper two classes of BTC, thus the landscape heterogeneity (τ) presents a medium value. Consequently, the landscape metastability LM = 35.6 (see Sect. 11.2.2) is very good, but not exceptional. The transformation of a natural landscape unit like this, step by step, cutting the forest and developing a rural area, reaches the maximum LM = 36.6 (an increase of 3%) at about 10% of the HH$_m$ presence. In a non-fanatic environmentalist view this fact is not astonishing, because the landscape biodiversity passes from 4.18 to 4.72 and the BTC decrease is still limited. We have to observe that the increase of LM up to 10% of HH$_m$ is possible only if $\sigma \geq$ 1.2; if this ratio results in $\sigma < 1.2$, the process changes. It is simple to calculate this change:

$$LM = LM' \times (\sigma/3 + 0.6) \quad \text{with} \quad LM = \tau\ BTC_e$$

where BTC$_e$ is obtained reducing the BTC$_m$ of the mosaic through the % difference (HH$_e$ – HH$_m$). When σ= 0.9 and HH= 10%, there is only a (–7.5)% decreasing effect of LM in comparison with the starting situation (HH= 0%), which becomes (– 40.6)% at HH= 50%.

This fact confirms the influence of population density on the natural balance: when people living in the HH area reach an habitat per capita not sufficient for their normal needs, they exert a pressure on the surrounding environment which decreases the BTC of the entire landscape unit for instance, improper use of a proportional part of the forest patches (for herds, for wood, etc.) even if protected. The effects of the population density become maximum when $\sigma < 0.2$ (SH corresponding to a city) but, generally, the values do not go under the limit of 60% of LM', otherwise it is a direct enlargement of the HH.

In any case, the model shows that after HH$_m$ = 18.5% we always have a decrease of the LM, even of LM'. When HH$_m$ = 55% the entire landscape unit becomes agricultural and the first three classes of standard BTC contain 54% of the ecotopes. Note that the entire landscape unit remains semi-natural until HH$_m$ = 27.5%, when the three highest classes of BTC contain 65-70% of the ecotopes. We can see also from Fig. 8.7 the rapid decline of LM after 20% of HH$_m$, even if the maximum τ = 5.33 is reached at HH$_m$ = 45%.

Over HH$_m$ = 85% we can find a tendency toward the urbanisation of the entire landscape unit. At HH$_m$ = 90% the situation is generally that of a suburban landscape, which becomes urban soon after. Reaching the HH$_m$ = 100% we may have 90.5% of the ecotopes in the first two BTC classes.

For a correct human habitat control, it is indispensable to check the HH relation with the LM of the landscape unit and the SH/SH* ratio. It is also useful to follow the dynamic of these parameters on a graphic like that of Fig. 11.5. Monitoring proper values must be done even in comparison of the landscape types resulting from the survey.

The territorial planning for conservation aims may be particularly reinforced by the application of a model like this.

9 Landscape Criteria of Evaluation and Diagnosis

9.1
Criteria of Evaluation and Diagnosis

9.1.1
Diagnostic Methodology

A clinical-diagnostic method is indispensable to studying a complex living system. On the other hand, the differences between analysis and diagnosis are often fuzzy, as are the differences between structure and functions.

Therefore, it is not possible to solve any diagnostic problem by focusing all our interest on a single component only; by contrast, it is necessary to jump from one part to another, enriching our knowledge of each part. Even if we can give and receive information only in a temporal sequence, because of the structure of our language and logic, rigid logical sequences are incompatible with a complex system. We need other criteria.

In fact, a provisional scheme of the entire system is inevitable, because a researcher may understand a single sub-system or a part of it only after having understood all the others. That is why we mentioned (see Sect. 7.1.1) the necessity for a biologist (or a physician or an ecologist) to follow the method of a painter: a correct methodology must proceed from the system as a unit toward its parts. These types of methodologies can be defined as *gestaltic*, being based principally on *gestaltic* perception (which means form and form-ness, pattern and pattern-ness). According to Lorenz (1978), in this system of knowledge even the observation of a single complex of symptoms one time only may acquire a scientific meaning (this is in conflict with statistical methods).

We need to follow an iterative method, based on a sequence of: (1) observation, (2) construction of models and/or (3) model-free evaluation, directed toward the extraction of the essential information from a complex system.

Contemporarily, we have to compare these data with a series of "normal behaviours", derived from the capacity to evaluate in a system behaviour that is considered correct from that which is not. As a matter of fact, the physiology-pathology relations are very difficult to study, because to remove an alteration we need to know the normal process, but viceversa, it is the alteration which in general permits the normal process to be understood. So, any therapeutic intervention on a living system will remain without success if we do not know its normal behaviour. Therefore, the application of models to understand the relationships between a normal process and an alteration can be very useful.

Nonetheless, remember that even models have intrinsic limits. To extract the essential information from a complex system we have to balance realism, precision and generality. It is very well known that they are not compatible: thus we need to establish our limits for each different aim. As expressed by Zadeh (1987), the increasing complexity of a system leads to the decreased significance of precise statements. It is a principle of incompatibility: increasing precision leads to decreasing reality.

In addition, it may seem that objectivity is often only an opinion, but it is not true. The problem of the "judgement of value" is intrinsically possible and necessary because it is the base of all life processes, in the sense that a correspondence between values perception and the passage from what is disordered and probable to what is ordered and improbable exists. The scale of values derived from this fundamental process can be applied, wrote Konrad Lorenz, to an animal species, to a landscape, or to a cultural work like an artistic creation.

This premise is essential for elaborating diagnostic models, the methodology of which must be iterative and "split shaped", avoiding the myth of a unique, complex decisive model.

9.1.2
A Split Method for Models

Normally one begins by observing the ecological system, as a landscape unit to be examined, describing its main subsystems and the set of properties which pertain to them and the ones which pertain to the entire system. Analysis and ordinations of their interrelations are often needed. Then, a set of hypotheses, the most complete in relation to our point of view, have to be formulated. This leads to the construction of what is called a preliminary model. A series of successive refinements are requested for any model (Fig. 9.1). Briefly, the main steps are the following:

1. Description of the objectives
2. Research of the main linkages among the components
3. Provisional assemblage (preliminary model, pM)
4. Research of the functioning of pM
5. Observations of the differences with the real system and eventually returning to precedent points
6. Construction of the functional model

This last point is particularly delicate and difficult and generally needs a split method to reach a correct formalisation. All that we described in Chap. 2 may obviously be very useful for the construction of models. The observations have to be organised through the system theory, remembering all the content of the structure and dynamics of the landscape and having to be iterative and often redundant to assure the perception of their information even in the presence of disturbances.

A split method is required because of the high degree of complexity of living systems, which present hyper-space variables and branching functions and have to be evaluated through projections and sections necessary to understanding their components.

The semantic rightness of the method has to be controlled by the observer, whose ability to manage the set of split models is essential. That is why a good diagnostic physician is compared with an artist.

The sequence of analysis and evaluation models, their number and the linkages among their information have to be adjusted with personal experience and with a vast knowledge of the principles of the disciplines regarding living systems, in particular for us the landscape.

The syndrome categories presented in Chap. 5 can be useful as a set of check-up sections to be tested.

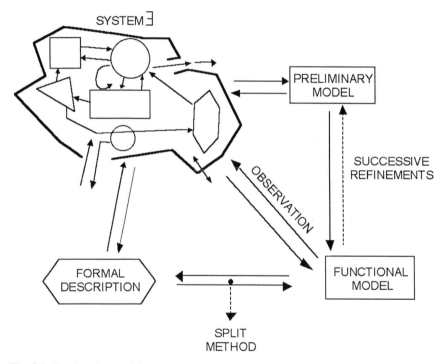

Fig. 9.1 Simple scheme of the position of a model, built with a split methodology, in a general scheme of the modelling process

9.1.3
Fuzzy Logic: Model-Free Evaluation

The relationships between physiology and pathology are in many cases irresolvable. Relatively few (and simple) subsystems may be described through mathematically formalised models. In fact, no one knows how to write the equations of the majority of complex living systems. Therefore, we need other criteria for the evaluation and control of our systems.

The freedom from mathematical model evaluation is possible. It comes from fuzzy systems (Kosko 1993). Fuzzy logic has a simple but powerful basis: it includes bivalent logic (A *or* non-A) as a borderline case of polyvalent logic (A *and* non-A).

In fuzzy logic all is a question of measure. Even numbers are fuzzy: numbers very near to zero belongs to the zero set 80%, 50%, 10%. We may represent the number zero fuzzy as a bell curve or a triangle centred on zero. Also the observation that a part contains the whole of a certain measure is a fuzzy concept: if the part is 2/3 of the whole, and the whole is coherent, then, surveying the part, we collect information on the whole with a probability of 66.7%. Thus the measure of the sub-setness (how a set contains another set) is equal to a conditional probability. Moreover, it is possible to define a fuzzy entropy. In fact, entropy signifies disorder or indeterminateness in a system. In a fuzzy system each element belongs to the system only to a certain extent. Entropy indicates the measure of this indeterminateness.

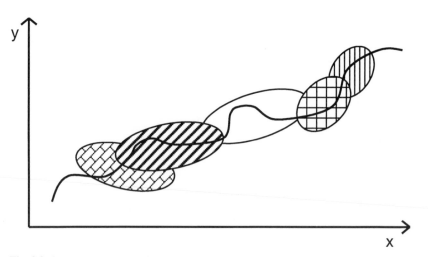

Fig. 9.2 A system represented by a non-linear curve is approximated with fuzzy patches

The FAT (fuzzy approximation theorem) states that any complex system can be approximated by a fuzzy system. The idea is simple: it is possible to create a non-mathematical model covering a non-linear curve (representing the system) with a series of patches (Fig. 9.2). Each splinter of human knowledge fixes a rule in the form "if this, then that", so defining a patch. Patches and grey gradations are the two key rules of fuzzy logic. To develop a fuzzy system we have to follow three steps. The first consists of defining the variables A and B as the input and output of the system, respectively. The second finds the limits of the fuzzy subsystems of A and B. They can be traced using a triangle instead of a bell curve. Note that the wider sets are less important: for a more refined control we need more reduced sets. The third step consists of defining the fuzzy rules. In this phase we have to associate the A sets with the B sets.

A simple evaluation of an agricultural landscape unit can be used as an example. In a regional park of Lombardy (Italy), the Adda River Park, we studied a unit of 320.5 ha, about 85% of which is composed by agricultural fields (Table 9.1).

Table 9.1. Agricultural landscape unit of 320.5 ha in the Adda River Park, near Corneliano (Lombardy): synthetic land use data

Landscape characters	1920	1960	2000
Urbanised (ha)	2.3	3.5	4.0
Cultivated (ha)	257.6	272.5	267.2
Hedgerows (ha)	9.8	7.0	5.2
Forest and woods (ha)	40.5	27.0	33.1
Waters and canals (ha)	10.3	10.5	11.0
HH/NH	4.00	5.25	4.75
Hedgerows/fields	0.0306	0.0218	0.0162

Being a natural park, it was important to control the ecological state of this landscape unit. Among other analyses, it was useful to check the dynamic of two related variables: the HH/NH ratio (A) and the hedgerow/field ratio (B).

The range of A might vary from 1.2 (if less, the matrix could not be agricultural) to 6 (if more, no available natural tessera could be found). This range was divided into five subsystems: semi-natural, agricultural-natural, agricultural, agricultural-rural, rural-technologic. The optimal subsystem was the agricultural, centred on the value 3.6. The range of B might vary from 0% (no corridors) to about 12% (all fields enclosed by hedgerows). This range was divided into five subsystems: very low, medium-low, optimal, medium-high, very high; it was centred on the value 3% because of agricultural economic reasons. The following fuzzy rules were consequently fixed:

1. If A is semi-natural, then B is very high
2. If A is agricultural-natural, then B is medium-high
3. If A is agricultural, then B is optimal
4. If A is agricultural-rural, then B is medium-low

These rules were derived from good sense and are fuzzy because their variables are a question of measure and represent fuzzy sets. The patches formed by the correlation of these two ranges of A and B subsystems represent the projection of the mathematical product of the triangles related to the fuzzy rules. These patches are overlapped (Fig. 9.3).

Note that in a fuzzy system the rules are always activated in parallel and are excited in a partial way. They follow an associative memory and the result is a fuzzy weighted mean. In addition, remember that a triangle is not a number but it has a mean number, the centroid, or the mass centre of the output set. Thus we can "de-fuzzyficate" the set if we need a precise number (for instance to size some landscape recovery).

As plotted in Fig. 9.3, within the fuzzy system it is possible to evaluate the movement of the transformation of the studied unit of landscape, thus giving a sense of it. In fact, referring also to Table 9.1, there was the possibility to design the best way to restore this agricultural landscape.

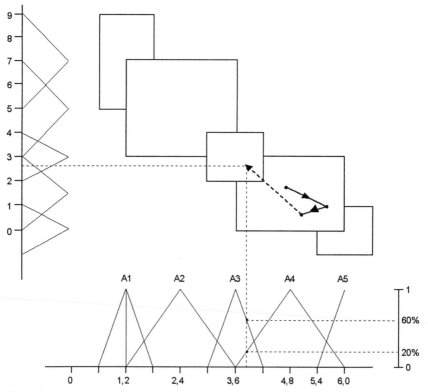

Fig. 9.3 The representation of the fuzzy system and comparison with the movement of the transformations in the studied landscape unit

Suppose we want to restore the ecological state of the agricultural unit by the year 2015. As shown in Fig. 9.3, this purpose activates two fuzzy rules (3 and 4) on A: rule 3 operates at 60%, rule 4 at 20%. Now we have to work on two sets, as in Fig. 9.4. Kosko demonstrated that the output set (in our case B) can be obtained summing the two triangles (20% and 60%). Thus the output set is an additive fuzzy system, the x_m value of which is the mean of the segmented curve (Fig. 9.4): that is, the x value corresponding to the y arithmetic value in the field between rules 3 and 4. When we calculate it, the result is 3.98%.

The quantification of the total hedgerows needed is $x/320.5 = 0.0398$, that is 12.76 ha, of which only about 41% now exist; thus about 7.56 ha of new hedgerows, half simple (6 m wide), half double (12 m wide), are needed. Therefore $6,300 + 3,150 = 9,450$ m of new vegetated corridors.

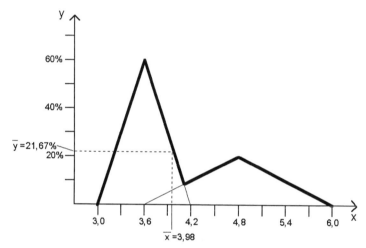

Fig. 9.4 Example of an additive fuzzy system, related to Fig. 9.3. x (mean) is related to the complex figure composed of the addition of the two triangles of 20 and 60%

9.2
General Landscape Diagnosis

9.2.1
Main Methodological Criteria

The evaluation of a landscape unit needs to be referred to the concept of ecotissue, as explained in Chap. 3. Remember that the ecotissue is the result of three main processes of integration, to which we may add a further operative one: (a) a range of spatial scales, from regional to local, (b) a set of thematic mosaics and arguments, (c) the range of temporal scales, which allow us to reconstruct and forecast the evolutionary dynamic of the landscape. This result can be configured in (d) an operative chart of integration.

The study of a landscape is usually led by thematic mosaics (geomorphology, vegetation, zoology, agronomy, land use, human needs, etc.) which have to be integrated for a certain purpose. But this integration is a sort of *a posteriori* process, following traditional multidisciplinary criteria, because the landscape is ultimately viewed as a *support* for biological and human systems.

By contrast, having defined the landscape as a living system, we must refer to a more complex structural model in which the integration is made *intrinsically*. This is possible utilising the proper and exportable characters of the landscape and their related ecological principles, analyses, indexes and evaluation criteria.

We affirmed that it is possible to study a self-organising system only through projections and sections of an hyper-space, as a way to gather information to be integrated, step by step, in a hierarchic correlation on the operative chart of integration, interpreting the structural components and the dynamic processes in coherence with the ecotissue model.

In Table 9.2, the analysis, evaluation and diagnosis of a landscape unit are shown. The integration processes have to be made within each scale and then among different scales, both at present and in the past. These results have to be again integrated and evaluated, arriving at the design of therapy and control.

Table 9.2. Analysis, evaluation and diagnosis of a landscape unit. The integration processes have to be made within each scale, then among different scales, both at present and in the past; these results have to be again integrated and evaluated, arriving at the design of therapy and control

Ecological scale	Analysis and evaluations: present	Idem past	Diagnosis	Therapy and controls
Regional 1:250,000 to 1:1,000,000	- Main components - Transformation dynamic - Ecological state			
Landscape 1:25,000 to 1:250,000	- Structure and delimitation of landscape units - Transformation dynamic - Ecological state - Comparison with regional data		Integration of the correlated mosaics into the operative chart	Design of operative criteria (therapy)
Landscape Unit 1:5,000 to 1:25,000	- Geotope analysis - Biotope analysis - Human system analysis - Basic mosaic characters - Changing trends - Operative objectives			
Detailed < 1:5,000	- Species control - Samples survey - Problematic spots		Ecotope diagnosis	Ecological controls on future order

The integration of the correlated mosaics of an ecotissue model into the basic one is not easy to achieve. The considerations derived by the mentioned mosaics of vegetation, fauna and agriculture are very important, like those derived from the inter-scales linkages. The distribution of local disturbances, the different roles of the groups of tesserae, the presence of possible urban patches at the boundaries, the networks of canals and ditches, the local points of connection with the surrounding parts of the landscape, the distribution of source and sink patches and corridors, and the difference in the landscape matrix help to localise the ecotopes of the operative chart of integration.

9.2.2
Ecotope Diagnosis

In fact, the basic mosaic of an ecotissue must be a synthetic one, because it is the integration among many other correlated mosaics. Generally it is the vegetational mosaic which facilitates this integration, but we may have exceptions, for instance in suburban landscapes. In cases like these, where in Europe the vegetation has no connected pattern and is widely invaded by e.g. North American species (*Robinia pseudacacia* L., *Populus x canadensis* Moench, *Phytolacca americana* L., *Solidago canadensis* L., *Erigeron annuus* (L.) Pers., etc.), the synthesis is more difficult.

In any case, our basic mosaic results in a group of ecotopes (e.g. 12 to 20 in a European suburban landscape unit) on which it is possible to make an ecological diagnosis and then to propose planning and design criteria. The problem is the complexity of an ecotope diagnosis, because an ecotope is defined as the smallest landscape unit with a given meaning. A survey on physiotope, biotope and landscape roles for a given research aim indicates only a rough preliminary method of delimitation of an ecotope.

The methodological basis of landscape evaluation was discussed in the first section of this chapter (see Sect. 9.1). Furthermore, to facilitate the diagnostic process a check-up through a schedule could be useful (Table 9.3). Partially analogous to the evaluation schedule of vegetated tesserae, this method considers the three main groups of parameters of an ecotope:

- A: Functions of the ecotope structure
- B: Functions shared with upper scale
- C: Some inferior scale characters

The four classes of evaluation are weighted through scores derived from a very general model. It is based on the observation that landscape evolution is a complex phenomenon, one following asymptotic curves, frequently exponential-parabolic shaped. In this case, the scores have simply been derived from the equation: $y = k \sqrt{x}$. Note that many parameters imply analytical and evaluative processes and some of them can be replaced by other variables, depending on the local ecological problems.

Table 9.3. Schedule for the preliminary diagnostic evaluation of an ecotope (or small landscape unit) with some human presence

Ecotope (Et)	1	4	9	16	score
A. FUNCTIONS OF THE ECOTOPE STRUCTURE					
A1- Shape/function corresp.	lacking	poor	near good	fitting	Sect. 3.3.2
A2- Ecotonal alterations	general	medium	partial	none	gradient
A3- Connectivity network	absent	minor	partial	normal	Fig. 7.10
A4- Geomorphologic alter.	heavy	few	marginal	none	
A5- Standard BTC classes	1-2	3-4	5-6	7	presence
A6- Diversities	too low	poor	mean	normal	H and τ
A7- SH / SH* (= σ)	< 1	1.01-2	2.01-3	> 3	HH
A8- Land use functions	intensive	mixed	tradition	mutualist.	mean
A9- Source elements	0	1	2	> 2	presence
A10- Sink elements	> 2	2	<1	0	presence
A11- Weighted average BTC	< 1.2	1.21-2.2	2.21-3.2	> 3.21	Mcal/m^2/yr
A12- Plant biomass (w. avr.)	< 2	2.1-6	6.1-14	> 14	Kg/m^2
A13- Movement dysfunctions	heavy	limited	marginal	none	flux etc.
A14- Faunal sensibility	critical	relevant	mean	low	see Sect. 8.3
A15- Historical rests	0	trace	1	important	artif. or nat.
A16- Transformation deficit	high	partial	none	positive	Sect. 4.8.2
A17- Contrast	none	too high	partial	normal	tesserae
B. FUNCTIONS SHARED WITH UPPER SCALE					
B1- Boundary connections	0	< 10	11-50	> 51	% perimeter
B2- Network participation	none	possible	minor	evident	natural
B3- Technical networks	many	few	marginal	none	artificial
B4- Difference with BTC(r)	< (-10)	(-9)–(0)	0 – 9	> +10	%
B5- Role in the landscape	reduced	minor	evident	important	
B6- Geo-physic instabilities	evident	partial	risk	none	physiotope
B7- Landscape apparatuses	0-1	2–3	3–4	> 4	types
B8- Permeant fauna interest	indifferent	medium	near good	attraction	key sp./ Le
B9- Pollution contamination	heavy	not much	marginal	none	from outside
B10- Positive margin attribute	none	rare	few	many	surround. Et
B11- Negative margin attrib.	many	few	rare	none	surround Et
B12- Landscape pathology	extremely serious	near chronic	easy to recover	none	landscape (L)
C. INFERIOR SCALE CHARACTERS					
C1- Plant η diversity	< 30	31–60	61–80	> 80	Sect. 11.2.2
C2- Plant forms	< 4	5–6	7–8	> 8	Table 7.3
C3- Exotic species	> 10	10–4	< 4	0	%
C4- Faunal microhabitats	few	common	uncomm.	rare	key sp.
C5- Critical singularities	danger	risk	neutral	none	artif. or nat.
D. RESULTS OF THE SURVEY					
D1- Total score Y (a+b+c+d)	a=	b=	c=	d=	
D2- Quality of the ecotope			Q=Y/544		

A1, shape/function correspondence; *A6*, H is the general diversity (see Sect. 7.2.1) and τ is the landscape diversity (see Sect. 11.2.2); *A7*, see Sect. 8.4; *A12*, weighted average among the PB of all the different tesserae of the ecotope; *A16*, referred to ecotope dynamics; *B4*, difference between the measured BTC of the studied landscape unit and the regional BTC; *B5*, functional role of the ecotope in the entire landscape unit; *B7*, no. of landscape apparatuses to which the functions of the ecotope belong; *B10-B11*, in relation to surrounding ecotopes

9.2.3
Evaluation of the Best Therapy

Another aspect of the evaluation of landscapes is related to the necessity to check the best solution for ecological compensation designs or similar therapeutic operations on the territory. An example seems to be the optimal way to explain our criteria regarding this delicate theme.

The study area is located at the northwest part of Milan, near the Ticino River. The Lombardy Regional Authority has requested a study for the ecological evaluation of a new railway-motorway exchange area. This area is going to transform a wooded territory, once characterised by the heath-land of Gallarate with many English oak woods (*Ornithogalo-Carpinetum* Marincek, Poldini e Zupancic 1982), today degraded, with the presence of *Robinia pseudacacia* L. trees. In 1856 the local landscape mosaic was composed of: heath 15%, woods 14%, coppice 55%, agriculture 16%. Today the same mosaic is: woods 5.6%, coppice 40%, agriculture 39.5%, urbanised 14.9%.

The Administration of the Ticino River Park has proposed a vegetated belt (about 100 m) around the transformed area, a rectangle of about 1.25 x 0.38 km. By contrast, we demonstrate that if we restore the mosaic, with about the same surface of reforestation as in the previous belt proposal, the ecological parameters should be enhanced.

In the first project, whose name could be "Belt Restoration", the coppice woods would be replaced by mature woods, replanted with central European trees. However in our project the name of which could be "Mosaic Restoration", the coppice woods are replaced by mature woods, replanted or integrated with central European trees like *Quercus robur* L. and *Carpinus betulus* L., and with patches of *Betula pendula* Roth and patches of restored heath-land. But the project is based on the ecological restoration of the core area of the local basic mosaic, out of the central part of the exchange area.

The mentioned demonstration is based on the simple synthetic model of Fig. 9.5 (besides it is an example of a split of a more complex split model). In the x axis we measure the fractal dimension D, on the y axis the median BTC of the same ecological mosaic.

We see the movement, that is the transformation dynamic, of this landscape unit during time: from 1856 to 1995 and from 1995 to 1998. We have plotted two bands and one rectangle, which represent:

- The BTC of Lombardy, as the median BTC of the region, about constant during the last century, the value of which is 1.9-2.0 (Mcal/m^2/year);
- The BTC of the natural habitats (NH) of the Lombardy region, which is obviously higher: 2.6-2.8 (Mcal/m^2/year) ;
- The optimum field of existence for a regional wooded park

At present, the study area has completely departed from its historical level of naturalness, especially after the beginning of enlargement of the exchange area.

The two mentioned proposals of restoration are plotted:
1. The Belt Restoration may increase the BTC, from 1.9 to at least 2.1 (Mcal/m^2/year), but the D remains 1.3.
2. The Mosaic Restoration increases the BTC close to 2.3, but the fractal D passes from 1.3 to 1.4, just in the direction of the optimum field. This direction is very important, because the growing of the restored tesserae toward their maturity could lead the landscape unit at least to the bottom of the optimum field. Moreover, the cost of the restoration will be cheaper, because of the cheaper price of the land far from the railway.

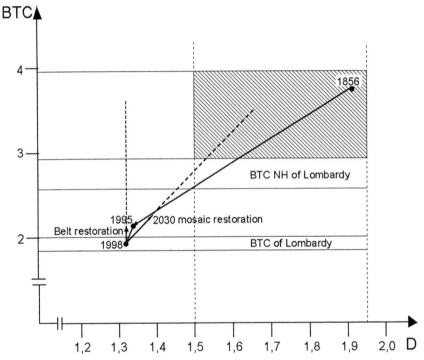

Fig. 9.5 Control of the transformation dynamic in the landscape unit of the Gallarate woods (Milan, Italy) and evaluation of the best solution for a therapeutic compensation of a newly built exchange area. See also Fig. 5.10

10 Landscape Ecology and Sustainability

10.1
Economy, Ecology and Sustainability

10.1.1
Economy and Ecology

Sustainable development is defined as the satisfaction of present necessities without compromising the possibilities for future generations. It should be clear that environmental sustainability is both an economical and a biological field of study. But a "weak sustainability" (in which nature and economics are interchangeable) is usually preferred to a "strong sustainability" (where critical natural systems must be preserved). It must be clear that environmentally sustainable development has to be defined as development that does not destroy the natural support function of natural living systems.

That is why *ecological economics* tries to evaluate the ecological services through a sustainability-based value (*S-value*): but ecological economics seems to find difficulties in understanding the right relationship between ecology and economy, principally because it refers only to traditional (out-of-date) concepts of ecology. Thus, the study of the fundamental relation between ecology and sustainability needs a deeper framework for economy and ecology, enhancing their peculiar limitations. Today it is not so difficult, even for a non-specialist, to recognise many obvious contradictions in the dominant economy:

- Natural evolution always makes an attempt to optimise levels of metastability, organisation, information, learning and beauty, maintaining the efficiency of the energy flux at a low level, in contrast with economical aims which maximise energetic-productive flux. The productive-consumption process in the ecosphere is based on recycling, while the economic-productive process seems to require a permanent expansion. Therefore, the ecosphere as a limited system can not sustain a permanent economic expansion plundering it, even putting aside demographic growth;
- Production is presented by economists as a function of labour and capital only, while resources are totally ignored. But it is so evident that work and capital are

not at the base of productivity: only nature is fully able to produce, because only nature possesses ontogenetic capacity. So, it is really non-sense what neo-classical economists affirm without any reasonable doubt: that ecological systems can be considered subsystems of the economy!

The true problem is that all the economists sensitive to ecological principles do not have a sufficient ecological background to propose alternative actions of wider cultural perspective. The relationship between ecology and economy is difficult because both are generally misunderstood and separated. For instance, in economy:

1. Exchange-value is thought to be the measure of richness and to be independent from nature. By contrast, we note that the value of something depends on its role.
2. Our culture rejects any ethical evaluation of means and ends. So the end justifies the means. By contrast, ends and means are strictly linked in nature, like a seed to its plant. Therefore they must not be separated and we have to choose among different ends.
3. Today absolute scarcity is not seen as a limitation, because the low entropy is abundant for our necessities. But these necessities must contain the proper functionality of all the ecological systems.

But also in ecology we have many problems:

1. Man is considered to be outside the ecological processes and in contrast with nature (e.g. E.I.S. environmental impact statement). By contrast, human ecosystems co-evolved and linked with all the others.
2. The concept of ecosystem prevails over all the other levels of biological organisation but it is now considered as ambiguous. General ecology must consider the entire biological spectrum.
3. Pollution problems prevail over all the other environmental questions. By contrast, the degradation and alteration of ecological systems (and of human health) depend first of all on the structural and dynamic dysfunction of systems of ecocoenotopes.

Our capacity to destroy the natural systems of the planet or to choose to save them (and ourselves) makes us a very special generation. We are veritably at a crucial turn of our history: we hope to change but not only because of disasters and wars.

Some scientists claim that economy is impossible to integrate with ecology, essentially because of the presence of the paradigm of growth. This radical position leads to a black pessimism on the man-nature relationships and it could be a dangerous position. Moreover, it seems not to be scientifically correct.

Although the neo-classical economic paradigm permits growth forever, we agree with the economist Herman Daly (1999), who underline that it is not the economic paradigm per se that mandates growth. Historically, the mandate arose because growth was the answer to the major problems raised by Malthus, Marx and Keynes. In fact neoclassical economists were thought to have found the answer to the problems of overpopulation (Malthus), unjust distribution (Marx) and unemployment (Keynes). Therefore, it is possible and urgent to develop a correct ecological-economic integration.

10.1.2
The Two Ecological-Economics Paradigms

The relations between ecology and economy show at present two major paradigms, the neoclassical one and the ecological-economics one. It is easier to explain their deep differences by referring to Fig. 10.1. The ecological systems, considered as subsystems of the paradigm of economy, postulates that ecology represents merely the extractive and waste disposal sector of economy. Even if these services become scarce, growth may continue without limits, since technology allows us to "grow around" the ecological sector by the substitution of natural capital with human-made capital. It results in the dictates of market prices, if and when prices of natural capital rise. Thus nature is nothing but a supplier of economy (Fig. 10.1B). The only limit to growth is technology, which is supposed to have no limits.

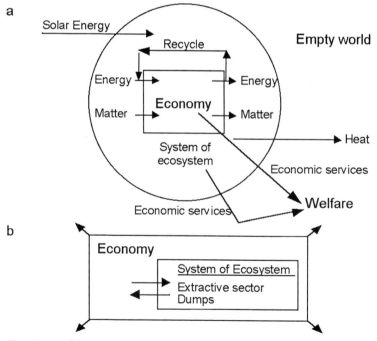

Fig. 10.1A,B. The scheme of the two main paradigms on economy-ecology relationships. (From Daly 1999: redrawn) **A** Ecological-Economics **B** Neoclassical Economy

Since economy is the whole, the growth of the economy is not at the expense of anything else: there is no opportunity cost to growth. Growth does not increase the scarcity of anything, rather it diminishes the scarcity of everything!

The other paradigm, the steady state at optimal scale, is at the opposite end (Fig. 10.1A). In the wide economy-ecology of nature, economy is a subsystem of ecology. Natural living systems are dissipative, not-growing as a whole, and materially closed. The flow of solar energy which sustains ecological systems is itself finite and not-growing. The natural environment physically contains and sustains the economy by regenerating the low-entropy inputs that it requires and by absorbing the high-entropy wastes that it can not avoid generating, as well as by supplying other systemic ecological services.

The two paradigms are per se logical: each is logical within its pre-analytic vision and absurd from the viewpoint of the other. Thus the true problem is not to integrate ecology and economy, but to overcome the discussed dichotomy.

10.1.3
Uneconomic Growth

Recently (1999) Herman Daly observed that the faith in economic growth can be tempered by the demonstration that growth does not imply only benefits but also costs. Uneconomic growth exists and it is defined as growth that, at the margin, increases environmental and social costs by more than it increases production benefits.

Economy, like ecology, can be divided into macro-, meso- and micro-economy, because the disciplines which have to do with the natural systems are scale dependent. Thus, some concepts pertain only to a particular level of scale. But some basic concepts have an universal logic: for instance, optimal scale, cost-benefit balance, system metastability and metabolic-antithermal functions. Remember the intrinsic and exportable characters of ecological systems (see Sect. 1.6.3).

Incredibly, in this perspective, macro-economy presents an evident contradiction. Cost-benefit balance is not considered! Let us refer to the GNP (or the GDP): we may think of it as the "built environment"; broadly conceived it is the set of all economic activities. Unique among economic magnitudes, GNP is supposed to grow forever. This basic index (GNP) increased about $ 85-90 per capita per year in the last half century, but the economic benefits are not comparable with its huge growth.

Note that Fig. 10.1A shows two general sources of welfare: services of human-made capital and services of natural capital. As the economy grows, natural capital is transformed into human-made capital. The law of the diminishing marginal utility of income and the law of the increasing marginal costs tells us that we satisfy our *most* pressing wants first and we sacrifice the *least* important ecological service first. Therefore, growth of economic services takes place at a decreasing rate, while natural services increase at an increasing rate. This process can be expressed in Fig. 10.2.

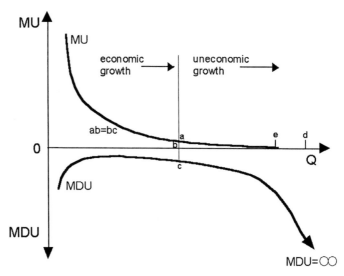

Fig. 10.2 Marginal utility (*MU*) and marginal disutility (*MDU*) vs quantity of produced goods and services (*Q*) (from Daly 1999, modified). See text for explanation

The MU (marginal utility) curve reflects the diminishing additions of marginal utility to the stock of human-made capital from consuming produced goods and services (Q).

The MDU (marginal sacrifices) curve reflects the increasing marginal cost of growth, as more natural capital is transformed into human-made capital. These costs can be for instance: sacrified natural capital services, disutility of labour, disruption of community, pollution, congestion, environmental destruction, sacrifice of leisure, etc.

Two further limits are noted in the diagram: point (*e*), where MU = 0 and further growth is futile even at zero cost; and point (*d*), where an ecological catastrophe is provoked (MDU goes to infinity). The figure shows that growth out to point (*b*) is economic growth (benefits > costs), while beyond point (*b*) there is uneconomic growth (costs > benefits). Thus, beyond point (*b*), GNP acquires another meaning and its growth does not reflects benefits but costs.

Moreover, we can observe that human-made capital and natural capital are complementary, not interchangeable; neoclassical economics is again wrong. This is an important statement. Substitution is reversible, complementation not. Another analogy with the scientific paradigm summarised in Chap. 2: both economy and ecology must follow these new paradigms.

There would be no point in expanding human-made capital beyond the capacity of remaining natural capital to complement it. What good are more fishing boats when the fish populations have become the limiting factor?

10.2
Landscape Ecology and Sustainability

10.2.1
A Crucial Ecological Discipline

In summary, we have noted that economy is clearly distorted but it is impossible to eliminate the distortions without the contributions of ecology and ethics. A more advanced ecology, meaning that it is necessary to pass from traditional to *unified ecology*. As we know, unified ecology studies all the hierarchies of the biological levels of life organisation (from the organism to the ecosphere), considering both the typical and the transferable properties of each level at all temporal and spatial scales. Landscape ecology is at the foundation of this ecology but it is important for sustainability even for other aspects:

- First of all, the landscape is the principal organisational level in which ecological systems and human systems fully interact.
- Therefore, it is the unique level within which actions consistent with great decisions necessary to sustainability can be concretely put into practice.
- Many of the exportable concepts of landscape ecology are indispensable to the evaluation of the state of the environment including at the scale of ecosphere (e.g. spatial patterns, biological territorial capacity, standard habitat, etc.).
- Many principles and criteria of application on nature conservation and territorial planning are based on landscape ecology.
- The strong space-orientation of this discipline and its articulate dynamic allow, through objective methods and results, the limits and the possibilities of the sustainability proposals to be demonstrated.
- The capacity to measure through ecological criteria both natural and human ecosystems interacting in a landscape or in landscape systems overcomes the apparently incommensurable complementary services.

The key importance of landscape ecology in the study of sustainability was recently enhanced in a scientific congress of German and Italian ecologists in the Villa Vigoni Centre (Lake Como), *Die Nachhaltigkeit in der Landschaftsoekologie* (Landscape Ecology and Sustainability), coordinated by Rita Colantonio Venturelli and Wolfgang Haber (Colantonio Venturelli 2000). In this congress the most recurrent argument was the possibility given by landscape ecology criteria to plan, design and control sustainable environments.

Creating sustainable environments is probably one of the major possibilities of landscape ecology to contribute to sustainability. Moreover, as underlined by Forman (1995), surprisingly, the role of spatial arrangement of the land is commonly overlooked: little literature exists on sustainability at the landscape and regional scales. But these aspects will be reviewed in the chapters on applied landscape ecology. We will show in the next section that even the control of the environment at world scale is presented, with the help of many landscape ecological concepts, in a different way compared to the common description of the ecological state of earth.

10.2.2
The Ecological State of the Earth

The World Bank (1992) stated that more growth was the automatic solution to the environmental problem! The answer of economic ecologists like Daly is: it could be, if that growth was economic, making us richer rather than poorer. But growth is becoming uneconomic. An uneconomic growth will not sustain the demographic transition and cure overpopulation, neither will it help redress unjust distribution, nor cure unemployment.

In Fig. 10.2 it is easy to understand that beyond point (b) growth becomes uneconomic: it is not able to balance the costs due to depletion of resources, pollution, disutility of some labour, sacrifice of leisure time, and especially disruption of life-support services and destruction of communities with the interest of capital mobility, take-over of habitat of other species and serious impoverishment of the inheritance of future generations.

The problem is: has point (b) been reached yet, is it going to be reached, or do we have some more time? The answers are at the opposite ends: environmentalists, ecologists and some economists are convinced we are already beyond point (b), but neoclassical economists are not. The response of the dominant economy, still led by the neoclassical paradigm, is global economy. Many national problems, still relatively tractable at landscape scale, are converted into one intractable global problem in the name of "free trade" and in the interest of trans-national capital.

We agree with the ecological economics paradigm: the neoclassical paradigm permits growth forever, but does not mandate it. As underlined by Daly, the mandate was the answer given to the problems raised by Malthus, Marx and Keynes. But the demonstration that the ecological state of the earth is really in peril, that we are passing beyond point (b), is still too weak to be indisputable, because of a non-scientific alarmism. Let us consider, for instance, some of the major ecological problems:

- *population growth*: the highest world's population growth rate was reached in 1975 (2.0%), but today it is 1.4%, the same in 1912 (Ives 1998). The tendency is today reversed, even if at least for one generation we will again have growth because of the parallel decrease of the mortality rate. Moreover, the carrying capacity of the ecosphere is not at the upper limit. In the case of a desirable redistribution of resources, we can calculate (see Sect. 8.4.1), considering a security coefficient of 1.8 on the weighted theoretical SH* mean, a need of about 2.500 m^2 per capita of human habitat (HH). Thus, the total HH being about 25.2% of the total emerged lands (1.49 x 10^{14} m^2) we can calculate that its value will be 3.755 x 10^{13} m^2 with a population of 6 x 10^9: so we can theoretically double the population without increasing the human habitat of the Earth. Note that in any case the increase of the ecosphere HH has been no more than ½ of the population growth rate.

- *Carbon dioxide*: the Earth is now at the peak of its inter-glacial period, at the beginning of the descent toward a new glacial phase. This was demonstrated studying the history of atmospheric CO_2 back to 420 Kyr ago, recorded by the gas content in the Vostok ice core from Antarctica (Sigman and Boyle 2000). During peak glacial periods, atmospheric CO_2 is 80-100 p.p.m.v. lower than during peak interglacial ones. Each period shows a cycle of about 100 Kyr, but at the end the rise of atmospheric CO_2 is very rapid, about 1 % per year (from 200 to 300 p.p.m.). The process is again not well explained, even the deep ocean currents are surely implied. In addition, at present, the sun is showing a strong activity, the influences of which are not completely known. In any case, the additional CO_2 given by human activities must be considered (it is at least 16-18% of the total), but it is not the principal agent responsible for the high increase.

- *Biodiversity decline*: the most important ecological process of this evolutionary phase of the ecosphere is the expansion of human population and its habitat. In the period 1850-1950, the human population passed from 1 to 2.9 billion, increasing its rate from 0.8 to 1.6%, while the habitat nearly doubled, from about 10 to 19% of terrestrial ecosystems. The need of natural resources obviously increased in a similar way. Therefore, the decline of biodiversity was in general simply inevitable. The rise of the species extinction rate in the same period was sensible, from 0.1 to 0.3% for birds and mammals (Primack 1998). But we have to observe that this increase was only ¼ of the increase of the human population. Moreover, we have to consider not only the species biodiversity but all the main types of biodiversity, first of all the landscape ones.

It must be clear that we *strongly agree* with the exigency of population control and redistribution of resources, with atmospheric gases control, with the principles of conservation biology for species preservation. We do not agree with incorrect scientific information. We fear that too much alarmism regarding arguments like the above mentioned ones may distract from other, more critical ecological questions: first of all, the evaluation of the transformations in the ecosphere's main ecosystems and landscapes.

10.2.3
The Evaluation of the Transformations of the Ecosphere

Uneconomic growth has been recently reached, the mentioned point (*b*) bypassed. Despite the changes in population growth and the increase of HH, the ecological state of the ecosphere was mainly stationary until about 1950. As shown in Table 10.1 and Fig. 10.3, the mean biological territorial capacity of the ecosystems and landscapes of the emerged lands changed (from 1882 to 1950) from 2.56 to 2.54

Mcal/m^2/year. It was a good ecological state, comparable with an open forest landscape, the most representative living system on the continents despite 2.9 billion population and a growth of human habitat of about 1.46% in 70 years.

Table 10.1. Evaluation of the transformations in the ecosphere's main ecosystems and landscapes in the past century, and estimation: (a) of the biological territorial capacity at that scale, and (b) of the human habitat. Note the increase in population

Ecosystems and landscape types	1882 10^6 km^2	BTC	1950 10^6 km^2	BTC	1985 10^6 km^2	BTC
Tropical rain forest			17.0		10.3	8.7
Tropical seasonal forest			7.5		8.0	4.0
Temperate evergreen			5.0		3.0	5.8
Temperate deciduous			7.0		3.0	5.4
Boreal forest			12.0		9.0	5.0
Woods and coppice			8.5		6.0	3.0
Forest	54.6	5.80	57.0	5.60	39.3	5.35
Graminoid savannah					6.0	1.2
Temperate grassland			9.0		12.5	0.5
Shrubby savannah			15.0		13.0	1.9
Grassland	25.0	1.29	24.0	1.25	31.5	1.21
Tundra			8.0		9.5	0.4
Swamp and marsh			2.0		3.2	1.9
Freshwaters			2.0		2.0	0.08
Other Wild	21.1	0.70	12.0	0.69	14.7	0.68
Tropical cultivation					9.5	1.0
Temperate cultivation					6.5	0.9
Cultivated	11.6	1.00	14.0	0.96	16.0	0.95
Semi-desert and urban			18.5		23.0	0.4
Extreme desert			8.1		8.2	0.05
Desert	21.2	0.28	26.5	0.29	31.0	0.30
Iceland	15.5	0.01	15.4	0.01	15.3	0.01
Total Terrestrial	149.0	2.56	149.0	2.54	149.0	1.90
Coastal seas	29.0	0.3	29.0	0.3	29.0	0.3
Open oceans	332.0	0.01	332.0	0.01	332.0	0.01
Total Ecosphere	510.0	0.76	510.0	0.75	510.0	0.58
Population (10^6)	1.400		2.900		4.800	
Human Habitat(%)	13.0		19.3		23.0	

The data on areas have been taken from: E. Goldsmith, R. Allen (1972) for 1882; R. H. Whittaker (1975) for 1950; Kimmins (1987) for 1985

By contrast, the successive time period shows a dangerous situation. The mean BTC decreases from 2.54 to 1.90, losing about 25% of its value in only 35 years! A simple projection to today leads to a BTC of about 1.60 (or less) Mcal/m^2/year, comparable with a mixed shrub-rural landscape as the present most representative living system at continental scale. The growth of human habitat can today (2001) be about 25.2% of lands, a quite limited increase if confronted with the increase of population: 133 vs 207% in 51 years. The mean BTC of the ecosphere, 0.76 to 0.75 in the first period (1880-1950), declined to 0.58 in 1985 and could today be only 0.49 Mcal/m^2/year, a value representing a decrease of 35% in 51 years!

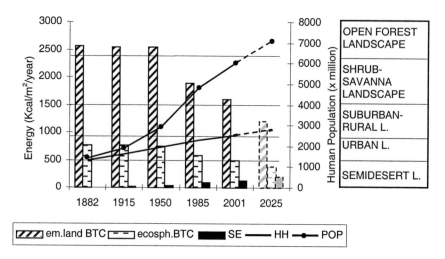

Fig. 10.3 Evaluation of the ecological state of the ecosphere starting from 1880 and projections up to 2025. Land mean BTC (em. land BTC) and ecosphere BTC (ecosph. BTC) in Kcal/m²/year; *SE* subsidiary energy in percent of natural energy (ecosphere BTC); *POP* population in millions; *HH* human habitat (% of emerged lands × 10^5). The decline of BTC is compared with the major types of landscapes: with a trend like this, in 2025 the HH mean BTC level will be reached, beginning a dismantling of the same *HH*. Note the exceptional increase of *SE*

Note that the BTC is proportional to the flux of energy needed by ecological systems to maintain their metastability. Therefore, we can utilise this "enlarged respiration" to compare the amount of subsidiary energy of human populations. In fact, we suspect that the relative limited growth of human habitat vs population indicates that what is mainly responsible for the decline of the BTC of the lands is the abuse of subsidiary energy. A study of Penner (in Ulgiati 1998) on energy used by man in 1952 and 1991 showed an increase of 517.4% in fossil oil, 284.9% in coal, 704.5% in natural gas, and an energy equivalent mean increase of 333.4% in only 39 years! This energy has been calculated as 16.71 (10^9 Gcal) in 1952 and 55.7 (10^9 Gcal) in 1991. A Gcal is 10^3 Mcal, thus we may compare the two energy fluxes, the natural and the human.

Observing that in 1950 the total BTC of the ecosphere could be estimated as about 3.775 x 10^{14} Mcal/year (the Earth surface being 5.10 x 10^{14} m² and 0.74 the mean BTC in 1952) and the total subsidiary energy 1.671 (10^{13} Mcal) we have a ratio of 4.4%.

Similarly, we can estimate that in 1882 we had about 0.6% of subsidiary energy in relation to natural energy. In 1991 the calculation results in 5.10 x 10^{14} x 0.54 = 2.805 x 10^{14} Mcal/year, vs 5.57 (10^{13} Mcal), thus in a ratio of about 20%.

Today (2001) following this trend we will reach at least 26%. After yearly mean increase of the subsidiary energy of 10.6% in the period 1882-1952, we pass to yearly increase of 11.6% in the period 1952-1991. A value like this is six times higher than the population increase (1.88 %) in the same period.

As shown in Fig. 10.3, the most critical period for ecosphere sustainability will be the decade 2020-2030. The projection of the decline of the mean BTC of the lands should reach about 1.20 Mcal/m^2/year; this limit is very important, because continuing in this direction means to destroy our own habitat (HH) resources. The landscape types representing the mean ecosphere ecological condition will probably be at that time (20-25 years from the present) a continuous sequence of cultivated, urban and grassland landscapes, surrounded by wide semi-desert, desert and shrubby territories and with scarce remnants of forested landscapes.

It is clear that any ulterior decline of BTC signifies destroying our human habitat. Note that from the present 25% the HH will grow until about 28%; it is not a huge increase, but we have to remember that the lands account for 29.4% of extreme deserts, ice-lands and waters and, we can add, impervious mountain landscapes: thus about 1/3 of the territories will be off limits. Another third (at least) is necessary to sustain natural systems (to prevent their ulterior dangerous destruction); thus even the HH is not far from its limits.

Moreover, the impressive increase of the subsidiary energy, reaching in 20 years about 50% of the yearly ecosphere mean BTC, enlarges the interference with the normal functioning of natural ecosystems and landscapes. The possible use of low cost energy could thus be at this moment a danger, even if it is clean, because of the entropy increase and the difficulty to control its use at the ecosphere scale.

In summary, we have only one generation (about 25-30 years) to begin to reverse the present trend. It must be clear that the ecosphere can support a wider destruction: the true problem is humanity itself. We can not destroy our own habitat. We have to underline again the danger of the tremendous increase of subsidiary energy. Remember that only when energy and technology are limited, the human dependence on nature is assured.

10.3
Principles of Landscape Ecology for Sustainability

10.3.1
Principles Related to Landscape Structure

Making a landscape or a region sustainable needs a brief review of previous chapters and of general principles. This will help to better insert landscape ecology into the context of proposals, discussion and elaboration of biological conservation designs and territorial planning.

1. *Landscape ecology should be a discipline like medicine.* In fact, since the landscape is a biological level, it is the physiology (ecology)/pathology ratio which permits a clinical diagnosis of a landscape, after a good anamnesis. It is necessary for landscape ecology to develop not as a simple predictive science, but also as a prescriptive one in order to contribute to the sustainability. This is just like medicine, and for similar reasons inclusive in the new scientific paradigms.

2. *Ecocoenotope, landscape, region are the three levels of life concerned.* An ecocoenotope is an ecobiota composed of the integration of the concepts of community, ecosystem and microchore. A landscape is a system of ecocoenotopes, in a recognisable repetitive configuration, the ecological model of which is the ecotissue. It is the basic element in a region, that is in the next broader scale composed of a non-repetitive, high-contrast, coarse-grained pattern of landscapes. Since the landscape is a living entity, socio-economic, technical and aesthetic disciplines are not sufficient to study it properly.

3. *The concept of ecotissue has a fundamental importance.* It can be found in the integration of elements and processes in a landscape. But this integration is not a sort of *a posteriori* process, obtained with traditional multidisciplinary criteria, in which the landscape is ultimately viewed as a support for biological and human systems. By contrast, defining a landscape as a living system, one has to refer to a more complex structural model in which the integration is made *intrinsically*. This is possible utilising proper and exportable characters of the landscape and related ecological principles and indexes, and interpreting the structural components and the dynamic processes consistent with the ecotissue model.

4. *Ecotope must be identified as follows.* The suggestion to survey the so-called physiotopes and biotopes is not sufficient. Recurrent (topographical) and genetic (geomorphological) relationships in fact are not always enough to locate an ecotope. It is necessary also to consider its configuration and its role within the landscape.

5. *Grain size has a great effect.* A coarse-grained landscape containing fine-grained areas is optimum to provide for large-patch ecological benefits, multi-habitat species including humans, and a breadth of environmental resources and conditions.

6. *Human and natural habitats are generally mixed.* It is difficult to find a completely HH tessera or ecotope. In many margins of human tesserae and ecotopes we may find natural species. Therefore even in urbanised landscapes generally it is possible to find NH patches and corridors. On the other hand, in natural landscapes many semi-managed patches may be found. Hence, the mosaic of NH and HH in a landscape is different from the land use mosaic.

7. *Landscape apparatuses are the best expression of the functional configuration.* Note that many landscape functions are typical of natural habitat (e.g. source, stabilising, geologic), others are typical of human habitat (e.g. productive, residential, subsidiary), and others may be in common (e.g. protective, resilient, connective). Note also that even human habitat apparatuses are intended in an ecological sense, not in a geographical or urban planning sense.

10.3.2
Principles Related to Landscape Dynamics

Here are reported the main landscape ecological concepts to be considered.

1. *Habitat standard per capita.* For permeant species the need for space is defined as a collection of adapted resource patches within a landscape. Its importance is based on the possibility to use SH per capita also for human populations, and on its coupling with landscape apparatuses and dynamics. Each type of landscape is characterised by a level of SH and its variations acquire ecological significance.

2. *Biological territorial capacity.* The BTC evaluation, associated with statistical data on the landscape, permits recognition of the regional thresholds of landscape replacement over time, and especially of the transformation modalities controlling landscape changes, even under human influence.

3. *Transformation in a human habitat.* The transformation modalities follow generally a cluster of parabolic functions crossing a series of thresholds, from semi-natural protective agricultural landscape type to the most urbanised one. The mosaic sequence remains that expressed by Richard Forman. Even if the opposite is theoretically possible, in human landscapes these transformations tend to be unidirectional.

4. *The change from one level of metastability to another.* The maximum metastability of a landscape generally does not correspond to the sum of the maximum metastabilities of its elements. This is an important statement, because it signifies that an ecological succession must not be viewed as simply orientated towards a climax. Therefore, in a landscape in which each component is fully at its climax, the maximum of metastability is not usually reached.

5. *Coevolution.* The history of the interactions among the elements of a landscape in a given area shows a particular dominion, characterised by the coherence of their reciprocal adaptation. This process leads to a stabilisation of the different homeostatic and homeorhetic capacities of a landscape, and to the formation of a landscape biodiversity which may be expressed with a particular degree of metastability of the entire system.

6. *Ecological alteration.* The first signs are expressed at the ecocoenotope level and are generally correlated with the most sensitive populations. If there is a sufficient redundancy, other ecocoenotopes may occupy the landscape degraded niches, but when the alteration influences the entire ecotissue level, then a degradation process is full in action.

7. *Disturbance incorporation.* If landscape disturbances are incorporated on a regional scale, the biological territorial capacity of the region remains almost constant during a very long period of time, even under strong landscape changes. The incorporation of disturbances on a broad scale is a sign of a good metastability capacity, but it can hide serious problems on lower scales, especially when some ecological perturbations present a discontinuity of

incorporation and they are incorporated only on large scales.

8. *Complementarity of transformations.* The complementarity of the ecological transformations in a landscape is due to the dislocation of the disturbances from one type of component to the other. If the entire range of perturbations does not increase too much, or if it does not present "out of scale" disturbances, changing the distribution of the disturbances changes the transformations of the ecological system without destroying its capacity of incorporation.

9. *Dysfunctional landscapes.* Compared with functional landscapes, dysfunctional ones will have less patchiness, and any remaining patches will have lower concentrations of soil nutrients, lower water infiltration rates, lower levels of biological activity and lower production cycles (Tongway and Ludwig 1997). We may also add: higher transformation deficit, decreasing BTC, NH loss, incorrect ratio of HH and NH, decreasing correlation between heterogeneity and information, higher fragmentation, loss of connectivity, not congruent landscape apparatuses, landscape resistance, etc.

10. *Semiotic character of a complex system.* This is given by a specific set of dominant and rare signs, both important in finding the sense of the structure. Moreover, an altered state of a system is not always perceived through its clear disfigurement. This kind of degradation may be not visible, but the syndrome remains serious. The alteration may be perceived noticing a peculiar lack of information in comparison with that expected, or noticing the beginning of disorder where not expected.

11. *Interactions among ecocoenotopes.* All ecocoenotopes in a landscape are interrelated, with the movement or flow rate of objects dropping sharply with distance, but more gradually for species interactions between ecocoenotopes of the same type.

10.4
Environmental Ethics

10.4.1
The Need of Ethics for Sustainability

The greatest possibility to begin to reverse the present trend of ecosphere alteration depends on ethics. The relation between man and nature is at the basis of this dangerous process. Survival depends on knowledge, because life is a process of knowledge. An evolutionary law of gradual increasing consciousness was postulated by Teillard de Chardin (1955). Thus, we may hope to overcome this crucial challenge. However, man fears carrying out the prescription derived by the observation of nature and puts himself instead outside, over nature. His systems and disciplines are thought to be able to enclose and control nature itself.

Thus, in fearing to carry out the prescriptions derived from nature, we limit our knowledge and our conscience, we therefore limit evolution. We prevent the integration of acts, because we can not perceive the unity in the variety of things. The conceptual separation and the failure of integration among knowing, believing, thinking, wanting, choosing and acting lead the culture to a dangerous non-sense.

The relation between man and nature is principally a question of justice. As asserted by Gandhi (1937) "Justice is the truth in our acts", therefore harmony with nature is a question of justice. This has been confirmed ever since the roots of our western civilisation, in the first book of *"De Legibus"*, by M.T. Cicero:

> Thus once [man] will have studied heaven, earth, seas and the nature of all the things and will have seen their origin and their possible end, when and how they will perish, what in them is mortal and transient, what divine and eternal, and almost will have touched the Being who governs them, and once [man] will recognise himself as not closed by walls in a limited site, but citizen of the entire world such as a single city: in this magnificence of things, in this spectacle and knowledge of nature, immortal Gods, how much he will have known himself such as the precept of Apollo Pytius! How much he will condemn, how much he will despise and consider as nothing what mass reputes as important. (Chap. XXIII)

"Truth is as old as the hills and the entire universe", wrote Gandhi. And all the great religions sustain this vision. From vision like this emerges the role of man as a ruler: the *munus regale* (regal gift) of Christians, the *raj* of Hindus. The instinct of territoriality is evident in managing behaviour, very strong in human populations, which – reinforced from ethics - implies also the conservation of the ecosystem mosaic and the landscapes, the components of his unique habitat. Remember that in the Bible the verbs for work and care (*avàd* and *shamàr*) are the same used for the service to God: *"lavòd et haadanà"* to serve the soil (Genesis, Beresìtt 2.5) and *"lavòd et Iod Elohènu"* to serve our God (Exodus, Shmot 10.26).

10.4.2
Truth and Non-violence

The totality set of living communities has evolved a component, man, endowed with the same characters of the cognitive and creative process which govern them. Therefore, the major threat to biological conservation and sustainable economy is the betrayal of man's own role in nature. This is violence against life as a whole. Thus, the ethic of sustainability must be the ethic of non-violence. Non-violence is the other face of the truth.

Remember, wrote Gandhi, that to approach the truth it is necessary to follow a precise direction. And Popper (1994) underlined: if the truth were not absolute and objective, we should have the possibility to err. We create our theories, to which belong even our *gestaltic* perceptions and our hypotheses. Continuously we try to disprove these theories with reality, so that we can approach the truth. That is why we can not impose our theories through violence, because it signifies to refuse to approach the truth: in practice to fear this approach.

The search for truth through the only possible way, non-violence, shows that the first law of man is the law of sacrifice, as a consequence of living testifying to the truth. Lorenz (1978) wrote that a deeper vision of the causal linkages between the laws of nature can not eliminate our will, but it can change what we want. Thus we refuse to follow the prescription derived from nature: it is the main reason of our escape from responsibility. And the main excuse is the faith in false gods.

The first of these gods is "growth": the faith that activity overwhelmingly reflects benefits. This is not a scientific statement, but it helps to make all men richer: thus the richest fifth can continue to receive 86% of the total world income, the poorest fifth 1.4%. The price of this ideology is paid for by ecological systems.

Another false god is in fact the unimportance of nature, the "faith in human technology". Of course, we have excellent technology. But it is not omnipotent and can not go against the laws of nature. As shown in Fig. 10.1, economy includes ecology, there is no environment outside itself. The economic animal has neither mouth nor anus, only a closed-loop circular gut, the biological version of a perpetual motion machine!

But even the second economic paradigm has its false gods: "green fundamentalism", man as one of the many species of animals. All what is man-made is incompatible with nature protection. The faith in false gods obviously distances us from the truth and represents in this case an insurmountable obstacle to the integration between economy and ecology. Thus we have to turn toward the truth. There is no other choice.

Note that ethics may coincide in many cases even with the simplest wisdom, as considered by Sir Goldsmith (1994): what an astonishing thing it is to watch a civilisation destroy itself because it is unable to re-examine the validity, under totally new circumstances, of an economic ideology.

It is interesting to observe that the ideology of global capitalism has many aspects in common with the Marxist economy, albeit violent aspects. David Korten (1996) pointed out that, while there are clear differences, in both economies:

- There is a concentration of economic power in unaccountable and abusive centralised institutions (state or trans-national companies).
- There is a destruction of ecosystems and forms of life in the name of progress.
- There is an erosion of social capital by disempowering mega-institutions.
- There is a narrow view of human needs by which community values and spiritual connection to the earth are eroded.

Therefore a modern economic system based on the ideology of free market capitalism and growth forever is destined to self-destruction, for many of the same reasons that collapsed the Marxist economy in Eastern Europe and the former Soviet Union. To act in a way such that this modern economic system could collapse soon is an ethical consideration for all of us, before the collapse of our civilisation and of the entire ecosphere result.

11 Landscape Ecology and Conservation Biology

11.1
Naturalness and Parks

11.1.1
The Concept of Naturalness

In dictionaries the word "naturalness" is defined as the quality and/or condition of what is natural, but conservationists generally define it as the absence of human artefacts or influence. This is again a preconceived contrast between man and nature, forgetting that man is part of nature. What should be discussed is the modality through which man administrates nature rather than the right to manage nature.

Among the most important revolutions of the history of life were the emergence of prokaryotes, the emergence of eukaryotes, the development of multicellular organisms, the development of superior vertebrates and the emergence of man. Nevertheless, man does not realise all that. Trying to underline and preserve only the most natural habitats, in the sense of areas not influenced by human activities is not sufficient.

What is necessary is a concept of "diffuse naturalness", which implies management of landscapes in the sense of a re-balancing, following ecological principles, and including the human habitat. Natural conservation must be diffused in the human habitat, for example, the conservation of historical or archaeological artefacts has to be exerted even in natural habitats.

In addition, the present concept of naturalness must abandon every Cartesian reference, in which nature, intended without history, is mechanistic and consequently can be dominated by man, who becomes the engineer of nature. Instead reality has to be intended as historical and evolutionary, in which order and disorder play equal roles. Nature is not determined once and forever, because it is open to unpredictable creative perspectives.

An agricultural landscape can be considered natural or not. The discrimination is not the presence of man or his changes on wild ecological systems, but man's type of management, his abuse of technology, his differences with ethical principles and the disjunction between ecology and economy. Even an urban landscape can be considered as natural: this is a rare thing, limited to old villages in some mountains, but in theory it is possible and historically it happened.

For this purpose we can refer to an indicative episode. In an agricultural landscape near Parma (northern Italy) local ecologists and environmentalists protested against the building of a sugar factory which used many hectares of fields to contain the warm water expelled from the technical plant. This protest had no consequences. A few years after, the Italian Association for Bird Protection (LIPU) opened a site there for bird watching, because of the return of a relative rare species of stilt bird (*Himantopus himantopus*) exactly there, attracted by warm water!

We know that admitting the possibility to consider an agricultural or even an urban landscapes as natural can be source of discussion and/or disagreement with many conservationists. But, as we have seen in the last section (9.3) on ethics, we must separate the abuses of man which lead to nature destruction from the transformation of parts of natural landscapes following natural laws, in which the species *Homo sapiens* takes part.

This is not in contrast with the concept of naturalness such as the prevailing definition of nature, derived from the concept of wilderness according to Thoreau (1854) who wrote: "in wilderness is the preservation of the entire World", or according to Muir (1901): "Wilderness is a necessity; mountains parks and reservations are useful not only as source of wood and water for irrigation, but such as sources of life".

In fact, for instance, our Alps can be considered in many cases as good examples of diffuse naturalness able to include also human populations in extensive landscapes of forests and grasslands which preserve rare species and many predators in the valleys not disturbed by tourism and technology.

11.1.2
Natural Parks and Reserves

After the efforts of Troll in the 1950s and Zonneveld in the 1960s, nature conservation organisations initiated a change in their environmental strategies. But still today, many principles and concepts of landscape ecology seem to be in contrast even with international conservation associations like the IUCN (International Union for the Conservation of Nature), linked to the UNESCO. The definition of national park given by IUCN (Jongman and Smith 2000) refers to a relatively large area, where:

- One or more ecosystems have not been changed fundamentally by human exploitation and habitation, where plant and animal species, geomorphologic objects and biotopes of special value occur or that contain a natural landscape of great beauty.
- The highest authority in charge of the country took steps to avoid potential exploitation as soon as possible, to reduce settlement in the whole area and to stimulate effectively the conservation of ecological, geomorphologic, and aesthetic characteristics that led to the initiative of its foundation.

- It is allowed to visit the area under special conditions to inspire educational, cultural and natural values.

The concept of landscape seems to be reduced to beautiful scenery! Moreover the role of man is considered as negative. But we can not forget the result produced by the application of these rules in the Abruzzo National Park (in the Apennines, central Italy). In one of the core areas of the park the sheep herds were removed to improve the naturalness of the brown bear (*Ursus arctos marsicanus*) habitat; as a consequence, many bears followed the herds, going outside the park. In fact, many centuries of co-evolution between bears and shepherds can not be cancelled by a preconceived idea of nature without men.

Episodes like this could enlarge the different positions of American and European conservationists, considering wilderness areas vs cultural landscapes. But ethologically speaking (Mainardi, personal communication), co-evolution is a fundamental process even between man and nature.

If some conservationist insists on underlining the contrast between man and nature, that can be dangerous especially when the aim of the study is a conservation plan. In a vision like this, the ecotopes or the landscape units to be protected, such as parks and reserves, are generally considered as islands. This fact may be regarded as an alibi, leaving the surrounding landscape free from any bond to them, with a consequent further degradation.

What is more, the destruction of crucial habitats (both natural and human), which constitutes the first cause of biodiversity decline, is normally considered less important than pollution problems.

The increase of protected areas in the world (emerged lands) had a positive growth impulse after the 1970s.

- From 1880 up to 1970, parks and reserves increased from 19 to 2,213 x 10^3 km^2 (24.38/year).
- From 1970 to 1993, they reached 8,981 x 10^3 km^2 (294.26/year). A good result. But if we compare these values with the decrease of forest landscapes (see Sect. 9.2.3):
- from 1950 to 1985 we note a decrease of 505.71 x 10^3 km^2 per year.

In 1993, only about 6% of the Earth (emerged lands) was protected. In the same period, in the European Union and in the United States of America this amount increased to about 11%.

The protected areas do not always correspond to landscape ecological principles. At least in Europe, many of the regional or national parks need a strong rehabilitation of ecological characters.

That is why a concept like diffuse naturalness becomes important and a model like the one of Erz (1980) is more effective than preservation through isolated park areas. This well known model is founded on the presupposition of the necessity to attend to an entire territory, using different degrees of protection.

11.2
Biodiversity and Landscapes

11.2.1
Considerations on Biodiversity

According to many well known authors (Wilson 1992; Primack 1998; Meuffe and Carrol 1999), although the total number of species and families of organisms has increased during time, we can register five episodes of mass extinction over the eons, in which a large quantity of these families and groups disappeared.

In the Ordovician (500 millions year by present) about 50% of animal families became extinct; in the Devonian (345 millions year by present) about 30%; at the end of the Permian period (250 million year by present) the most dramatic loss of groups occurred and involved at least 50% of families, including about 95% of marine species, many trees, amphibians and all trilobites. The extinction of dinosaurs was in the Cretaceous (65 millions year by present), while the last extinction is recent, in the Quaternary, a few millennia ago.

One of the main agents of the last extinction was man and there is no doubt that this extinction is still in process. We have underlined that the most important ecological process of this evolutionary phase of the ecosphere is the expansion of the human population and its habitat. In the period 1850-2000, the human population rose from 1 to 6 billion, while the habitat area rose from about 10 to 25% of terrestrial ecosystems (see Sect. 10.2). The need for natural resources obviously increased in a similar way, therefore the decline of biodiversity was in general simply inevitable.

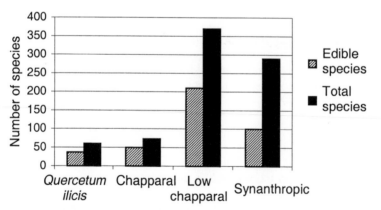

Fig. 11.1 Biodiversity of Mediterranean vegetation (from a study of Pignatti). Note the numbers of edible species

Thus, following the principles of landscape ecology, we observe that the problem of biodiversity decline must be reconsidered. If we strictly identify the biodiversity increase with the stability increase in ecological systems, forgetting the spatio-temporal scales, we may distort the reality. The basic importance of biodiversity requires no discussion, but each level of biological organisation, from genetic biodiversity to landscape biodiversity and up to the entire ecosphere has to be characterised by a proper degree of diversity, and not exclusively the species diversity in an ecosystem, in a landscape or in an ecoregion. What is more, the equation greater biodiversity equal greater stability is not applicable in every ecological system.

For example, in the Mediterranean landscapes the most mature forest (*Viburno-Quercetum ilicis*) generally is formed by 50-60 plant species, while the low chaparral (*macchia* and *gariga*) may reach 200-300 (350) species. Similarly on the alpine prairie (*Caricetum curvulae* Rübel 1911) we note a sequence of about 8-12 species in the pioneer phase, 40-45 species in the intermediate phases, declining to only 15-20 in the most mature state. It is clear that, if we limit the examination to the amount of specific diversity, we conclude that the most mature ecological systems are degraded ones: it is a nonsense, obviously.

On the other hand, evaluating the specific biodiversity at landscape level, it is possible to understand the importance of an ecotissue in which chaparral and forest ecotopes coexist. As shown in Fig. 11.1, the edible species can reach in the *gariga* as many as 200, therefore sustaining many more people than in a forested landscape, and in fact man transformed large forests into chaparral patches. The diversity of a mixed landscape formed by man is higher than in many natural conditions.

Here we have to recall the Prigogine process of order through fluctuations (see Sect. 2.3.6). The fluctuations of disturbances decide the evolution of living systems, in the historical sense. A history of the interrelations among the components leads to a higher order, therefore to a larger fitness and to the necessity to decrease the redundancy within a complex adaptive system. That is why the specific diversity is not to be emphasised. Rather, other ecological diversities have to be considered, remembering that the fundamental question of biological conservation is reducible to avoiding the decline of biodiversity, but considered in all its form of expression. At landscape scale we need heterogeneity and order, thus interacting ecocoenotopes with high, medium and low diversity, in the creative and evolutionary capacity of life.

11.2.2
Landscape Diversity and Metastability

One of the crucial problems in the evaluation of biodiversity is the incapacity of traditional ecology to measure it, aside from the species. Whittacker (1975) proposed his famous α, β and γ diversity indexes, but all based on species. [See an effective theoretical example used by Primack in his text (1998); consider three hill ranges in three landscapes (Fig. 11.2)].

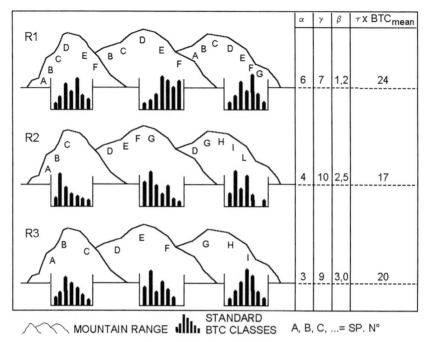

Fig. 11.2 Biodiversity indexes for three landscape units, each with three hills. The first case has the highest α diversity, the second the highest γ, the third the highest β. If we add the distribution of standard BTC values of the components of each ecotope, it is possible to measure also the landscape diversity and add the value of the landscape metastability (LM). The first case has the highest value, thus it seems the best to protect, even if the highest γ index is in the second case

Landscape 1 has the highest α diversity (per site), with more species per hill on average (6 species). Landscapes 2 has the highest γ diversity (per range), with a total of 10 species. Landscape 3 has the highest beta β diversity ($\gamma/\alpha = 3.0$), because all of its species are found on only one hill each. If we had to assign a priority, the optimum would be in landscape 1 in the case of having the possibility to protect only one hill; the optimum would be landscape 2 in the case of protecting a hill range. This scale follows classical criteria.

An attempt to link the diversity indexes with landscape ecology has been proposed by Brandmayr (1990): an η index relating the number of species to the ecotopes. But also this diversity is based on species.

To change these criteria, in order to make the right choice, it is necessary to add the index τ based on landscape ecological theory and applicable to measuring the diversity of landscape element types related to standard classes of BTC.

We know that landscape ecology considers also the heterogeneity of the components, emphasising that a very high heterogeneity of patch types is not per se a good index of the proper complexity of an ecological system. A question like this is always inherent to the Shannon index of diversity H (see Sect. 7.2.1), substituting the term species for the term landscape element. Remember that the equation of Shannon diversity expresses a measure of entropy and its maximum value is reached with the equiprobability of the distribution of the attributes of a system.

All this leads to considering that in complex adaptive self-organised systems the diversity of their components must consider both heterogeneity and information, therefore the proposed landscape diversity index is:

$$H (3+D) = \tau$$

where H is the Shannon diversity, D the dominance, τ the synthetic landscape diversity.

Looking at Fig. 11.2, the τ index is useful to distinguish between landscapes 1 and 2, because the latter has a higher γ index, but it is more probable that the optimal choice would be landscape 1, because its γ index is 30% lower, but its τ index multiplied per the mean BTC is 40% higher than in landscape 2.

This signifies a higher fitness among the components of the landscape and a higher level of organisation. In fact, both τ and BTC indexes are linked with the concept of metastability and that is why we prefer to call their product LM (landscape metastability), even if this index also measures a landscape complex diversity. Hence:

$$\tau \times BTC_{mean} = LM$$

Table 11.1. Standard temperate landscape types (see Table 7.7) and their landscape metastability index (LM), a measure proportional to the landscape diversity

Main temperate landscapes	Normal mosaic of standard BTC classes (d = dominant, r = rare)	Main range of LM	
Semi-desert	I(d) +II	0.1	0.8
Closed urban	I(d) +II(d) +III	0.4	2.7
Grassland	I +II(d) +III(r)	0.7	3.0
Open urban	I(d) +II(d) +III +IV	1.2	5.7
Suburban	I(d) +II(d) +III +IV +V(r)	1.8	7.0
Reed thicket	I +II +III(d) +IV(r)	2.0	8.0
Agricultural	I + II(d) +III +IV +V	2.5	11.0
Bush-land	I +II +III(d) +IV(d)	3.2	12.0
Rural-orchard	I +II(d) +III +IV(d) +V +VI(r)	4.5	17.0
Open forest	I + II(d) +III +IV +V(d) +VI	8.0	23.0
Closed forest	I +II +III +IV (d) +V +VI(d) +VII	15.0	37.0

Now we provide another example: for a more concise effect, it can be useful to collect the different types of tesserae for each standard BTC class (see Sect.7.5.2), seven within a temperate landscape, as shown in Fig. 11.3. Note that a balanced and general measure of complex diversity is very useful especially in designing a correct scenario to improve the biological conservation of an examined landscape unit.

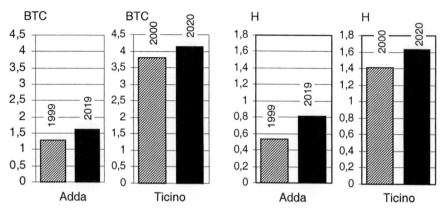

Fig. 11.3 Different conditions of the standard BTC spectrum for which it is easy to evaluate the parameters needed for the measure of a complex landscape diversity index. An agricultural landscape unit in the Adda River Park (see also Fig. 3.5.) vs a forest landscape unit in the Ticino River Park (see also Fig. 7.2). Proposed improved plans are added. See text for data

On the basis of Fig. 11.3, we may calculate the complex diversity of two cases of study:

A. Adda River Park (Lombardy), agricultural landscape unit

- Year 1999: Mean theoretical BTC = 1.30 Mcal/m^2/year
Shannon diversity H = 0.534; dominance D = 1.412; τ = 2.36
$\tau \times$ BTC = landscape metastability = *LM* = *3.06*
- Year 2019 (plan): Mean theoretical BTC = 1.62 Mcal/m^2/year
Shannon diversity H = 0.809; Dominance D = 1.137; τ = 3.35
$\tau \times$ BTC = landscape metastability = *LM* = *5.42*

B. Ticino River Park (Lombardy), forest landscape unit

- Year 2000: Mean theoretical BTC = 3.81 Mcal/m^2/year
Shannon diversity H = 1.418; Dominance D = 0.528; τ = 5.00
$\tau \times$ BTC = landscape metastability = *LM* = *19.06*
- Year 2020 (plan): Mean theoretical BTC = 4.14 Mcal/m^2/year
Shannon diversity H = 1.632; Dominance D = 0.314; τ = 5.41
$\tau \times$ BTC = landscape metastability = *LM* = *22.39*

Comparing the results of the case study with Table 11.1, we can see the correct perspective of the proposed environmental restructuring. Case A presents an agricultural landscape unit with a LM similar to a suburban area; the plan suggests to plant from 6 to 7 km of hedgerows and about 40 Ha of forest; the result of such a transformation leads to a more correct LM. The increase of 25% in the mean BTC and of 42% in τ produces the needed effect.

Case B shows that increasing by only 8.7% the mean BTC with an increase of 8% in τ leads to a 17% increase in landscape metastability.

Note that the new criteria suggested by landscape ecology are useful also in comparing regions of the world. The presence of a rare species changes in its importance, depending on the relation with the state of the ecotissue in which it is found.

11.3
Landscape Conservation Criteria

11.3.1
General Considerations

We agree with Myers (1996), who followed landscape ecological criteria and affirmed that in conservation planning it is better to preserve the possibility of species radiation than to focus on single rare, even if endemic, species.

Forman (1995) indicated the paradox of management. One can more likely cause or create an effect at fine scale, whereas success is more likely to be achieved at a broad scale. Thus, managing and planning for conservation and sustainability at an intermediate scale, the landscape or small region, appears optimum.

According to Forman, pattern and processes described in a book like this primarily apply to trees, shrubs, herbs, mammals, birds, reptiles and other conspicuous organisms. Like landscapes and regions, they are at a more easily surveyable scale, and a reasonable amount of scientific evidence is available on their species richness, dispersal patterns, edge-vs-interior preferences, and so forth. Conversely, little evidence exists for small organisms. Therefore, to increase the conservation success and natural sustainability, i.e. a sustainability not so strictly dependent on today's economy, the obvious solution is to make sure that a coarse-grained landscape contains fine-grained areas. This provides environmental resources and conditions for almost everything.

Concerning this argument we have to emphasise what was discussed by Sanderson and Harris (2000) on multilevel species evolution, because the distribution and abundance of small organisms may depend on keystone species and ecosystem engineers or top carnivores, thereby on species which affect ecosystems in a disproportionate way. Even humans can be considered in this way. For instance, top carnivores often act to organise prey populations in space and so also affect local resources disproportionately. Prey populations such as deer eat vegetation and can in turn affect the distribution and abundance of insects and hence insectivorous birds. As already underlined, evolution works on multiple ecosystems and hence on the landscape.

Another important consideration regards the necessity to refer conservation studies to a correct landscape unit, refusing administrative limits. The methodology of conservation planning should be elaborated in at least three principal phases:

1. *Diagnostic studies*: for instance the state of vegetation, the relationship between natural habitat and human habitat, the range of disturbances, the modality of transformation of the ecotissue, the state of ecotonal networks, the source-sink patch distribution, marginal problems and emergencies, etc.
2. *Therapeutic proposals*: strategies to eliminate landscape pathologies, such as new corridor formations, enlarging patches, substitution of exotic species, buffer zones design, alternatives for adapted human infrastructures, restoration ecology applications, strategic areas design, bio-engineering applications, etc.
3. *Ecological control and final design*: controls on ecotissue transformation, check on regeneration territorial niches, disturbances incorporation auditing, interference between natural and human networks, monitoring areas, etc.

According again to Forman (1995), measuring of *ecological integrity* is one of the most important attributes to check, especially for nature conservation purposes. A system like a landscape unit presents an ecological integrity when it remains near to natural conditions with four broad characteristics: productivity, biodiversity, soil and water, as discussed in Sect. 6.4.2.

Just as in the case of ecological integrity, basic human needs can not be forgotten even in biological conservation planning. That is why we have insisted on considering the basic human apparatuses in the landscape and the standard habitat per capita. But it is obvious that any nature conservation planning must be interrelated with territorial planning methods, as we will see in the next chapter.

11.3.2
Ecological Networks

One of the first and largest set of ecological protection criteria was elaborated on and proposed by the Society of Conservation Biology (SCB), in the United States of America. Since its foundation (1987), Michel Soulé, Reed Noss and their collaborators have been working on the "Wildlands Project", with the aim of preserving the entire continental territory of their nation. The modalities of this wide-ranging conservation plan are based on the concept of ecological network (Noss 1992), in parallel with a similar concept proposed by Bennett (1991) for the European Union.

As clearly reported by Jongman and Smith (2000), spatial transition from one natural reserve or park to another has attracted the interest of ecologists, geographers and land planners for several decades. Ecotones, "buffer zones" and natural corridors are concepts relying on the idea of transitional zones among ecological units.

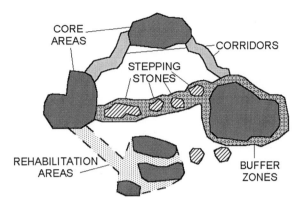

Fig. 11.4 The main structure components of an ecological network. With these components (core areas, corridors, buffer zones, stepping stones, rehabilitation areas) it is possible to put in evidence many aspects of a landscape, but not its entire complex structure

These concepts for nature conservation have recently been enriched by recognising their value regarding biodiversity maintenance and control of flows across the landscape. Especially in Europe, the concept of ecological network has had a recent wide spread affirmation, an incredible success.

The landscape is viewed as a network of patches or habitats connected by fluxes of air, water, energy, nutrients and organisms. Interactions among habitats are thus defined by these landscape fluxes and their function for certain habitat conditions.

Habitat quality is the main factor in determining the species occurrence in an area. The most important habitat conditions are: water supply, buffering of nutrients, energy and human impact, and dispersal of species.

An ecological network is composed of core areas, when possible protected by buffer zones and connected through landscape corridors and/or stepping stones. Buffer zones and ecological corridors preserve nature specifications of multifunctional landscapes. These landscapes support both natural and human functions.

As proposed by the wilderness project of SCB, there is a need to design some rehabilitation areas to implement the ecological network. A scheme of the structure of an ecological network is shown in Fig. 11.4.

One of the goals of the ecological network is based on the assumption that, by changing human influence in multifunctional landscapes, habitat conditions in the core areas may regain their former optimal status. Thus, proposals were made to establish networks on local, regional and national scales. The response to these new criteria of natural conservation is at present very positive in many European countries.

11.3.3
Merits and Limits of Ecological Networks

No doubt that ecological networks present clear merits. First of all, their methodological criteria are simple and the comprehensibility of the applied conservation principles gives territorial managers, especially public authorities, the possibility to implement the protection of nature. The territorial surface commitment is generally quite small and the political effects seem to be very much appreciated.

Some aspects of landscape ecology are transferred even to people with limited knowledge of ecological principles, or to people used to confusing ecology with pollution. That is why the Habitats and Species Directive of the European Union (EC 92/34) more or less includes the concept of ecological network, although it leaves much of its realisation to national governments. The development of the "Natura 2000" project included a network of special areas for conservation (SAC).

On the other hand, ecological networks present also many limitations. It is necessary to summarise the main questions, because we think that the first positive contribution to nature conservation will be cancelled if we do not change the methodology.

- The scientific basis of the model is excessively simplified: we have the impression that an easy solution to be applied in densely inhabited territories has been transformed into a scientific methodology.
- A landscape is a complex ecological system (see Chaps. 1 and 2), that can not be defined simply as a network of patches connected by fluxes of energy, matter and species.
- If the structural model of a landscape is an ecotissue (see Sect. 3.1.2), we have to study its processes, many of which can not be reduced to a network! See, for instance, the spread of exotic species which does not follow a network model.
- There is the problem of landscape diversity (τ), which depends on the presence of all the standard classes of territorial capacity of the vegetation, included fields, orchards and gardens.
- The distribution of natural habitat (NH) within a landscape is not distinguishable per landscape elements, but can be imbedded in each element.
- In ecological network planning, the core areas can not be only those having a high ecological quality, because of the source-sink theory.
- The distribution of the minimum resistance of a landscape may indicate a network pattern, but it does not represent the only possible pattern (for instance, new corridors can be designed even in high resistance zones).
- A landscape network can become, in some cases, too rigid with respect to the continuous transformation of many landscape units.
- For certain groups of animals a corridor can be of little or no utility, and even have negative effects: predators, parasites, etc.
- Consequently, landscape ecology risks appearing as an easy to apply trans-discipline, therefore allowing many professionals without an integrated natural background to plan and design networks.

It is not difficult to find other limits beyond these. A reference to a more complex methodology is needed at least for technicians and consultants. We have to consider, for instance, the relationships between the human ecological networks and the natural ones (see road ecology); or to make plain the hemi-networks following multi-purpose logic. In many situations of small or mean scale, many considerations "from the field" can be much more important than the ones we try to synthesise through ecological networks. We suggest the use of methods based on fuzzy logic (Kosko 1993, 2000) because of the impossibility to formalise mathematically complex systems, even if sometimes it is possible to use methods coming from auto-catalytic cycles and ascendancy (Ulanowicz, 2000). What is not scientifically correct is to plan and design ecological networks without having before, or in parallel, carried out a diagnosis of the ecological state of the landscape units involved, following criteria and methods of landscape ecology.

11.4
Conservation Biology, Bioengineering and Restoration Ecology

11.4.1
Bioengineering

The necessity to bypass many types of human barriers in densely inhabited territories requires a strengthening of ecological network applications together with bioengineering designs. Avoiding railways and highways constitutes the main problem, together with the re-naturalisation of channels. An example of green bridge design is shown in Fig. 11.5.

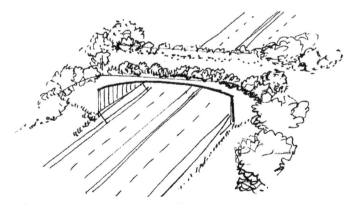

Fig. 11.5 An example of a green bridge crossing a road. These bridge have to be enlarged in their connections with the territory

Structures like these can be very expensive, thus they should be built only at crucial cross-points and combined with pedestrian paths. Note that barriers and patch fragmentation creates difficulties even in human populations in most cases.

Bioengineering represents an application method and a technique that can be considered complementary to landscape ecology in many aspects. We will summarise the differences and complementary aspects of these disciplines, because each conservation plan or design generally needs knowledge of detailed interventions. Let us compare what each discipline gives, allows and requires.

Landscape ecology:
- *Gives*: theoretical principles and methodological criteria to study a landscape and to check its state of ecological health; this is indispensable before interventions on territorial units (basin, river, landscape unit, etc.) if we want to contribute to their ecological effectiveness.
- *Allows*: to focalise on different objectives on different spatio-temporal scales; to apply therapeutic criteria on a detailed scale; to choose possible non-interventions in certain areas.
- *Requires*: bioengineering (and restoration ecological) techniques when overcoming barriers or other limiting factors and/or shortening of temporal scales for some processes (see transformation deficit).

Bioengineering:
- *Gives*: construction (and rehabilitation) techniques using primarily living materials, sometimes coupled with dead materials (even synthetic if forced by safety reasons).
- *Allows*: to reach diversified goals on various spatio-temporal scales (e.g. a punctual restoration of a connection may have very large influences); to realise landscape ecological therapies.
- *Requires*: a priori systemic study, in order to know all the characters of the landscape; a hierarchical scale of interventions.

Bioengineering can be a complementary discipline to landscape ecology even on a territorial scale, when its techniques are coupled with rehabilitation ecology methods.

11.4.2
Rehabilitation Ecology

When a landscape is affected by a serious pathology, each therapy must be based on a series of interventions that will restore its structural and functional behaviour. It is important to distinguish between the concept of rehabilitation and the one of restoration. The latter term means to bring back an ecological system to a supposed original state: this is theoretically impossible, as suggested by non-equilibrium thermodynamics. Thus, the correct possible operations can be:

- *Rehabilitation*: bring back to a normal life a subject (a landscape or its units) altered by a pathologic disease.
- *Reclamation*: bring back a wasted landscape unit or an ecotope to a useful condition, a site of cultivation or a marsh vegetation from heavy pollution, etc.
- *Reconstruction*: construct again an ecotope or a landscape unit completely devastated.
- *Recovery*: return to a former state of health, leaving an ecological system free from out of scale disturbances.

In projects of ecological rehabilitation we can insert also the compensation needed by a deep transformation of some ecotopes. A good methodology for a rehabilitation process can be expressed remembering the clinical-diagnostic sequence. The main phases are:

1. Analysis, evaluation and diagnosis of a landscape or of one of its parts
2. Prognosis of the corresponding pathology
3. Development of a realistic possible therapy, with its main objectives
4. Development of methods to reach the objectives
5. Incorporation of these methods in a territorial plan and its management
6. Control of the success of the rehabilitation project

A large quantity of landscape reconstruction is still designed by engineers, architects and landscape architects; thus, ecologists have a great deal to offer in this regard (Bradshaw 1983; Urbanska 1997). The methods of point 5 can be very numerous and each site generally presents a particular requirement. The aim of this section is not to describe technical methods, nonetheless it is useful to mention at least the importance of population dynamics in the vegetation rehabilitation.

According to Urbanska (1997), the safe sites and dispersal of plants have to be underlined. A safe site is defined as an environment immediately surrounding a seed which is favourable to its germination and establishment. Dispersal is the first component in the sequence of events which ensures the viability of a species in the long run. A simple model (Fig. 11.6) shows that the success of dispersal (i.e. deposition of propagules in safe sites) is influenced by two factors, namely (1) distance from the diaspore source to possible safe sites and (2) availability and density of safe sites.

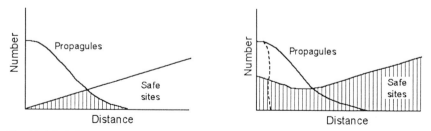

Fig. 11.6 Relations between dispersal and safe site abundance. (From Urbanska 1997)

11.5
Preservation of Traditions Rooted in the Landscape

11.5.1
Human Traditions Rooted into the Landscape

Forman (1995) wrote that since a sustainable environment maintains both human and ecological characteristics by definition, exploring the human dimension is essential. On the other hand, we underline (see Sect. 4.6.2) that even human culture is not outside of nature. Therefore, nature conservation has to consider also human components of the landscape, when non-altered by the breaking of traditions.

Traditions have a crucial role, because they may represent the information resulting from the interaction (and integration) between biological and cultural evolution. The tradition of human populations contains many cases of symbiosis, the most important of which is the agricultural one (Ingegnoli 1980; Frankel and Soulé 1981). It began in Neolithic times and represented one of the greatest revolutions in natural and cultural history. The agricultural symbiosis led to other important integration processes between man and nature, first of all mutualism with the landscape, and second the evolution of land races.

Mutualism with the landscape
- In the beginning of this book (see Sect. 1.1.1) we emphasised that man was obliged to gather information on the entire ecological mosaic which formed his territory, in order to be able to plan successfully his fields, orchards and settlements. Even defensive aspects, regarding both natural and human disturbances, played an important role in this sense, and it becomes clear that the emergence of the concept of landscape had the significance of an event of consciousness, regarding new linkages between man and nature.
- The landscape in which man decided to build a village or a town conditioned its construction in manifold ways: the settlement orientation and its roads, the form of the buildings, the local materials (woods, rocks, colours etc.), the wild as well as domestic plants and animals. Thus man changed the landscape and he himself was changed.

Evolution of land races
- The symbiosis between edible plants pre-adapted to cultivation, like wheat (*Triticum* spp.), was in fact based on a non-forced productivity, rather than on strength and easy cultivation. Thus, domestic species deeply influenced their selectors: cultures based on rice were very different from cultures based on wheat. While advanced cultivars present few differences among them, the local races show very marked differences, but they are not easily found. The importance of the landscape in influencing local races is evident, for instance, in wine types: a single tessera can be peculiar enough to produce a rare taste.

- Similar things can be noted for domestic animals. Many local races of cows, sheep and goats are going to be extinct in a few years, many others are already extinct. The same is true for chickens and pigs and their edible products.

The aspects of traditional culture that must be preserved are very numerous, and we have to observe the crucial importance of the landscape (Fig. 11.7). It is impossible try to conserve only single details or races without conserving their context and the tradition which led to them. Trying to save vernacular architecture in old towns and villages is not sufficient: to let their landscape be altered is already a sign of non-culture. A similar argument is still more evident in the case of old farms in their agricultural landscapes.

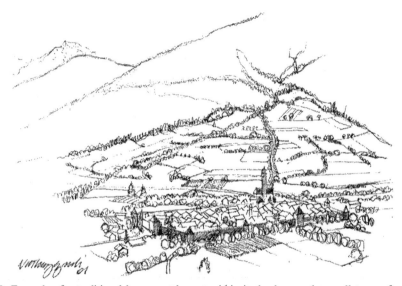

Fig. 11.7 Example of a traditional human settlement within its landscape: the small town of Glorenza-Glurns, South-Tyrol. Note the position at the bottom of the valley, along an important commercial road. The town walls act as a cellular membrane, the productive landscape apparatus is exactly localised on an alluvial fan, the protective and stabilising landscape apparatuses around (drawing by the author)

11.5.2
Landscapes and Archaeology

Archaeological areas are quite frequent over the entire world and they can even be large areas. Frequently these areas have lost many relations with their own landscape. However, to try to preserve these areas as isolated patches in a rather alien landscape is not a good cultural choice. An attempt to re-construct the surrounding environment must be done, at least in the most important cases.

The reconstruction can not be a true restoration, for the reasons we discussed before, but rather a rehabilitation leading to a characterisation of the same type as that of the past environment.

What is more, the conservation of monuments and archaeological areas can be reinforced by the biological conservation of the surroundings and vice versa. The occasion to preserve a very ancient constructed area and to rehabilitate the surroundings restoring for instance, a semi-desert to a planted landscape unit (e.g. with olives and pistachios in North Africa) may become an important intervention. Note that in many cases it is possible to find an ancient temple or theatre accompanied by prevalent exotic plants (Fig. 11.8). In Sicily, for example, it may be common to admire a classical Greek temple surrounded by:

- *Pittosporum tobira* (Thunb.) Aiton fil. (white flowering shrubs from Japan)
- *Eucalyptus globulus* Labill. (tall tree from Australia)
- *Eucalyptus* cfr. *pauciflora* subsp. *nana* (shrub from Australia)
- *Opuntia ficus-indica* (L.)Miller (thorny Indian figs from Mexico)
- *Agave americana* L. (agaves from Mexico)
- *Acacia dealbata* Link (yellow flowering shrubs from Australia)
- *Acacia karoo* Hayne (thorny shrub from South-Africa)

In cases like these the reconstruction of some patches and corridors with "macchia" shrubs and typical Mediterranean trees is truly important in a landscape in which very few remnants of natural vegetation are found.

Fig. 11.8 The importance of coupling archaeological areas with nature conservation: a negative example from the Valle dei Templi of Agrigento (Sicily, Italy)

12 Environmental Design and Territorial Planning

12.1
Design and Planning: An Ecological and Ethological Process

12.1.1
The Process of Design

From an ecological viewpoint, the role of man in nature should be principally managerial, because he should be conscious of possessing many attributes of the process which has evolved him, such as creativity, capacity to reach a goal, abstracting capacity of perception, capacity to organise space, etc. (Lorenz 1983; Ingegnoli 1980). In antiquity, man was conscious of this basic role: all religions emphasised it. We know that, even in science, an evolutionary law of gradually increasing consciousness was postulated by Teillard de Chardin (1955). We also know that survival depends on knowledge, because life is a process of knowledge.

Therefore, environmental design and territorial planning acquire a certain ecological and ethological significance (Ingegnoli 1975; Godron 1984). But we *have to recover it*, because even if the role of man is not in conflict with nature, as we underlined in Chap. 10, man fears carrying out the prescription derived by the observation of nature and tends to put himself outside, over nature.

Landscape ecology represents the best way to recover that basic significance of planning and design. In fact, to follow its principles means to design with nature, in the sense stated by Ian Mac Harg in 1969 and summarised by E.P. Odum in 1971, as we will see later. But also a deeper knowledge of the process of design can be useful, because it demonstrates that strong dominance of man in designing choices is an abuse, not compatible with a correct methodology. Let us summarise a design method, omitting its many feedbacks, and useful also for planning purposes:

1. *Preliminary phase.* Illustration of the theme and collection of main data on it; research of the significance of the objectives
2. *Setting out.* Main requirements of the project and their implications on the environment; trans-disciplinary linkages; preliminary project criteria and

structure typology
3. *Investigation*. Project functionality and its compatibility with the environment and local traditions; evaluation of the behaviour of the project and possible trends
4. *Proposition*. Choice of a compatible model and its development; reciprocal relationships with the environment; times, alternatives and controls
5. *Resolution*. Final creative elaboration and executive details; execution of the work and completion of details

Note that a correct design methodology can not avoid being confronted with the environment and with local traditions, especially in the first phases. But the main reason for the need to couple this method with nature is dependent on culture.

According to Konrad Lorenz (1978) the evolution of human culture occurs following particular processes able to develop and to conserve information which can not be genetically transmitted. This information is complementary to that of the genome and, together with it, permits man to be conscious of acquired knowledge through his central nervous system. These processes are governed by antagonistic bio-mechanisms, the harmonic development of which leads to a widening knowledge, helping in ordering disorder. This fact signifies that traditions can not be completely altered without receiving a dangerous answer from the system in which we operate; similarly for culture, because it is not outside nature. Culture needs nature as its principal substrate and reference, even if man can not is not always conscious of this.

Note that the culture which produced the paradigm of neo-classical economy (see Sect. 10.1.2) does not follow these criteria. Consequently its design method is different from what has been described here, because it is based on technology, efficiency, industrialisation and a strong simplification of reality, with no regard for the environment or, better, with a too strong simplification of it. The name of its design method is "rationalism". For instance, architectural rationalism is seen as being opposed to nature, while organic architecture is not.

Rather, in architecture, the discipline in which design is the essential method, classicism and non-classicism, rationalism and organicism, should coexist. What is more, architects have been interested in nature conservation and in man-nature relationships since ancient times. In the first important theoretical book of western civilisation on architecture (Vitruvius, first century) we can read:

> "*Preterea de rerum natura, quae Graece phyysiologia dicitur, philosophia explicat. Quam necesse est studiosius novisse, quod habet multas et varias naturales questiones*". (Besides, philosophy explains the nature of things, which Greeks name physiology. This [discipline] needs to be studied deeply, because many and various questions depend on nature).

The first explicit approach coupling architecture and natural sciences started with the foundation of the university course in landscape architecture at the Graduate School of Harvard, the same course which Forman teaches today, more than one century ago. This course derived from the actions of Frederic Olmsted, who was both biologist and engineer, and a well known American planner and designer, for instance, the creator of the Boston park system. European universities came later, but a good school programme arose especially in the Netherlands and in Germany.

The parallel process to couple nature and urban studies in the scientific faculties was much more recent (see Sect. 6.4). About 25 years ago, Sukopp founded the discipline of urban ecology (*Stadtoekologie*) in Berlin. He underlined that an urban area is not an homogeneous habitat, rather it is a mosaic of diverse biotope types (for us, it is better to say landscape element), many of which can be described as new types of environments, mainly created by design and planning applications.

But the process of design and planning has deeper roots than the theoretical one and it represents an innate human character. That is why we can learn from history if we want to recover the true significance, that is ecological and ethological, of these activities as we noted before.

12.1.2
Vernacular Design and Planning

Historical examples show that, since ancient times and up to the first half of the nineteenth century, only large cities and their suburban landscapes and some parts of the agriculturally most intensive landscapes were planned with a high degree of human technology and economical needs in a "rational" way, even in conflict with natural laws. Some parts of the territorial infrastructure could be added, such as bridges, main international roads, harbours, aqueducts, first railways, etc.; however the formation of a complete technical network is very recent, and particularly highway systems, oil and electrical powerlines, large super-markets, concentrated industrial areas, inter-ports and airports.

Let us observe the "spontaneous" settlements: they have nothing of spontaneousness in the sense of casual; they were planned with great attention and with much more attention to the natural characters of the locality than is given today. The old motto "genius loci" enhanced this capacity of our ancestors to understand and to follow the "spirit" of the landscape unit in which they wanted to settle. We have to underline that an urban ecosystem can be formed only through the mutualism and the coevolution of all the other ecosystems. *Civitas* (city), even if it was a small one, was never an isolated concept for Romans, rather the integration between the concepts of *urbs* (town) and *territorium* (territory). An example of an old settlement within an alpine landscape can be rich in significance (Fig. 12.1).

The localisation of each small town or village is accurately planned in relation to the geomorphology, the exposition of the slope, the heterogeneity of the ecological mosaic of vegetation, the presence of water resources, etc. The possibility to find some ecosystem amenable to being transformed without destroying or seriously damaging the landscape structure and function is another important factor. A village shows the form of an organism, with the characters just described for a single house. Local tradition emerges from the smallest scale (details) up to the large one (village-fields configuration).

AGOSTO 92   Vittorio Ingegnoli

Fig. 12.1 An example of an old alpine settlement (Mascognaz, Vallée d'Ayas, 1900 m above sea level, AD 1400 - 1500, northern Italy), showing the capacity to understand the "genius loci" of its landscape unit. Note the localisation on a grassy side of the mountain, near the margin of a forest of spruce and larch, the exposure, the morphologic adaptation, the building typology and materials, the agriculture, the vegetable gardens, the sheep farming: all are perfectly assembled in this anthropic ecocoenotope, demonstrating a real "organic" structure (drawing by the author)

Vernacular design does not follow the method of rational design, rather the organic one. The similitude with the evolutionary criteria, through which an organism is designed is incredible. In fact, as pointed out by Lorenz (1978), the structure of an organism can never be compared with that of a building designed by an experienced architect in a unitary project, but it is very similar to a country house built by a peasant.

The hut, built to protect from cold, wind and rain is enlarged (Fig. 12.2) proportionally to reflect the wealth and the family composition without destroying the old hut, rather changing it into a storeroom. In this way each room is enlarged and changed through time. The historical remnants are well recognisable and they are conserved because the building can only be restructured, not destroyed, since it is continuously inhabited.

The planning methodology was multifunctional, integrating and comprehensive of the management of common natural resources, first of all the forests. The distinction between the human habitat of each town or village and the natural habitat of the entire community has always represented an important limit to human artefacts. Note that the freedom of people too was maintained, even in difficult times, such as during the Middle Ages, in proportion to the capacity of planning: a typical example is the "Magnifica Comunità di Fiemme" (Trent, Italy).

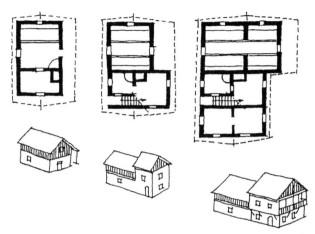

Fig. 12.2 An example of a vernacular architecture (map and section) from South Tyrol. Note the design method, planned step by step, to fit the house to the changing exigencies of the family

12.2
Requirement of a New Planning Methodology

12.2.1
Large Scale Changes

In January 1993 the international magazine Time published an important article titled "Megacities". It underlined the incredible increase of urbanisation in the last decade, openly speaking of a change of epoch, because the human population was going to live mostly in urban and suburban landscapes rather than in the countryside, as previously. As shown in Table 12.1, the increase is very large. Since 1950 (50 years) the percentage of people living in urban residential and industrial landscapes increased from 28.6 % to 54.2 % !

The reason for this process is only secondarily economic. Landscape ecology may demonstrate that it is principally due to the functioning of agricultural landscapes. In fact a productive landscape has a drastic limit on population density, the ecosphere mean SH* being equal to 1390 m^2/inhabitant, applied with a confidence coefficient of about 1.5-1.8. Thus, it seems impossible for an agricultural landscape to sustain a population with a SH<6250-7500 m^2/inhabitant, because such a landscape needs a potential stock of at least 5000-6000 m^2/inhabitant for the exportation of agricultural products. Theory and estimate converge very well, as shown in Tables 12.1 and 12.2.

Table 12.1. Development of the main types of human landscapes at ecosphere scale (million of inhabitants) (from Ermer et al. 1996, modified)

Human landscape type	Inhabitants							
	1950		1975		2000		(2025)	
	no.	%	no.	%	no.	%	no.	%
Urban, residential	210	7.2	1100	25.6	2180	36.3	4090	51.1
Urban, industrial	620	21.4	800	18.6	1090	18.2	1060	13.3
Rural, industrial	410	14.1	350	8.1	280	4.7	270	3.4
Rural, agricultural	1660	57.2	2050	47.7	2460	41.0	2580	32.2
Total population	2900	100	4300	100	6000	100	8000	100

Table 12.2. Development of agricultural landscapes at ecosphere scale and comparison with standard habitat per capita

	Surfaces			
	1950	1975	2000	(2025)
Human habitat (10^3 km^2)	28,800	32,600	37,200	(41,400)
Agricultural landscapes (10^3 km^2)	14,000	15,400	16,800	(18,200)
SH (prd) per capita (m^2)	8430	7510	6830	7050
SH per capita (m^2)	9930	7580	6200	5170

SH(prd) productive standard habitat per capita

Table 12.3. Population change within new large suburban landscapes in typical central European cities

	Population change %	
	Cities	Districts
HAMBURG (1970-87)	-11.2	+19.5
FRANKFURT (1970-87)	-11.6	+13.1
MUNICH (1970-87)	-8.4	+37.9
MILAN (1975-91)	-20.0	+12.2

The population increase is compelled, even in absence of specific laws, to migrate, gathering in abnormal suburban and urban landscapes and producing enormous environmental alterations, such as pollution, habitat fragmentation, temperature and rain increase (city heat bubbles), field abandonment, destruction of ecological networks, etc.

Moreover, this process tends to create new large suburban landscapes, even decreasing the urban population of the cities, not only transferring people from countryside. In fact, the quality of life in large towns has been altered by overcrowding, the formation of industrial peripheries and the increase of tertiary centres. Contemporarily, the relationship between city and agricultural landscapes has been broken. Thus, even advanced central European cities show this behaviour, as seen in Table 12.3.

Note that all these ecological processes can not be limited by administrative boundaries. Therefore, even in Europe, planning becomes more difficult, and without the contribution of landscape ecology it is rather impossible to plan properly these deeply changing territories.

12.2.2
Urban Ecology Problems

The first human settlements were more or less multifunctional in every component; by contrast, the emergence of cities produced a clear differentiation among their components (urban districts): residences, gardens, markets, administration, religious centre, defence, transport exchange areas, etc. Therefore, the city, a recent type of environment, can not be properly reduced to a single ecosystem, being in fact a typical landscape unit. These landscape units were produced by human populations that colonised a landscape, after having transformed some other landscape units into agricultural ones. The city is an heterotrophic system: this is one of the most evident characters of a city. It is dependent on a landscape, first of all through its agricultural resources (see Sect. 12.1.2).

Sukopp (1998) also suggested that urban areas are very heterogeneous, thus creating many unusual ecological conditions. Nevertheless, studies in urban ecology have been made principally on the basis of the traditional (and ambiguous) concept of ecosystem. But today it has become increasingly evident that landscape ecology plays an indispensable role in this field of science.

An urban system of ecocoenotopes differs from the surrounding landscape in a number of ways. Cities have a particular climate, generating a so-called warm island, which can have an average annual air temperature of 0.5-2.0 °C warmer than the normal one for the landscape. The column of warm air in the centre of the city attracts cooler air masses from the country, hence the suburban and peripheral districts have different mesoclimates. The increase of the temperature of a city is proportional to its enlargement: for instance, the annual mean temperature of Tokyo increased in the period 1953-1982 from 14.8 to 15.9 °C, while in Milan (1835-1999) the mean temperature rose from 12.4 to 14.2 °C. The relative humidity decreased in similar proportion. Research of the Physical Faculty (Historical Climatology) of the University of Milan showed a logarithmic correlation between the increase in air temperature and that of the radius of the city

$$\Delta T = 0.39 \ln R + 0.87$$

where: ΔT is the increased temperature of the surroundings, R the radius of Milan. The city's radius was in 1860 about 1.5 km, and today it is about 9-10 km, the difference of temperature ΔT changing from 1 to 1.75.

Material inputs into a city exceed normally the output; after centuries the ground level in old districts may rise several metres. As a result, the reduced infiltration capacity of water reduces the groundwater level. Urban soils are subjected to eutrophication and are over-compacted. In some cases (industrial zones) soil can be very polluted.

The urban environment is particularly rich in non-native flowering plants. The reasons for this are the physiologic attributes of many neophytes, demographic strategies, or degradation of soils (e.g. ruderal, eutrophic, etc.). But also other reasons can be found in very contrasted ecological mosaics, or in non-natural forms of patches, or in the absence of margins, and thereby in landscape ecological attributes. Data from Poldini (University of Trieste, Italy, personal communication) show that in the north-east of Italy, we may consider that from 3 to 4% of neophytes are in rural landscapes; in suburban units this value rises to 8% with an urbanisation of 20% and to 12% (or more) when the urbanisation reaches 60%. In large towns districts the percentage of neophytes is higher.

In the composition of the flora of Berlin, 839 native species and 593 non-native species have been found, of which 167 are archaeophytes and 426 neophytes, hence about 30% (Kowarik 1990). In Rome this value decreases to about 20% of its 1300 species (Celesti 1995). In the central section of the suburban South Milan Agricultural Park, neophytes reach 27%, but cosmopolites are very frequent (Ingegnoli 1999b).

Changes produced by man are in some cases favourable to animals, especially of small-medium size. New associative relationships can emerge, thus new communities. Urban waste discharges become a resource for some animals, influencing their behaviour, as for gulls, many populations of which have abandoned the sea coast for towns. Urban parks, especially because of their island effect, can host unusual species, like kingfishers (*Alcedo atthis*), or night heron (*Nycticorax nycticorax*).

Urban buildings also have ecological effects. They affect microclimate behaviour, geomorphology characters and urban topography, and resemble natural rock faces. We may have diverse types of districts characterised by many typologies of buildings: from the closed courtyards of a historical centre to scattered cottages with small gardens, to high towers in recent tertiary zones, etc. Many of the recent residential districts are too overcrowded and lack social local centres, remaining mainly as large dormitories.

The largest cities have incorporated many peripheral villages or towns, thus the growing process generates many contrasts to urban ecotopes: we may find an industry near a garden or a big condominium near an old villa. These contrasts produce in general ecological nonsense and traffic difficulties. The differences between the structure of the city and the structure of their peripheral suburban belts becomes increasingly fragile. The few suburban settlements that were strictly near the main entrances of the city (even only a century ago) have been incorporated into the growing city and the recent peripheral ecotopes show a confused structure, which contaminate also the rural landscapes.

An idea of the drastic transformation process of a growing city can be given by an example concerning the west district of Milan, which absorbed the peripheral town of Baggio about 70 years ago: more precisely the small district of the Gravel Pits Park of Baggio (135 ha) (Fig. 12.3). Using the BTC standard classes (see Sect. 7.5.2) we can see in the last 50 years a doubling of the first class (urbanised tesserae) from 19.3% to 43.0%, the decrease of field tesserae (II class) from 47.5% to 36.8%, the disappearance of hedgerows, orchards and coppices (III and IV class) from 30.1% to 3.3%, the increase of woods (V class) transformed into an urban park from 3.5% to 16.7% (see also Sect. 13.4.4).

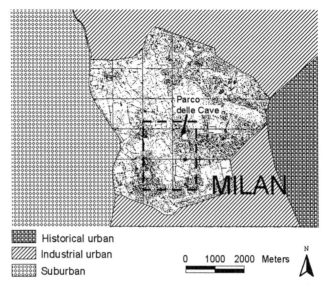

Historical urban

Industrial urban

Suburban

0   1000   2000   Meters

N

Fig. 12.3 The district Gravel Pits Park of Baggio (Parco delle Cave) and the landscape unit of the western side of the city of Milan (Italy)

Note that in this case study we had a rather contained change in the mean BTC, from 1.8 to 1.2 Mcal/m^2/year. But the landscape unit which contained this park district has been transformed from a rural landscape to an urban one with a change of BTC from 1.76 to about 0.50 Mcal/m^2/year between 1850 and 2000.

The cities are characterised by a very high human habitat, generally >90%, and the main landscape apparatuses show a standard habitat per capita varying from about (100) 120 to 500 (600) m^2, that means an ecological density of about (100) 84 to 20 (17) inhabitants per ha. Consequently, the productive standard habitat per capita (SH$_{prd}$) may reach only a small percentage of the theoretical SH (SH*). For example, let us summarise some data on Milan (Table 12.4): the urban district of Milan includes about 4,000,000 inhabitants (officially the Province of Milan is reported in the table, but we should add the southern parts of Como and Varese provinces), and in this large urban landscape SH$_{prd}$ contributes to less than 1/4 of its agricultural needs.

Sukopp (1998) reported that urban utilisation of subsidiary energy usually amounts to 25-50% of the natural energy; but we have evidence to assume a larger amount of it, at least 50-100% of the natural energy or more. In summary, any sustainable solution to planning urban landscapes, with their so complex and uneasy problems, needs a new planning methodology on a new and enlarged scientific basis.

Table 12.4. Human habitat, landscape apparatuses and SH of the city and district of Milan (limited to the Province of Milan), 1999

	HH	Total SH	SH subdivided into the main landscape apparatuses			
	(%)	(m^2)	RES	SBS	PRD	PRT
Milan, city (1,304,200)	96.8	142	69	15	40	10
Milan, district (3,753,900)	88.8	469	189	37	235	8
Lombardy (9,065,500)	64.4	1700	175	70	1305	150
SH*	(70-80)	1427	105	79	1043	200

See Sect. 3.4.1; *SH** is referred to Lombardy Region

12.2.3
Changes in Agricultural Landscapes

In the previous paragraph, we underlined that cities were founded by human populations that colonised a landscape, after having transformed some other landscape units into agricultural ones. Thus, heterotrophy is one of the most evident characters of a city: a rural-urban symbiosis of mutualistic character is indispensable to survival. Furthermore, as well expressed by Wolfgang Haber (1998), this fundamental ecological role of agriculture is misjudged or neglected by the majority of our society, who are city dwellers. Instead they are inclined to accuse farmers and economists of damaging or ruining important life-supporting components of the environment, even nature in general.

In western civilisation, agriculture developed following scientific suggestions of the Romans (Columella, first century): regular fertilisations, organised water supply canalisation, hedgerows, rural roads, well organised rural farms, crop rotation, selection processes, even the use of the first (horse-driven) machinery. Many agricultural landscapes in Europe still show some features of these sophisticated farming practices, e.g. the wine tradition of the Rhein valley, the remnants of *centuriatio* in France, Italy and Spain, etc. But the European population seems to have not exceeded 75 million people (third century) thus many forest patches and corridors remained undisturbed.

In the Middle Ages, empirically based agriculture dominated again almost everywhere, with the only exceptions being the most populated parts of France, northern Italy and south-west Germany. At that time the main agricultural landscape elements (cultivated fields, meadows and pastures, forests, uncultivated lands) were distributed without a real dominance of the fields. The exploitation of forests (from heating to building) without regard for their many functions or without replanting became commonplace for the majority of regions. The few fields, intensively cultivated, suffered from soil erosion, acidification and decreased productivity.

Wider rural landscapes emerged during Renaissance times, when Europe returned to about the population level of the third century; but the exhaustion of forests and soils continued in many European regions and at the beginning of the eighteenth century agricultural productivity had dramatically decreased. The need for modernisation was pressing, because the population began to re-grow after the terrible epidemics and the cruel wars of the century before. Forage meadows improved the nourishment of cattle and the production of more manure. Crop rotations practices were reintroduced in every region, sylviculture procedures and the wide use of coal for civil energy supply revitalised the forests. The introduction of mineral fertilisers (about AD 1750) resulted in a greatly enhanced productivity of crops. Drainage of swamps and irrigation of arid soils enlarged the agricultural landscapes, which in 1850 sustained about 250 million Europeans.

The twentieth century led a new type of transformation in agricultural landscapes, especially through the diffusion of machinery and technical devices, and the introduction of chemical crop protection. Intensive monoculture industrial farming enhanced habitat destruction and the spoiling of beautiful scenery, finally resulting in an excess of production. The EU Common Agricultural Policy has taken measures to contain surplus production and market failures, for instance with set-aside and promoting extensive agriculture. And, in the last four decades, agricultural landscapes have even been used for industrial buildings, expelled from the too dense cities, creating a sometimes dangerous mixture.

Interesting research on agricultural landscape transformation (Caravello and Ingegnoli 1991) was carried out in the Po valley, comparing two territories: Veggiano, 1642 ha (Padua), and Colorno-Torrile, 8950 ha (Parma). The main components of the agricultural landscape were compared after a reconstruction of their territories in the late Middle Ages (respectively 1421 and 1352), in the nineteenth century (1857 and 1808) and recently (1980 and 1987). An estimation of the mean BTC allowed a synthetic ecological diagnosis of these landscape transformations. See the analogies between the two landscape units (Fig. 12.4).

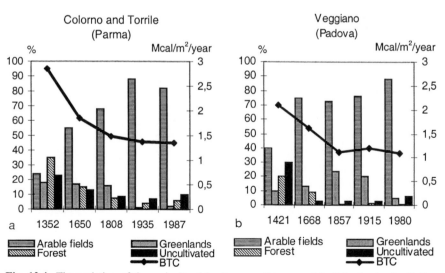

Fig. 12.4 The variation of the agricultural landscapes of two municipalities in the Po Valley (Italy). Colorno and Torrile (Parma), Veggiano (Padua). See the text for explanation

In the late Middle Ages, arable fields, meadows and pastures, forests and uncultivated lands were quite equally distributed; crops and grasses reached 50% in Veggiano, 41.5% in Colorno-Torrile and forests covered 19% and 35% of their territories. In the nineteenth century, crops and grasses increased to 97% in Veggiano and 83.2% in Colorno-Torrile, and forests were reduced to a few remnants, about 1% and 7%. The BTC decreased from 2.21 to 1.34 Mcal/m^2/year in Veggiano and from 2.69 to 1.64 in Colorno-Torrile; this was not so bad because of the presence of arable fields with trees and hedgerows (near 80% of landscape elements). The decrease of the BTC continued until today: 1.16 and 1.38 Mcal/m^2/year respectively, with a higher degradation due to the disappearance of grasses and hedgerows.

This example show the transformations of an agricultural landscape on very fertile soils. We know that it is ecologically not justifiable that protection of biodiversity should have priority over farming in cases like these; but this does not mean that it should be excluded altogether! It is not so difficult to change the ecological state of an altered agricultural landscape unit, as discussed in Chap. 11 for the Adda River Park, at present with a BTC = 1.30 Mcal/m^2/year and a LM = 3.06; by only decreasing the crop and grass area 10% (from 85% to 75%), we should have BTC = 1.62 and LM = 5.42. But a change like this needs a new planning methodology through landscape ecological principles and indexes.

12.3
Design and Planning with Nature

11.3.1
Basic Methodology

As already discussed, the study of a territory is traditionally led by thematic mosaics (concerning geomorphology, vegetation, zoology, agronomy, land use, human needs, etc.) which have to be integrated for a certain aim; but this integration is a sort of *a posteriori* process, when the landscape is ultimately viewed as a *support* for biological and human systems. By contrast, we must refer to a more complex structural model in which the integration is made *intrinsically*, because the landscape has been defined as a living system. This is possible utilising the proper and exportable characters of the landscape and their related ecological principles, analyses, indexes, and evaluation criteria.

That means to study a territory through its structural landscape composition, in which the ecotopes, the ultimately meaningful parts of landscape, have a fundamental importance (see Sect. 9.2.2) in parallel with the evaluation of the main processes related to the structure of landscape units. Therefore, the basic methodology must be the clinical-diagnostic one.

In this framework, remembering also what was previously briefly presented (see Sect. 6.6), we may present a sequence of phases related to a planning method, always noting the iterative elaboration of the design process.

1. Delimitation of the main structural elements of the examined territory, e.g. a landscape complex unit and its simple sub-units; hierarchic levels of spatio-temporal scales become important and geographic-administrative divisions have to be related to the ecological structure. Climatic data, geomorphologic data, vegetation belts, broad human land use and structural networks (natural and human) become necesseties for this integration. This phase needs also the identification of the broader scale landscape system (region or sub-region), and a preliminary ecological analysis of the borders.
2. Determination of the main vegetation types and human land uses, identification of the physiotopes, biotopes, landscape role, thus first identification of ecotopes.
3. Research and reconstruction of at least two past states of the given territory comparable with the present ecological structure, with their vegetation types and human land uses (from a few decades to a few centuries, depending on available historical data and the scale of interest).
4. Analysis and first evaluation of the ecological states of the examined landscape units, using ecological considerations, particularly what was explained in Chap. 7 and 8, starting from the present and going back to past situations. This phase can be very complex and needs also accurate field studies, thus many iterative passages from detailed scale to the scale of interest.
5. Diagnostic determination, beginning from the study of the general dynamics of the examined landscape units, and compared with the regional one. The evaluation of the landscape alterations and pathologies (see Chap. 5). The evaluation of the ecotopes may be the conclusive part (putting in evidence also eventual detail or spot problems).
6. Pre-therapeutic study. Needed interventions derived from landscape ecological alterations. Limits are given by ecological integrity and basic human needs with respect to the considered ecological system, by the ecological adaptability of the landscape structures and functions, by exceptional cultural indications, by the definition of the main possible objectives.
7. First design strategy. Delineation of major areas for nature conservation, human activities, territorial infrastructures, main roads, ecological networks, etc.; their dimensioning, their attributes, preliminary synthetic ecological controls, controls of sustainability (even ecological economics) and possible alternatives and scenarios and controls.
8. Plan design. Choice of the best scenario. Development of the plan design with sub-controls before the detailed design. Prescriptions on managing the plan and on its main subsystems (water, cultivation, housing, biodiversity conservation, etc.). Executive examples and details. Priorities and successive sub-plan strategies. Plan illustration to the local professional society and to people.
9. Control evaluation of the effects of planning on the ecological state of the landscape unit and surroundings landscapes. Possible monitoring webs.

12.3.2
Principles of Landscape Ecology for Planning

It is essential that a planning methodology is continuously related to basic principles of landscape ecology in a synthetic way. Some of the general principles have been presented in Sect. 10.3 and are obviously applicable here; but now we also need to add some other principles, more strictly related to planning requirements.

1. *Structural congruence.* A good plan must maintain at any scale a structural congruence. This signifies that transformations can not cancel or alter the spatial relationships among geomorphology and vegetation, hydrology and agriculture, old settlements and their grain and contrast with the surrounding, etc. Instead, design criteria have to follow historical (both natural and human) signs.
2. *Patch shape.* To accomplish several key functions, an ecologically optimum patch shape usually has a large core, with some curvilinear boundaries and narrow lobes, and depends on orientation angle relative to surrounding flows (Forman 1995).
3. *Aggregates with outliers.* Land containing humans is best arranged ecologically by aggregating land uses, yet maintaining small patches and corridors of nature throughout developed areas, as well as outliers of human activity spatially arranged along major boundaries (Forman 1995).
4. *Local compensation.* Serious transformations which alter many characters of a landscape unit have to be compensated through an ecologically balanced therapy inside the same unit, not far from the degradation.
5. *Indispensable patterns.* Top-priority patterns for protection, with no known substitute for their ecological benefits, are a few large natural vegetation patches, wide vegetated corridors protecting water courses, connectivity for movement of key species among large patches, and small patches and corridors providing heterogeneous bits of nature throughout developed areas (Forman 1995).
6. *Attractors.* In the presence of a landscape element presenting a high attractor potential, it is necessary to insert constraints to avoid the formation of a barrier due to human activities.
7. *Source-sink and boundaries.* Any landscape element presents traits of its boundaries sensitive to the surrounding disposition of source and sink structures. Consequently it is necessary for planning to leave the wider possibility of connections and fluxes regarding those traits.
8. *Complementarity.* The main ecological law (not too much, not too little, just enough) underlines the necessity to avoid any excess; therefore each sub-system of landscape elements needs to have the presence of at least one complementary component.

12.3.3
Underlining Some Planning Problems

Remembering what was expressed above, we need to underline at least three arguments regarding the most common planning problems: the limits of administrative data, the design of development models (general and local) and human habitat quantification.

- *Limits of administrative data.* In our concise synthesis of the planning methodology, we wrote that hierarchic levels of spatio-temporal scales become important, and geographic-administrative divisions have to be related to the ecological structure. But in the majority of cases the administrative divisions do not coincide with the landscape ecological one. Transformation of the data into the most correct ecological divisions is not always possible, because of the rigidity of the survey districts. Thus we have two main possibilities: the estimation of that part of the data not comprising our ecotopes or landscape units, or comparison of the administrative divisions data with the characters of the ecological divisions related to them. In any case, consider with care this problem, especially when it concerns historical data.

- *General development model.* Natural tendencies and human needs have to be integrated only after evaluation of the general landscape unit diagnosis. Starting from the diagnostic indications, once the human activities compatible with a therapeutic plan are dimensioned, we need to elaborate a sort of meta-design model capable of following landscape ecological principles. For instance, we mentioned some of the most important principles such as aggregates with outliers, local compensation, indispensable patterns, attractors, etc. In the above mentioned first design strategy, the physiognomy of the vegetation and its natural distribution directions play, together with geologic information, an important role in characterising the general development model. Also, the pre-existing human settlements have to be considered. Here the main problem arises when some landscape architects want to impose their vision, giving too much importance to aesthetic design. On the other hand, there is no doubt that a correct architectural composition in many cases can be very useful even when the ecological problems have priority. Sustainability controls and possible alternatives lead to new scenarios.

- *Local development models.* Each cluster of landscape elements which characterise an ecotope has to be typified and its need for landscape rehabilitation has to be put in evidence. In this way, adding the limits of transformations for human activities, it is useful to design a development model. For instance, a trait of a river corridor to re-naturalise or a buffer zone to be planted with the proper vegetation can be presented in good accordance with the contents of the previous general model and used as a compensation for the part of the ecotope which has the possibility to receive new buildings or other infrastructures. An economical account of the cost of realisation of the model becomes indispensable for real sustainable development.

- *Quantifying human habitat.* One of the crucial problems in landscape ecological planning is concerned with the design of the right amount of human habitat, especially when the case study presents landscape units to be preserved. No doubt that there is a limit to the transformation of a natural forested landscape: but the question is how to find these limits. It may be true that a narrow rural ecotope in a natural landscape can enhance its heterogeneity, and sometimes also its landscape metastability, but the question is how to verify this supposition. To answer these questions see Sect. 8.4.

- *Controlling human habitat transformations.* The variation of landscape ecological indexes (BTC, SHproductive, SHprotective, SHresidential, SHsubsidiary) and its relation with the ecological density of population lets us find different combinations of them per landscape type as presented in Fig. 12.5

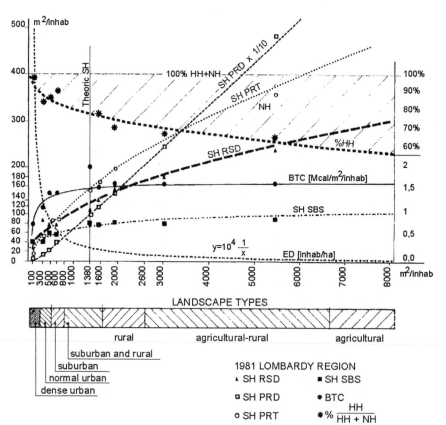

Fig. 12.5 Control of the principal functions of human habitat (HH) and their correlations with the different types of landscapes: an example concerning the Lombardy region. (From Ingegnoli 1993)

12.3.4
Planning Criteria Depending on the Remnant Patch System

Especially in large suburban landscapes systems consisting of remnant rural and/or semi-natural areas, derived from the strong fragmentation process of the previous agricultural landscape, can be frequently found. Such systems must be evaluated using a landscape ecological analysis, the constraint of which will form the basis for subsequent planning criteria. Otherwise, the landscape will become rapidly chaotic.

The structural pattern of the above-mentioned remnant areas is composed of many types of elements, characterised by their size, role in the ecotissue, biological territorial capacity, incorporation of disturbances, etc. Consequently, they can not be considered indifferently as free land patches available for every human activity and infrastructure.

In practice, after having defined the range of incorporable disturbances for the specific type of landscape under study (for instance, small field roads, small irrigation canals, small isolated rural buildings, medium-low power electrical lines), the first step consists of checking the characters of the remnant patches by defining the gradient of disturbance caused by their surroundings. This can be done by mapping a defined set of belt, the width of which depends on the characters of the landscape unit; for example, there will be a difference between an agricultural landscape unit and a suburban or open-forest one. In fact, one of the principal functions of a landscape element is its capacity to incorporate a disturbance. So, the width of a considered belt will be minor in the presence of a landscape element with a high BTC value and larger when the landscape element on which the disturbance is exerted has a lower BTC value (thus enhancing the extreme difficulty to incorporate the disturbance). In the case of the study for the municipality of Capannori (Lucca)(Fig. 12.6; see also Table 12.5), a set of belts, moving from the outside to the inside, has been defined as:

- Disturbance belt (Db), 10--100 m in width
- First buffer belt (B1), 60--100 m in width
- Second buffer belt (B2), 80--100 m in width
- Potential core area (PCa), 100--120 m in width
- Potential inner core area (PCi), >25 m in width

The resulting bounded area inside the third belt can be called the "potential core area". Thus, a remnant patch, even a large one, may lack a potential core area due to its shape or to the presence of disturbances that are too serious. Inside the "potential core area", it is sometimes possible to define a "potential inner core area" with better restoration possibilities.

Table 12.5. Example of a set of belts for the evaluation of remnant patches studied for the municipality of Capannori (Lucca). Note that the width of the belts depends on the BTC value of the landscape element through which the belt passes

Type of disturbance	BTC value of the landscape element close to the disturbance source	Belt type	Distance from the source of disturbance (m)
Residence buildings, small railroads, small roads, small industries, etc.	Low BTC	Db	20
	High BTC	Db	10
State roads, large industrial plants, main railroads, quarries, etc.	Low BTC	Db	40
	High BTC	Db	25
Airports, highways, dangerous industrial plants, etc.	Low BTC	Db	100
	High BTC	Db	60
Buffer zone perimeters.	Low BTC	B1 and B2	100
	High BTC	B1	60
		B2	80

The second step will be to give to each remnant patch a specific quality value (Table 12.6): the presence of a potential natural or semi-natural core area, a strategic linking capacity to connect other remnant patches, a good level of BTC of its vegetation, will lead to a high value of the patch. For example, two close "core areas" can have different values of BTC, thereby needing different types of evaluation. Analogously, a patch without intrinsic value could be very important as a stepping stone for an ecological network, so assuming a strategic functional role (SR).

Table 12.6. Evaluation levels of a mosaic of remnant patches characterised by a five belt set

Number of belts	$i<20\%$	$i>20\%$	$BTC_p>BTC_m$	SR
2 (Db+B1)	A	A	C	D
3 (Db+B1+B2)	A	B	C	D
4 (Db+B1+B2+PCa)	B	C	D	D
5 (Db+B1+B2+PCa+Pci)	C	C	D	D

Abbreviations: i is the most inner belt mapped; BTC_p the BTC value of the remnant patch; BTC_m the mean value of the BTC of the landscape unit; SR the strategic role of the patch within the landscape unit; A, B, C, D evaluation levels from worst to best

Obviously, the strategic roles of remnant patches have to be clearly expressed after a specific study of their distribution in a landscape unit. For example:

- Stepping-stone function for connection
- Meso-climatic mitigation of an intensively urbanised area
- Island for birds temporary refuge (e.g. along a flyway)
- Protection for hydrologic processes (e.g. a spring or resurgence)
- Necessity for maintaining good porosity of the landscape matrix

Depending on their evaluation, remnant patches present different constraints which are reflected in the different planning capacities. To measure these capacities, the main transformation processes need to be checked, for instance:

- N: Each transformation improving the naturalness of the patch (e.g. reforestation)
- M: Each transformation bringing the patch back to the original landscape matrix condition (e.g. from abandoned fields to cultivation)
- RG: The building of recreational green areas (e.g. public gardens)
- HS1: The building of light human structures (e.g. rural roads or canals)
- HS2: The building of heavy human structures (e.g. industrial roads or new technical plants)
- TU: Urbanisation of the entire patch

Table 12.7. Different planning capacities of the remnant patches.

Patch values	N	M	RG	HS1	HS2	TU·
A	+	+	+	+	+	+
B	+	+	+	+	(+)	[+/-]
C	+	+	(+)	(+)	[+/-]	-
D	+	(+)	(+)	[+]	-	-

Where: + Freedom of transformation; *(+)* transformation possibility only marginal and only with mitigation projects; *[+/-]* transformation possibility only marginal and only with mitigation and compensation projects; *shaded squares* no transformation possibility

Referring to the last two tables (12.6 and 12.7) note that the patch values (A,B,C,D) can be obtained in different ways: for instance, a 3D value can be obtained starting from a 3A or a 3C; analogously, a 4D value can derive from a 4B or it can be a simple 4D. Therefore the value degrees are generally more than four and they could be indicated, respectively, as 3D(A), 3D(C), 4D(B) or 4D(D).

These evaluation degrees give the possibility to map with different colours the system of patches of a landscape. A map like this can help to study an ecological network and to compare some case studies of transformation proposed by the urban planners of the municipality. Moreover it can be used to restore the landscape structure and to divide the landscape in different planning sub-units.

Fig. 12.6 The present situation of a portion of the territory of the municipality of Capannori (Lucca, Italy). Note the highway and the airport at the bottom and the high level of fragmentation of the landscape

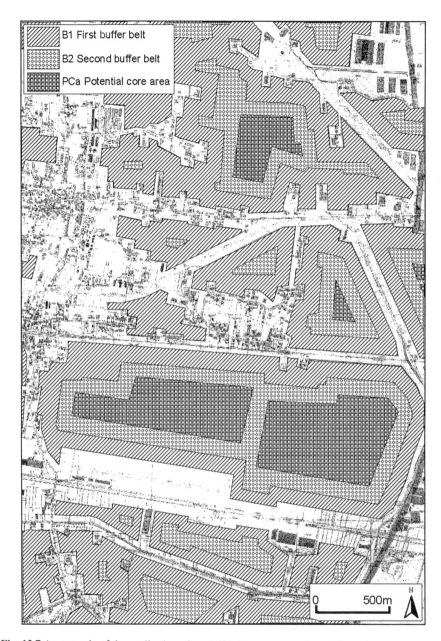

Fig. 12.7 An example of the application of a set of belts to measure the gradient of disturbances exerted on the remnant vegetated patches by their surroundings in the territory represented in Fig. 12.6. The potential core area is *darker*.

13 Examples of Applications

13.1
Introduction

A book like this is quite condensed, therefore it needs some examples of applications. The possibility to describe in detail a sufficiently various group of examples is, however, very reduced, hence we will choose only four study cases. They are different both in spatio-temporal scale and in latitude, but the main landscape ecological principles and indices are similar.

We decided to omit the multiple more traditional applications of landscape ecology, e.g. derived from the more known texts of Forman, Naveh, Gardner and Turner, Zonneveld, etc., and generally utilised also in our case studies, to concentrate at least on some of the principal concepts and methods proposed by this text. The references to the chapters of this book are obviously indispensable, even when non-explicitly cited.

The presented examples cover a field of study, the extremes of which go from central-northern Europe, i.e. north-east Poland (about 53° of latitude north and 23° 30' longitude east, first example) to central Burkina (Africa) (12°45' latitude north and 1°15'longitude east, fourth example). While this is not in proportion to the emerged lands of the ecosphere, it provides sufficient variation: from the margin of the boreal zone to the arid tropical one. This is the field experience of the author: other case-studies in Colorado parks, in Indian parks, in South African and Tanzania parks, are not yet completed.

Synthetic data on regional European dynamics have also been limited by difficulties, not only economical but also practical, in obtaining proper data in some regions, because they were not collected by cognizant authorities: a completed transect of Europe through its regions, from north to south, remains a challenge. Nonetheless we discuss a small transect from central Germany to central Italy (second example): Rheinland-Palatinate, Trentino- South Tyrol, Lombardy and Tuscany.

The third example concerns a case study of a suburban park in Northern Milan with some indications as to its planning criteria.

13.2
Studies on the Bialowieza Landscape (Poland)

13.2.1
Synthetic Landscape Description

The studied landscape belongs to the landscape system of the west Russian lowlands (eastern Europe). It is a forest landscape and situated in the north-east part of Poland and about a half is now in the territory of Byelorussia. The forest is localised, from an hydrologic point of view, on the periphery of the Vistula and Niemen catchment areas. The Białowieża landscape is one of the best forest landscapes in Europe and presents peculiar characters. It is the remainder of a larger forest territory of central Europe, from which it was separated in the fifteenth century. This territory has never been heavily populated and even today the geographic density of population in Bialystock Province (10055 km^2) reaches only 70 people/ km^2, one of the lowest of the European regions (e.g. Lombardy 379, Rheinland-Palatinate 202, Provence 143) because 40% of the population lives in the city of Bialystock.

The extent of this landscape is about 1500 km^2, 75% of which is covered by forest. The Podlasie-Byelorussia uplands geomorphology (134-202 m of altitude) is dominated by undulating plains of ablation moraine, 10-15 m above the level of the ground moraine. These plains consist of boulder clay of different thicknesses. Sandy and gravelly small hills occur in the area as remnants of denudation and erosion processes.

The landscape is mainly characterised by the climatic ecotone between the boreal zone and the nemoral one, that is between the central European climate, with more continental influxes, and the boreal climate. The annual mean temperature is 6.8 °C, −4.7 °C in January and 17.8 °C in July and the annual rainfall is 614 mm, mainly in the vegetative period. The period of snow cover is about three months.

According to Falinski (1994a) the Białowieża forest differs from west European forests above all by the absence of beech (*Fagus sylvatica* L.) and from the adjacent eastern European forests by the abundance of oak (*Quercus robur* L.) and hornbeam (*Carpinus betulus* L.) in the structure of the forest communities. This may be indicative of a strong influence of the continental climate, because in the warmer Brianza (Milan, Italy), with an annual mean of about 12.0 °C but with a sub-continental climate and moraine hills (0 °C in January, 23.5 °C in July), *Querco-Carpinetum* is the fittest vegetation and the beech is absent.

In the Białowieża forest, the surveyed phytosociological types of forest communities show quite close relations to morphological-genetic soil types: but the most variable and common forest community of *Tilio-Carpinetum* (linden-hornbeam) occurs on three different soil types (brown soils). In this forest numerous groups of species, particularly with western distribution, attain the absolute limit of their occurrence. The original differentiation of this forest vegetation (still observed in many parts of the landscape) started in the Atlantic period with the formation of forests of the *Querco-Carpinetum* type, and the present coexistence of deciduous and coniferous forests began in the sub-Atlantic period (500 BC-600 AD). Extrazonal forest communities, such as thermophilous *Potentillo albae-Quercetum,* are dispersed in the forest as are azonal ones, such as *Circeo-Alnetum* or *Vaccinio uliginosi-Pinetum* (see Falinski 1994, 1998).

This wide landscape (Fig. 13.1) is not very fragmented and the human habitat is rather limited. Traces of human use of parts of the forest go back to the tenth century (necropolia, small colonizations) but a strong pressure of anthropogenic activities arose in the seventeenth to eighteenth centuries, with agriculture, forestry, charcoal production and hunting. The last two centuries have presented an increase of human pressure, with larger communication networks, frequent hunting, new colonisations, forestry and tourism. All these pressures led to the institution of the national park in 1921, and even if the protected area is too small (8%), it was a good decision.

In any case, the human habitat is contained within about 7.8% of the ecological mosaic, hence about 15% of the ecotissue structure. Particularly dangerous seems to be the impact of forestry outside the park, which has caused the gradual disappearance of thermophilous oak forests in the western part.

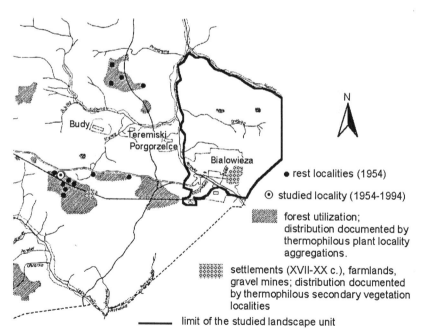

Fig. 13.1 The map of the Białowieża forest landscape, its surroundings and the borders of the studied landscape unit. Note the small proportion of human settlements, but also the relative lack of protected areas

The main objective of the research on this forest has concerned the landscape unit composed of the national park and the Polana Białowieżka: it is a mainly operative unit because of the presence of the border Poland-Byelorussia, but it presents a very notable contrast between a primeval forest and a human colonisation.

13.2.2
The Polana and the National Park

Referring to good maps of the vegetation (Falinski 1994b; Pedrotti 1983) and surveying the national park and the Polana, it was possible to carry out a synthetic study of this landscape unit. In spite of the detailed description of the vegetation communities and the synthetic scale of the maps (1:50000), some examples of landscape ecological measures on typical forest tesserae have been made in order to have a better basis on which to evaluate the BTC of the unit (Table 13.1). The table shows some very high values of plant biomass (accurately re-controlled) and also of BTC, difficult to find in Europe, and other more typical results.

The entire landscape unit measures 6100 ha, of which 4700 pertain to the park (77%). The evaluation of the national park has been done on the basis of the map of dynamic tendencies of the vegetation of Białowieża (Falinski 1994a), representing 85% of the park surface and of surroundings. That is one of the reasons why in Table 13.2 the ecological measures have been limited to 3925 ha. The mean BTC of 8.65 Mcal/m^2/year confirms the exceptional state of preservation of this forest and encourage further studies. The mentioned values can be extended, without large errors, to the entire park surface. The immediately contiguous Polana (an area of about 1400 ha, 23 % of the landscape unit) has been studied by Pabjanek (1999) from a geographic viewpoint (Fig. 13.2).

Table 13.1. Some examples of measures on vegetation tesserae in the Białowieża landscape

Examples of some characteristic tesserae	Veget. height m	Plant biomass m^3/ha	Diversity Sp. no.	Quality of tessera %	BTC Mcal/m^2/year
1. *Tilio-Carpinetum*, - national park (dominant tree: *Quercus robur*)	32.7	912	41	90.3	10.90
2. *Tilio-Carpinetum*,- national park (dominant tree: *Carpinus betulus*)	26.1	480	28	74.4	8.27
3. *Tilio-Carpinetum*, - national park (dominant tree: *Tilia cordata*)	29.5	770	36	95.1	11.07
4. *Tilio-Carpinetum*, - Polana (dominant tree: *Tilia cordata*)	26.0	620	24	64.3	7.70
5. *Vaccinio myrtilli-Pinetum*,- surrounding (dominant tree: *Pinus sylvestris*)	20.0	257	20	63.0	5.87

Note: the plant biomass has been measured with a Spiegel Relaskope, reducing the result x 0.9; the BTC has been estimated using the method presented in Chap. 8.3

Table 13.2. Measure of the biological territorial capacity of the national park (3925 ha out of 4700). BTC (estimated) in Mcal/m^2/year

Main landscape elements in their dynamic states	ha	%	BTC
Fluctuation			
- *Vaccinio myrtilli-Pinetum*	50	1.3	8.0
- *Calamagrostio arundinaceae-Pinetum*	120	3.1	8.5
- *Calamagrostio arundinaceae-Piceetum*	290	7.4	9.5
- *Tilio-Carpinetum*	1650	42.0	10.0
- *Circaeo-Alnetum*	330	8.4	8.7
- Other, (less dense)	120	3.1	6.0
Regeneration	80	2.0	8.0
Regression	75	1.9	5.5
Recreation			
- *Carici elongatae-Alnetum*	200	5.1	7.5
- *Thelypterido-Betuletum pubescentis*	780	19.9	8.2
- Other (Less Dense)	100	2.5	6.0
Natural clearances	70	1.8	2.1
Water	60	1.5	0.1
Total	3925	100	8.65

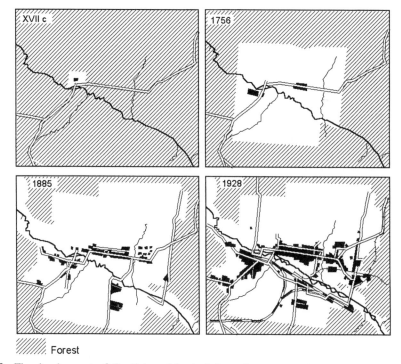

Forest

Fig. 13.2 The development of the Polana (clearing) from the seventeenth to the twentieth century (from Pabjanek 1999). Today the clearing is not much changed, but it is less inhabited.

Table 13.3. Landscape ecological dynamic in the Białowieża Polana. Data from land use categories (adapted from Pabjanek 1999)

Landscape element	HH	BTC	1885		1928		1953		1989	
	%		ha	%	ha	%	ha	%	ha	%
Arable fields	95	1.1	891.6	67.2	708.0	51.2	809.9	56.3	350.9	25.4
Grasslands	85	0.8	289.8	21.9	421.5	30.5	425.3	29.6	717.0	51.9
Forest + coppice (open)	20	4.5	7.5	0.6	2.0	0.1	16.2	1.1	61.9	4.5
Forest + coppice (closed)	10	7.0	22.8	1.7	2.2	0.2	5.3	0.4	13.2	1.0
Forest in manor park	25	7.5	20.0	1.5	20.0	1.4	21.2	1.5	24.1	1.8
Urban greenery etc.	75	2.3	16.7	1.3	30.7	2.2	29.2	2.0	27.9	2.0
Urbanised (scattered)	98	0.7	(9)	1.2	(25)	6.0	11.8	0.8	20.4	1.5
Urbanised (dense)	100	0.3	(34)	2.0	(114)	4.1	59.9	4.2	95.7	6.9
Roads and transportation	100	0.1	26.2	2.0	45.5	3.3	43.1	3.0	43.4	3.1
Other work areas	90	0.2	-	-	1.7	0.1	2.4	0.2	15.5	1.1
Waters	10	0.15	8.5	0.6	12.6	0.9	13.2	0.9	11.2	0.8
Total area			1326		1384		1438		1381	
Population			600		9000		2500		2600	
Human habitat HH %			89.3		90.0		89.0		83.7	
BTC Mcal/m²/year			1.22		1.04		1.11		1.19	
SH m²/inhabitant			19735		1384		5119		4446	
σ = SH/SH*			11.9		0.8		3.1		2.7	

Note: $SH^* = 1663$ m²/inhabitant; number in parentheses are estimated values

The reciprocal influences with the national park were checked. Until the end of the nineteenth century, Białowieża was a small forest village, the duty of which was to serve royal game hunting. A hunting palace with a manor-park was created in about 1880 following the gardening typology of the villas of Brianza (Ingegnoli 1987). Soon after a railway was built. During World War I Germans greatly developed the forest industry. Quickly rebuilt after the war, Białowieża increased its population, which reached about 11000 inhabitants before World War II. This overcrowding decreased rapidly after the war. In the last fifty years land use has changed greatly, following the abandonment of arable land, which decreased from 56 to 25% in the period 1953-1989. The urbanisation is still going on, even if it was never very extensive.

Table 13.3 shows the human habitat of the Polana decreasing in the last period (6 % in 36 years), as effect of the land abandonment. Consequently, the BTC has come back to the value of 1885. σ is good today, but it denotes a rural landscape, not an agricultural one as in 1953: in 1928, σ reached its worst value 0.8 (< 1.2, see Sect. 8.4), hence it created a dangerous situation for the newborn park.

Comparing these data with the forest ones, see (Fig. 13.3) the complementarity of the standard BTC classes distribution; for the old forest we have 89.2% of their ecotopes in classes VI and VII, while for the Polana we have 90.7% in classes I and II. Table 13.4 underlines that the landscape unit integrating the two components forest-Polana is not negatively influenced by this; its LM is not reduced but remains near constant. Some wild animals are even attracted by their complementarity. Thus, if the HH remains <20% of the total landscape unit, if σ remains >1.2 and if the human activities are not aggressive but protective, it is not per se a problem for a national park to border on a human settlement.

Table 13.4. Landscape parameters of the unit park-Polana.

Sub-units	Area %	Shannon H	Landscape diversity τ	BTC Mcal/m^2/year	HH %	LM
National park	77.3	1.048	4.085	8.65	0.5	35.34
Polana	22.7	0.759	3.178	1.19	83.7	3.78
Landscape unit	100	1.426	5.020	6.96	19.4	34.94

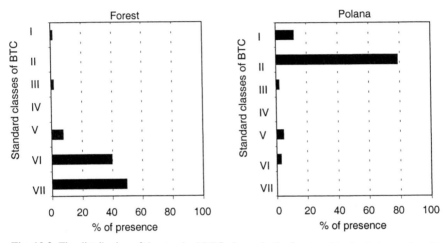

Fig. 13.3 The distribution of the standard BTC classes in the forest and in the Polana sub-units. Note the opposite symmetry and the incomplete distribution of each

13.3
Regional Transformations Across Europe

13.3.1
Objective and Method

As we know, landscape ecology needs to work at different spatio-temporal scales. If we have, for instance, to control and plan a protected territory, we need to compare its mean BTC and other ecological parameters to the regional ones. Otherwise we cannot know the real extent of the rehabilitation actions a park area should receive. It implies that we have to refer our studies not only to local and detailed scales, but also to regional ones. As already mentioned, this is quite a problem, because in general the administrative regions are indispensable to ranking territorial data, but they hardly coincide with ecological regions: or, at least the coincidence is only partial. On the other hand, especially in Europe, regional landscapes are strictly linked to regional traditions. In any case, we simply can not avoid referring our studies to the present regions, because the alternatives are practically impossible. So, the reported study was much more complex than it should have been. It was very difficult to find the correct statistical data and, moreover, they needed to be re-adapted. The main sources of information were: ISTAT (2000), De Agostini G. I. (2001), CIP. (1997), CORINE-Land Cover (1996), Statistisches Jahrbuch fur Reinland-Pfalz (1997).

Here we began to present some researches on Rheinland-Palatinate from 1951 to 1996, Trentino-Alto Adige (South Tyrol) from 1928 to 1998, Lombardy from 1878 to 1999, Tuscany from 1911 to 1998, mainly central European regions, from 42.5° to 50°of N latitude, and from 7.5° to 12.5° of E longitude (Fig. 13.4).

Following Bailey (1996) (see Sect. 3.6.2), all four administrative regions belong to the humid-temperate domain: concerning the first three, their mountains can be assigned to the Marine Regime Mountains (= 240M) and their plains in the Marine Division; the mountains of Tuscany can be assigned to the Mediterranean Regime Mountains (= 260M) and their plains to the Mediterranean Division.

Fig. 13.4 Localisation of the studied four administrative regions in Europe.

As shown in Table 13.5, the seven main groups of landscape elements were synthesised considering their sub-components, thus the BTC evaluation was accurately adapted, because at a wide spatio-temporal scale even a small difference may have a large effect. The human habitat (HH) was much more approximated, but always in proportion to the common use.

Four tables will be presented, each measuring the main landscape ecological parameters at regional scale: human habitat (HH, %), mean BTC (Mcal/m^2/year), SH (m^2/inhab.) and σ (SH/SH*).

Table 13.5. Synthesis of the ranges of BTC per region and per landscape element; mean HH

Landscape elements	Lombardy wBTC	Trentino-South Tyrol wBTC	Rheinland-Palatinate wBTC	All regions %HH
- Forest	5.0	5.2	5.1	0.15
(deciduous, coniferous, coppices)	(3.6 - 6.7)	(3.6 - 7.0)	(3.8 - 6.8)	
- Woody cultivation	2.25	2.0	2.0	0.90
(plantations, orchards, vines)	(2.0 - 2.4)	(1.9 - 2.3)	(1.8 - 2.4)	
- Arable land	1.21	1.15	1.2	0.96
(dry- wetted- sowable, hedgerows)	(1.0 - 1.4)	(1.0 - 1.3)	(1. 1 -1.4)	
- Grassland	0.79	0.8	0.8	0.85
(prairies and meadows)	(0.7 - 0.9)	(0.7 - 0.8)	(0.7 - 0.8)	
- Wild shrub	2.0	1.9	2.0	0.10
(heathlands, marshes, shrublands)	1.5-2.5	1.1-2.6	1.5-2.5	
- Natural unproductive	0.1	0.1	0.1	0.05
(Waters, glaciers, rocks, sands)	(0.0 - 0.2)	(0.0 - 0.2)	(0.0 - 0.2)	
- Urbanised and subsidiary	0.4	0.45	0.45	1.00
(Cities, industries, roads, pits)	(0.1 - 0.7)	(0.1 - 0.8)	(0.1 - 0.8)	

wBTC (Mcal/m^2/year) weighted average; number in parentheses are the most common range of values

13.3.2
Regional Dynamics

During the studied interval of time, *Rheinland-Palatinate* (Table 13.6) showed a mean HH of 54.125%, with a slight tendency to decrease, and a good mean BTC (2.63 Mcal/m^2/year), rather constant and equal to 1.64 times the BTC of the ecosphere emerged lands. In fact this is the most forested region of Germany. The SH is still abundant, even if σ diminished from 2.56 to 1.87 in 45 years, with a limited population increase. The main transformation is related to the strong decrease in arable land (- 22.5% in 21 years) and a symmetrically strong increase of urbanised and subsidiary elements (+ 336.8% in 45 year).

Trentino-South Tyrol (Table 13.7) shows an HH mean of only 42.5 in the examined period because of the alpine mountains, with a tendency to decrease about 10% in the last 30 years. The BTC mean (2.76 Mcal/m^2/year) is good, equal to 1.72 times the BTC of the ecosphere emerged lands and slightly increased. The SH is certainly abundant even if we consider about 130,000 beds for tourists: in any case σ diminished from 6.57 to 4.02 in 70 year. The main transformation is related to the disappearance of the arable lands (- 99.8% in 36 years) and a strong increase of urbanisation (+ 207% in 47 years).

Lombardy (Table 13.8; see also Fig. 5.5) is one of the regions of Europe most transformed during the last century: the decrease of arable land was −23.4% in 48 years and of grassland − 29.4%, while urbanisation augmented + 383% in 48 years. In this period, the data show a quite high mean HH of 66.0%, with a slight tendency to decrease (- 4.3% last 30 years): remember that this region has been one of the most populated in Europe since the fourth century; the SH value is too low. This region is comparable with Belgium or Portugal in terms of population dimension. In any case it recently reached the σ limit under which its territory should not be able to sustain the population, which in fact consistently diminished its previously strong increase after 1975. At the opposite end, the mean BTC is not high (1.89 Mcal/m^2/year, only 1.18 times the BTC of the ecosphere emerged lands), but it tends to a slight increase.

Tuscany (Table 13.9) shows a mean HH of 57.0% with a small tendency to decrease (- 8.8% in 30 years) The mean BTC is good (2.61 Mcal/m^2/year), i.e. 1.63 times the BTC of the ecosphere emerged lands, and near constant. SH is good, and also σ. The main transformation is related to the strong decrease of arable lands (- 40.0% in 36 years) and a very strong increase of urbanisation.

In summary, all these European regions highlight transformations in their territory, in some cases even deep ones, and more or less in the same direction; the field abandonment is a consequence of the strong arable land decrease; urbanisation is expanding even when the population does not or no longer; but the mean BTC is generally near constant and it means that, at regional scale, the incorporation of disturbances was possible (but not on a more detailed scale); HH is generally not so heavy, excluding Lombardy; the same for SH.

Table 13.6. Landscape transformation in the Rheinland-Palatinate region. Data on landscape elements in %

Rheinland-Palatinate	1950	1975	1985	1996
Total surface (km^2)	19 847	19 847	19 847	19 847
Forest	37.0	37.9	39.4	35.4
Woody cultivation	4.5	5.0	5.0	5.3
Arable land	31.0	25.8	21.6	20.0
Grass land	13.2	13.1	11.6	12.3
Wild shrub and marsh	7.4	7.1	8.1	11.0
Natural unproductive	3.1	3.1	3.2	3.2
Urbanised and subsidiary	3.8	8.0	11.1	12.8
Population (10^3)	2 909	3 678	3.619	3 880
(%) HH	55.3	54.3	53.1	53.8
BTC (Mcal/m^2/year)	2.62	2.68	2.68	2.53
SH (m^2/inhabitant)	3 770	2 930	2 910	2 750
σ	2.56	1.99	1.98	1.87

Table 13.7. Landscape transformation in the Trentino-South Tyrol region. Data on landscape elements in %

Trentino-South Tyrol (mountains 90%)	1928	1951	1968	1987	1998
Total surface (km²)	13 910	13 620	13 620	13 620	13 620
Forest	43.8	43.1	43.7	45.2	46.4
Woody cultivation	1.3	1.8	3.1	3.3	2.7
Arable land	6.1	5.5	3.7	0.9	0.7
Grass land	33.4	32.3	31.7	26.9	26.4
Wild shrub and marsh	2.4	3.3	2.4	8.0	7.3
Natural unproductive	11.2	11.2	11.1	11.0	11.0
Urbanised and subsidiary	1.8	2.8	4.3	4.7	6.0
Population (10³)	642	736	899	884	936
(%) HH	44.6	44.6	44.4	39.5	39.8
BTC (Mcal/m²/year)	2.71	2.68	2.71	2.83	2.86
SH (m²/inhabitant)	9 660	8 43	6 870	6 220	5 910
σ	6.57	5.73	4.67	4.23	4.02

Table 13.8. Landscape transformations in Lombardy region. Data on landscape elements in %

Lombardy (mountains 40.5%)	1878	1911	1928	1951	1968	1987	1999
Total surface (km²)	23 530	24 180	23 810	23 850	23 850	23 860	23 860
Forest	19.0	16.0	16.1	18.8	20.1	20.6	20.7
Woody cultivation	10.0	7.0	7.0	7.0	7.5	9.4	9.3
Arable land	39.8	43.0	44.6	44.0	39.5	33.6	33.7
Grass land	17.0	16.0	16.3	15.7	17.0	13.6	12.0
Wild shrub and marsh	4.6	8.3	6.1	4.2	3.4	6.9	7.4
Natural unproductive	7.9	7.8	7.8	7.8	7.7	7.7	7.7
Urbanised and subsidiary	1.7	1.9	2.1	2.4	4.8	8.2	9.2
Population (10³)	3 500	4 790	5 450	6 610	8 330	8 890	9 039
(%) HH	65.7	65.0	67.6	67.5	67.3	64.6	64.4
BTC (Mcal/m²/year)	1.94	1.82	1.81	1.89	1.90	1.93	1.94
SH (m²/inhabitant)	4 420	3 280	2 950	2 430	1 930	1 740	1 700
σ	3.09	2.29	2.06	1.70	1.35	1.22	1.19

Table 13.9. Landscape transformations in Tuscany region. Data on landscape elements in %

Tuscany (mountains 25%)	1911	1928	1951	1968	1987	1998
Total surface (km²)	24 099	22 934	22 991	22 991	22 992	22 992
Forest	38.3	40.3	35.6	37.7	38.7	38.8
Woody cultivation	1.2	1.3	4.0	4.8	7.9	8.8
Arable land	35.8	33.8	45.4	40.7	27.3	25.5
Grass land	19.1	17.8	6.1	6.9	9.4	6.1
Wild shrub and marsh	1.1	1.2	4.0	2.1	5.9	7.2
Natural unproductive	3.8	3.8	3.7	3.8	3.9	4.0
Urbanised and subsidiary	0.7	0.9	1.2	4.0	7.0	9.6
Population (10³)	2 690	2 760	3 160	3 460	3 560	3 530
(%) HH	58.4	56.0	59.4	59.3	54.9	54.1
BTC (Mcal/m²/year)	2.55	2.62	2.56	2.60	2.67	2.68
SH (m²/inhabitant)	5 230	4 650	4 310	3 940	3 550	3 520
σ	3.56	3.16	2.93	2.68	2.41	2.39

If we could continue this transect of European regions toward north and south, we might find more consistent differences in their dynamics. For instance Sicily, which has only 8.6% forests, cut by half its arable land (about 60% in 1950) in 45 years and has a mean BTC near 1.6, the altered mean of the entire ecosphere today.

13.4
Planning Criteria on a Suburban Park in Milan (Italy)

13.4.1
Introduction and Analysis

An urbanised landscape has to concentrate the ecological functions of the natural components in often isolated patches, which acquire a particular linkage with the surrounding ones, even over a large distance. Patches such as suburban parks are also very sensitive to the changing of their ecological mosaic (Ingegnoli 1993).

The urbanised landscape of Milan changed in the last century from about 500 thousand to 4 million inhabitants over an area of about 1500 km^2. Therefore, the importance of its parks has increased enormously. Four main types of parks may be found in the suburban landscape of Milan: historical parks (e.g. Monza), natural remnant parks (e.g. Groane), recently preserved agricultural parks (e.g. South Park) and new parks built over restored areas (e.g. Baggio).

Each of these parks has its own problems, some of which are difficult to solve without landscape ecology. The environmental evaluation needs comparisons with "normal" patterns of behaviour of a complex system of ecocenotopes. For that reason, the principal problem becomes how to determine the normal state of these types of ecological system and the levels of alterations of the same systems.

Applying the principles and methods reported in this book, the result is a quite severe situation. Monza Park currently of presents an ecological state lower than the one it had before its foundation in the eighteenth century. Groane Park, even if well managed, today has a low metastability and a high fragmentation. South Park (Ingegnoli 1998) is still partially a rural-agricultural landscape, not able to incorporate the high disturbances caused by urban structures. Gravel Pits Park of Baggio (Parco delle Cave) is practically a park which needs more planning.

The most interesting case study of these suburban parks is the one of Monza. This historical park has served as hunting reserve and agricultural research and improvement area starting in 1770 (Villa Reale di Monza), when the Vice-King ruled the State of Milan for the Augsburg Emperor of Vienna. Its area measures 730 ha and it once was the largest park in Europe surrounded by a wall. This park enclosed the old forest remnants of the medieval Bosco Bello, near the river Lambro, and enriched it with plantations, boulevards and farms. It was designed by good landscape architects and botanists (see an old map in Fig.13.5).

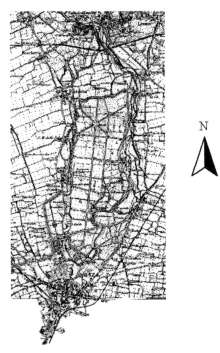

Fig. 13.5 An old map of the Parco della Villa Reale (Monza Park) in the nineteenth century. Note the relation with the surrounding agricultural-rural landscape, the complex design of the gardens, the forest remnants cut by hunting roads, and the farms

The park was situated in the south part of the Brianza, the district near Milan historically richest in villas and gardens. The park development had been based on two old gardens of the villas of the noble family Durini. Today, the park has been enclosed in the southern part of the River Lambro Park, north-east of Milan (of 6662 ha), but it is almost completely surrounded by the urban landscape.

After the murder of King Humbert of Savoy in 1900, the park was donated to the cities of Milan and Monza. In 1921, the Monza autodrome (motor-racing) and a large golf course were built, degrading the exceptional park.

Ten years ago, some studies on the transformation of the Monza suburban park, and the proposal of a new plan to restore it, were made by the Specialised School of Landscape Architecture of the University of Genoa, directed by A. Calcagno Maniglio (Ingegnoli and Gibelli 1992). More recently Ingegnoli and his Natural Sciences students (degree thesis) M. Cannata, C. Rezia Loppio, M. De Paolini, continued the research.

As we know, all the complex adaptive systems, formed by highly correlated configurations of elements, in a proper range of spatio-temporal scales can be studied only through an historical methodology, that is a reconstruction of the past landscapes. So, the local landscape unit containing the park, about 4000 ha, has been studied at four temporal thresholds: 1720, 1840, 1937, 1994. In this case, we put in evidence a list of seven major types of ecotopes, as shown in Table 13.10.

The transformation of this landscape unit has been very impressive: in 1720 the landscape matrix was formed by fields with trees (75.5%) and urban ecotopes covered only 3.8%; in 1840 the matrix continued to be constituted by fields with trees, but at 58.2%, and urban ecotopes showed slower growth, reaching 4.9%; in 1937 the matrix changed into simple arable fields (64.2%) and urban ecotopes reached 14.1%; in 1994 the landscape matrix became urban (45.6%) and the fields with trees disappeared.

Table 13.10. Control of the transformations in the landscape unit of Monza Park (Milan)

Landscape types	BTC Mcal/m^2/year	1720 ha	1840 ha	1937 ha	1994 ha
Forest	8.0	35.1	36	23.1	5.4
Woods	4.5	273.4	358.5	419	784
Fields and trees	2.2	2964.2	2289.7	238.9	5.6
Arable fields	1.0	212.8	714	2534.8	1468
Humid meadows	1.3	260.7	318.5	146.5	6.2
Urbanised and roads	0.45	151	192	555.7	1931
Barrens	0.1	27	28	30	30
Total area		3 924.2	3 936.7	3 948	4 230.2
BTC $_m$ (mean)		2.20	2.09	1.42	1.40
HS (m^2/inhabitant)		3004	1737	452	227
BTC$_{HH}$ (Mcal/m^2/year)		2.0	1.96	1.22	1.09
BTC$_{NH}$ (Mcal/m^2/year)		3.5	3.39	4	4.25
BTC$_{NH}$ / BTC$_m$ (%)		15.2	14.7	20.1	30.3

In fact, studying the change in the relationships with agricultural landscape units outside the park boundaries, we go from 14 double linkages in 1720 to only two double and two simple linkages in 1994 (Fig.13.6).

Remembering the ecotissue concept, which allows correlation of the previous mosaic with other information relative to the ecotopes, we estimated, the variation of the human habitat (HH), standard habitat per capita (SH) referred to the human population and biological territorial capacity within the landscape unit and then we integrated these data with other thematic data.

Surprisingly, the values of HH remained practically unchanged: HH_{1720}= 90.3%, HH_{1994}= 90.0%. By contrast, the SH variations are very high: σ= 2.18 in 1720, 1.26 in 1840, 0.33 in 1937, 0.16 in 1994. The BTC shows a lower, but significant, change: BTC_{1720}= 2.22 Mcal/m^2/year; BTC_{1840}= 2.09; BTC_{1937}= 1.42; BTC_{1994}= 1.40.

Other analyses have been made (Fig. 13.7). First of all on geomorphology, climate, vegetation, then on the river Lambro, on the BTC distribution, the role of the ecotopes, the disturbances, the connectivity, the fragmentation, the fauna, and on the farms, the recreational areas and monumental villas. It is impossible to describe all these studies here.

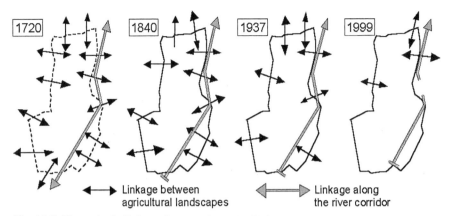

Fig. 13.6 Change in the linkages between the park of Villa Reale and the agricultural landscape units outside its boundaries. There is a decrease from 14 to two linkages in the last two and a half centuries

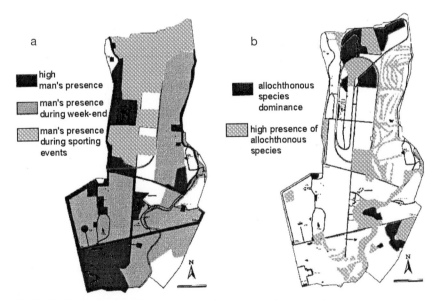

Fig. 13.7A, B. Examples of two thematic analyses concerning Monza Park. **A** Human presence impact. **B** Allochthonous plant species presence

In extreme synthesis:

Climate and geomorphology: Notwithstanding the short distance from the city centre of Milan, the park still exerts a positive effect on climate, which is not influenced by over-warming; the wide river bed of the river Lambro is divided in two terraced levels on which the park has grown, but the area at risk of flooding is quite limited, just before the English Garden of the Villa Reale.

Vegetation: In the northern part of the park some forested tesserae are remnant of the *Querco-Carpinetum boreoitalicum* (Pignatti 1953), and are therefore very interesting, with some oaks (*Quercus robur* L. and *Q. petreae* (Mattuschka) Liebl.) about 30-35 m high. Its characteristic species are also present, such as *Acer campestre* L., *Cornus sanguinea* L., *Brachypodium sylvaticum* (Hudson) Beauv., *Hedera helix* L., *Polygonatum multiflorum* (L.) All., *Vinca minor, Ornithogalum pyrenaicum* L., etc. But frequently these tesserae are partially invaded by Canadian oaks (*Q. rubra* L.), planted at the end of nineteenth century in nearby tesserae and in some cases even by *Robinia pseudacacia* L. and *Prunus serotina* Ehrh., all North American species. A wide portion of the forest patches in the park was coppice woods, no longer utilised today. Many trees are threatened.

River Lambro: This small pre-alpine river was very polluted, but now its conditions are better; the river functionality index, IFF (see Sect. 7.5.2), gave few tracts at level II-III (good-mediocre), the other being III-IV (mediocre-cheap) and only some small tract near Bad. The river tends to flood with a dangerous cycle of 20-30 years, more frequent than in the past because of the urbanisation of its surroundings.

BTC distribution and role of ecotopes: These are very unbalanced and fragmented; the survey measures (see Sect. Table 8.2) were not above 8.0-8.5 Mcal/m^2/year (compare with Białowieża values!) and with a BTC mainly about 4.5-5.5. The role of ecotopes is generally reduced because of the dismantling of the entire park.

Disturbances: The range of local disturbances is wide; the presence of motor racing induced air pollution and deafening noise and especially an abnormal overcrowding (more than 100000 people). The golf course also is a source of pollution (for greenery) and of fragmentation (for forest patches). Moreover, a road crosses the park east-west, and represents a problem also for humans.

Connectivity and fragmentation: The present connectivity shows α and β indices of urban level, while in the past they were quite good. The patchiness also, defined as the ratio between the mean grain and the park area (before and after the motor racing and the golf course), has changed from 12.6% to 5.5%.

Fauna: The park is today a peninsula (near an isle) in the metropolitan area of Milan: thus it is a possible refuge for many animals, especially birds. For instance, a study on the tawny owl (*Strix aluco*) gave a density approximately double that of the mean of Brianza. But the park is still mainly uninhabited by many mammalian species.

Monuments: The main monuments need to be restored (e.g. the Durini villas), and, moreover, some old boulevards planted with *Carpinus betulus* L. (hornbeam) and other garden species have been destroyed.

13.4.2
Diagnosis and Planning Criteria

Trying to extrapolate a correct diagnosis from all the analyses was not easy. Let us check up the main syndromes (see Sect. 5.3.3):

- Excess of road density
- Insufficiency of network density
- Lack of natural habitat
- Excess of landscape resistance for key species
- Too high boundary crossing frequency
- Critical thresholds of landscape habitat per capita (both in HH or NH)
- Presence of not-incorporable range of disturbances (out-of-scale)
- Invasion of exotic species
- Metastability level not compatible with structure and functions
- Break of connectivity networks in the landscape
- Excess of noise from the road network
- Suburban-industrial-rural chronic degradation of the landscape unit
- Drastic change in the surrounding landscape

No doubt that the syndrome is serious. After this framework, and an operative chart of integration (Fig. 13.8), we need to detail the most crucial dysfunctions.

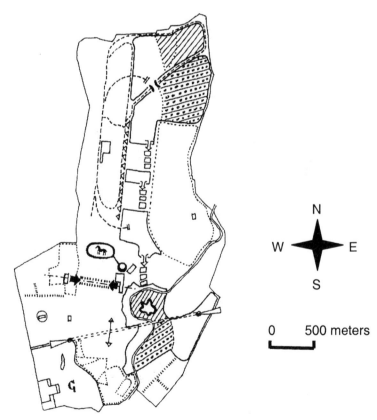

Fig. 13.8 Example of a simplified operative chart of integration prepared for the rehabilitation plan of Monza Park.

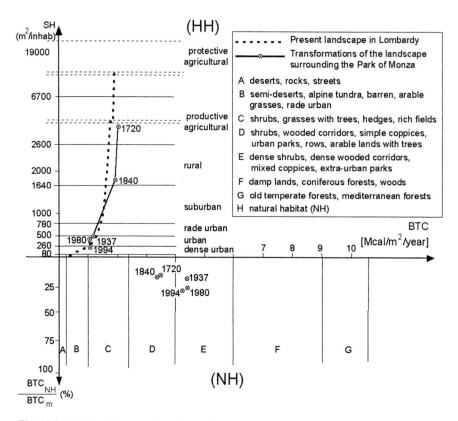

Fig. 13.9 Control of the transformations of the landscape unit of Monza Park. It changed from a rural-productive type of landscape to a dense urban one in only 270 years, notwithstanding the presence of the park (see also Fig. 5.4)

The significance of the transformation in relationship to landscape ecological criteria may be evaluated by comparing the movement of the complex system of Monza Park with the thresholds of metastability of the possible landscape types, which form the human habitat (HH) and the natural habitat (NH) (Fig.13.9).

In the field of NH, remember that the x axis reports the values of the BTC index and the y axis the percentage ratio BTC_{NH}/BTC_m. This figure shows the sharp change of the landscape unit of the studied park, which changed from a rural-productive type of landscape to a dense urban one in relatively few years, notwithstanding the presence of the park. The change in the NH shows a bigger importance of the BTC_{NH}/BTC_m, increasing from about 15% to 30%.

At the same time we compared this kind of transformation with the transformation in the regional system of landscapes, that is, with the upper level of scale, i.e. the regional one (Table 13.8).

As expressed in Table 13.11 we can now measure the variation of the importance of the park BTC.

Table 13.11. Simple but effective control of the importance of the park BTC in relation to the regional and the local BTC

BTC (Mcal/m²/year)	1720	1840	1920	1937	1994	(2030)
Park BTC (p)	2.48	2.70	3.00	2.15	2.38	3.80
Region BTC (r)	2.10	2.00	1.80	1.85	1.94	1.95
Landscape unit BTC (u)	2.14	1.95	1.65	1.25	1.19	1.15
Park BTC importance (p-r)/(p-u)	1.12	0.93	0.89	0.33	0.37	0.70

The ratio (p-r)/(p-u) is equal to 1.0 when the local landscape unit and the region have the same BTC value, thus it gives an idea of the influence of the change of the surroundings exerted on the park: its regional importance is reduced because of the need to re-balance the local transformation. This diagnostic result drives the first choice toward the criteria for territorial planning. In fact, the proposals to restore the park to its original design (e.g. the plan of the Vienna archives) is to be avoided, because the landscape conditions have completely changed: today the first need is an ecological balance of the landscape unit, not a disputable aesthetic restoration.

From Fig. 13.9 and Table 13.11 we can see that the BTC of the park should be planned to achieve values of 3.50-3.80 Mcal/m²/year in the next 20 years, if possible. So, to verify this hypothesis and plan the park we need to evaluate the ecological state of the park itself on a more detailed scale.

Using Table 13.12 and with the help of the standard BTC classes (not forgetting the suggestions derived from the field study) it is easy to verify that the design BTC will not be over 2.80 Mcal/m²/year, first of all because of the time limits, at least until motor-racing will is moved outside the park.

Table 13.12. Control of the possible planning criteria for the rehabilitation of the Monza Park

BTC class	Mean BTC	1994 %	2014 %	(2034) %
VII	9.8	-	1.0	5.0
VI	7.5	1.0	6.0	10.0
V	5.2	14.0	18.0	20.0
IV	3.4	32.0	25.0	16.0
III	2.3	2.5	3.0	4.0
II	0.9	43.0	39.5	38.0
I	0.2	7.5	7.5	7.0
BTC (Mcal/m²/year)		2.38	2.80	3.29
Shannon diversity H		1.335	1.536	1.677
Landscape diversity τ		4.82	5.24	5.48
Landscape metastability LM		11.47	14.67	18.04

Only in a second plan (2034) will it be possible to arrive at a BTC of 3.29 Mcal/m^2/year. Moreover, even in a plan for 2014, the ecological state of the park could be decisively better, reaching a LM = 18.04.

Note that for any planning purpose, it is crucial to have reference values compatible with ecological criteria. The use of landscape ecology becomes simply indispensable. Following the methodology and the principles discussed in Chap. 12 (see Sect. 12.3), starting from the values of Table 13.12 and from all the analyses, we can now propose a rehabilitation plan. Let us observe the present map of the park and compare it with the main planning proposals (Fig. 13.10). The park is today only about 50% available for people, because of motor-racing and the golf course; no doubt that their location outside the park remains the main

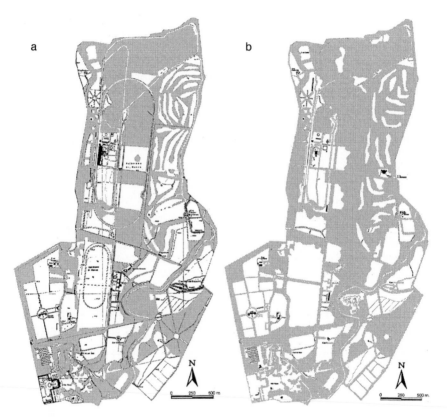

Fig. 13.10A, B. Maps of the Monza Park (Lombardy). **A** Present situation. **B** Rehabilitation plan. In the first period we can maintain the motor-racing. See text for explanation

need. But we also have to be realistic, so "Realpolitik" imposes for the moment the following plan:

- To dismantle the eastern part of the first (old) racing ring, leaving the bridge over the modern racing track, which can be transformed into a green bridge, to link the two main patches of forest
- To readjust the forest tesserae following sylviculture methods converging with landscape ecology and adding small patches of autochthonous reforestation
- To plant a new forested corridor between the golf course and the dismantled racing track, from north to south
- To create a new humid forest patch in a bight of the river Lambro, with a small lake useful also as expansion against flooding (Fig. 13.11)
- To restore the trees of the boulevards and to plant the hornbeam corridor linking the two villas ex Durini
- To reorganise the agricultural surroundings of the main historical farms
- To create a green riding ground for high school horseback riding
- To create an underground gallery for the central part of the traffic road crossing the park
- To restore the historical villas and the old monuments
- To restore the English garden of the Villa Reale

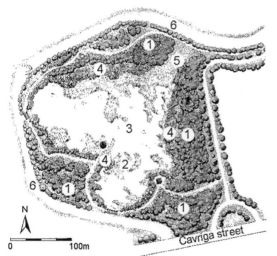

1. Never submerged zone
2. Submerged zone
3. Sheet of water
4. Submerged zone at intervals
5. Grass
6. Lambro River

Fig. 13.11 The creation of a small lake in the bight of the river Lambro: more detailed studies (from C. Rezia-Loppio, unpublished)

13.5
Agricultural Landscape Dynamic in Burkina (Africa)

13.5.1
Territorial Frame

The study area is located in the territory of Zitenga, province of Oubritenga in the northern part of the capital town of Ouagadougu in central Burkina Faso (Fig. 13.12). The main biogeographic character is the aridity but it is not so harsh as to impede the development of a savannah forest; the rain varies from 600 to 1000 mm/year with a mean temperature of 26-27 °C and the arid period is 6-8 months.

The study is part of a research project on the territory of two villages: Tanlili and Tanginna or Tanghen-Kossodo, 26250 ha; it was co-ordinated by E. Tibaldi (University of Milan, Department of Biology). The specific research was conducted for the graduate thesis of D. Paradiso, coordinated by the author and G. Soncini. The surveys were always made by a mixed group of Italian and Burkinabé students. The study was not only descriptive, but also oriented toward active nature conservation and territorial management.

A long transect from north to south (9 km) facilitated the description of soils, vegetation and human use. The disposition of the two villages is nearly continuous, interrupted only by small hills of lateritic soil cover.

Fig. 13.12 Burkina Faso map and the localisation of the study area

Fig. 13.13 An example of a so called *zaka*: a study of typical houses is essential to understanding the significance of the urbanisation of agricultural landscapes in Burkina, which consists mainly of a condensation of many zakas without a true village structure

Often the excessive exploitation of the soil led to a loss of its fertility. The Mossì houses (*zaka*) are built along a circular fence or wall (diameter: 30-40 m) around a well: three or four round thatched cabins for wives and children and a rectangular cabin for the husband (Fig. 13.13). They are self-sufficient.

Near the hills, systems of mammalian dens are still present, but limited mainly to *Orycteropus afer* (earth pig) and *Crocuta crocuta* (tiger wolf) which use the same dens at diverse periods of time. Other animals are *Genetta* sp., *Hystrix cristata*, *Viverra civetta*. The trees species found along the transect are: *Acacia albida, Acacia seyal, Acacia senegal, Adansonia digitata, Anogeissus leiocarpus, Balanites aegyptiaca, Butryuspermum parkii, Diospyros mespiliformis, Ficus platyphilla, Khaya senegalensis, Lannea microcarpa, Mytragina inermis, Sclerocarya birrea, Tamarindus indica;* the shrub species: *Combretum glutinosum, Combretum micranthum, Combretum nigricans, Guiera senegalensis, Piliostigma reticulatum, Ziziphus mauritana* (species following White 1986).

The degradation of the vegetation is due to the increasing aridity and to fires and cutting of the forested savannah. It concerns especially the species *Parkia biglobosa, Khaja senegalensis* and *Butyospermum parkii*. Pastoral and agricultural landscape units differ very little in trees species: pastoral areas have only three species more then the agricultural one, *Acacia seyal, A. senegal* and *Zyziphus mauritana*.

The FAO (1993) sustainability indicators for arid agricultural landscapes have been deducted from thematic maps. These indicators (Table 13.14) underline three main changes in the landscape: (a) the strong population increase, (b) the consequent increase of agricultural land to the detriment of pastoral land, (c) the great decrease of the forested savannah in favour of pastoral lands.

Table 13.14. Variations of the FAO indicators for agricultural landscapes

FAO sustainability indicators	1980	1994	Variation %
Houses/km^2 of total surface	1.88	3.61	+ 92.02
Houses/ km^2 of cultivated area	6.18	7.86	+ 27.18
Cultivated area/total surface	0.31	0.46	+ 48.39
Tree + shrub area/ total surface	0.69	0.52	+ 48.39
Tree area/ tree+ shrub area	0.33	0.14	- 24.64
Highly degraded area/ total surface	0	0.02	+ 2.00

13.5.2
Landscape Ecological Considerations

The FAO indicators were insufficient to understand the real ecological state of the examined territory. So, the use of landscape ecological principles and indices have been adopted. Among them we report the fractal dimension and the BTC.

The fractal dimension of the forested savannah went from $D = 1.4$ (1980) to $D = 0.97$ (1994), underlining that the degradation of this important type of local vegetation was not only due to its decrease, but also to its spatial distribution, which became evidently more fragmented (Fig. 13.14).

Observe that the landscape dynamic of this African territory was even more rapid than in a suburban European landscape. Besides, for climatic and other instabilities, the safety coefficient for SH* has to be at least double that in Europe.

We needed to calculate the percentage contribution of the BTC of NH to the mean BTC (BTC_{NH}/BTC_m). For this purpose, remembering that NH is the opposite of HH, we easily found the total amount of NH in 1980 in order to weight the other contributions: $\Sigma NH_{80}= 120$, and then the same for 1994 (see Table 13.15 too):

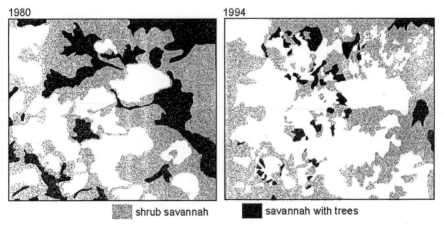

1980   1994

▨ shrub savannah   ■ savannah with trees

Fig. 13.14 The decrease and fragmentation of the forested savannah in 1980 and 1994

BTC_{NH} (Agricultural land)= ......$10/120 \times 1.4 = 0.1167$

BTC_{NH} (Shrub savannah)= ....$40/120 \times 2.3 = 0.7667$

BTC_{NH} (Forested savannah)= .....$70/120 \times 5.5 = 3.2083$

BTC_{NH} (Total) = 4.0917 Mcal/m^2/year (1980)

BTC_{HH} (Total) = 1.94 Mcal/m^2/year (1980)

$BTC_m = 0.3769 \times 4.10 + 0.6231 \times BTC_{HH}$ (Total) from which

$$BTC_{NH} / BTC_m = 1.49 \ (1980)$$

BTC_{NH} (Total)= 2.95 Mcal/m^2/year (1994)

BTC_{HH} (Total)= 1.72 Mcal/m^2/year (1994) from which

$$BTC_{NH} / BTC_m = 0.70 \ (1994)$$

Table 13.15. Recent landscape dynamic of the territory of Tanlili and Tanghen-Kossodo.

Landscape element	HH	BTC	1980		1994	
	%	Mcal/m^2/year	ha	%	ha	%
Urbanisation (scattered)	100	0.4	76.1	0.29	196.9	0.55
Agricultural land	90	1.4	7932.7	30.22	11935.9	45.47
Shrub savannah	60	2.3	12229.9	46.59	11649.7	44.38
Forested savannah	30	5.5	6011.3	22.90	1918.9	7.31
Degraded areas	50	0.2	-	-	601.1	2.29
Total			26250	100.00	26250	100.00
Population			7087		13571	
Human habitat HH %			62.31		71.44	
BTC Mcal/m^2/year			2.75		2.07	
SH m^2/inhabitant			23079		13818	
$\sigma = $ SH/SH*			7.59		4.54	

SH* = 1014×3 (safety coefficient) = 3042 m^2/inhabitant

Plotting these values in Fig. 13.15, we saw strong changes in the territory, from an agricultural-seminatural landscape to a near rural one in only 14 years. This territory, able to sustain seven times the number of its inhabitants in 1980, in 1994 might sustain only four the number of inhabitants. Even the natural habitat was reduced from a mixed trees-shrubs savannah to a simple herbaceous one.

On the same plot, we designed a way to direct the development of this territory, evaluating its limits and proposing criteria for a correct landscape planning.

Some criteria for the design of an ecological network were proposed in 1996 after this work, increasing a ministerial programme to plant 6500 *Acacia senegal* (producing Arabic gum). The preferential linkages are plotted in Fig. 13.16. At the western end of these linkages (arrows in the figure), in the shrub savannah, we need to plant some forest patches to rehabilitate the natural environment. It seems that the local people will act positively in order to save their environment. But the true problem is time.

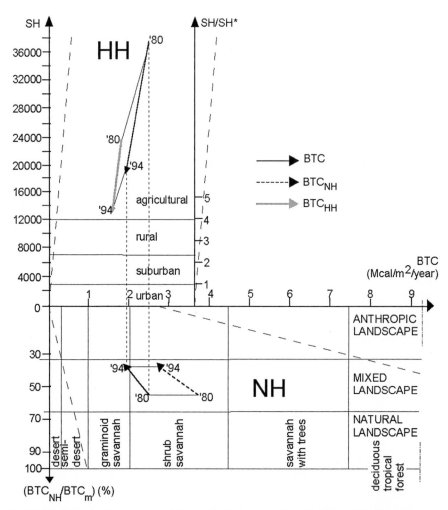

Fig. 13.15 Control of the transformations of the landscape unit of Tanlili and Tanghen-Kossodo from 1980 to 1994. In the upper quadrant we can measure the HH movements of the ecological system, in the lower quadrant the movements of the NH. This plot can be used to check the proposals of land management and population control (see also Figs. 5.4 and 13.9)

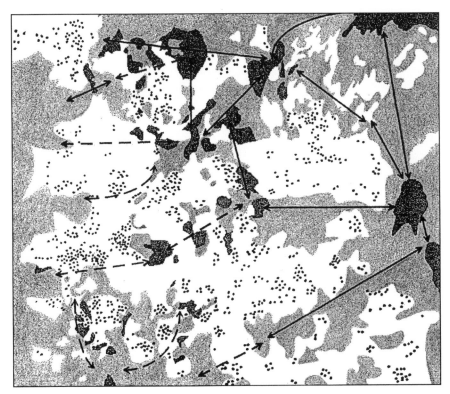

- - - → potential corridor

⎯⎯⎯→ existing corridor

Fig. 13.16 Proposal of the development of an ecological network in the landscape unit of Tanlili and Tanghen-Kossodo. *Pale grey* shrub savannah, *dark grey* forest savannah, *black spots* houses, *white* agricultural areas (with some degraded patches). *Dotted arrows* rural plantations, *plain arrows* forest plantations.

Glossary

ELENA GIGLIO INGEGNOLI

Adaptive system: a system able to modify itself to meet new requirements in response to changes in its environment. An active adaptive biological system is able to specify the admissibility of the environmental constraints, so as to define the domain of environmentally pertinent perturbations.

Alliance: in phytosociology, taxonomic level superior to association.

Allometry: the quantitative relation between the size of a part and the whole in a series of related organisms differing in size.

Archaeophyte: plant of old naturalisation in the studied land (before AD 1500).

Association, vegetational or plant: in the phytosociological approach, a plant community presenting a defined floristic composition (characteristic specific composition), not necessarily always identical, but fluctuating around a medium value. It represents a vegetation type corresponding to defined ecological conditions and its taxonomic name comes from the Latin name of particularly significant species with the suffix *–etum*.

Attractor: a geometrical object toward which the trajectory of a dynamic system, represented by a family of curves in the phase space, converges in the course of time. A fixed-point attractor consists of a set of points in the phase space and describes a stationary state of a system.

Auto-state: (eigenstate) a dynamic state the wave function, or state vector, of which is an auto-vector (or a proper function) of an operator, corresponding to a specified physical quantity.

Biocoenosis: the first definition of ecological community.

Biogeocoenosis: a part of the relief where, within a certain area, biocoenosis and the relevant parts of the atmosphere, lithosphere and pedosphere are homogeneous and are associated with each other in a homogeneous relationship, creating a single interdependent complex.

Biodiversity: there is no universal meaning of this word. The Earth Summit of Rio de Janeiro (June 5, 1992) defined biodiversity as "the variability of living organisms of every origin, included *inter alia* the terrestrial, the marine ecosystems and other aquatic ecosystems, and the ecological complexes which they are part of; it includes the diversity in the circle of the species and among the species of the ecosystems". As life is not only organisms or species, and ecosystems are part of landscapes (upper level of living systems) and not of generic 'ecological complexes', a more correct definition might be: "the variety and variability of living systems". So, a proper biodiversity exists for each level of the hierarchy of life organisation, from the small scale (e.g. genetic diversity) to the medium scale (e.g. specific diversity) to the upper scale (e.g. landscape diversity). Different indexes have been proposed to measure it.

Biological spectrum: the whole of the hierarchic levels of life organisation on Earth.

Biological territorial capacity: a magnitude representing the energy flux per square metre per year ($Mcal/m^2/year$) that the system needs to dissipate to maintain its equilibrium state and its organisational level. It expresses the latent capacity of a landscape to return to its metastable equilibrium state. It can be estimated through a proper methodology referred to the vegetation.

Biomass, plant: the dry weight or the volume (above ground) of the individuals of a plant species population in a given area.

Biome: the largest land community unit which it is convenient to recognise. They are formed by the interaction of regional climates with regional biota and substrate. The biome is identical with major "plant formation" of plant ecologists, except that the biome is a total community unit and not a unit of vegetation alone. It is the same thing as "major life zone" as used by European ecologists (Odum, 1971). In a given biome the life form of the fittest vegetation is typical, although the species of dominant plant communities may vary in different parts of the biomes.

BTC: an index measuring the biological territorial capacity of a landscape or of a part of it.

Chaotic system: a system, complex or simple, the behaviour of which depends so sensitively on the system's precise initial state that the behaviour itself and the final state are in effect unpredictable and can not be distinguished from a random process, even though they are strictly determinate in a mathematical sense. The reason is that even the most sophisticated approximation can not precisely coincide with absolute exactness: these systems amplify the initial differences exponentially in time, acting as a zoom. So, even if in theory two initial positions would reproduce the same trajectory, in reality the starting point will never be exactly the same each time and the trajectories consequently will have to diverge.

Chaos theory: the theory studying randomness in system, that is, how systems during their movements are able to acquire degrees of freedom that they did not own previously, and going toward a chaotic behaviour even if these systems can be described by deterministic equations. Chaos theory concerns even simple systems because the variety of factors and causes or the complexity are not the unique source of randomness in systems.

Climax: in classical Vegetation Science and General Ecology it indicates a mature, relatively stable community in an area; a community which will undergo no further changes under the prevailing geo-climatic conditions, and which represents the culmination of ecological succession (from Ancient Greek κλιμαξ 'scale').

Complex landscape unit: an interacting arrangement of simple landscape units, the origin of which is ascribable to the same group of geomorphologic processes. It could represent a good proportion of the entire landscape.

Configuration: an arrangement of juxtaposed or intersected landscape elements and the way in which they are interconnected. It can be functional and/or spatial.

Corridor: scale independent concept indicating a linear portion of a territory, the aspect of which differs from the surrounding environment on both sides. It can be linear or stream-like or a strip.

Determinism: the principle that the specification of the dynamic state (or of the dynamic variables) of a system at a given time, and of the interactions of the system itself with its environment, completely determines the dynamic state of the system (and the values of the dynamic variables and their probability distribution) at later times. Also known as causality.

Disciplinary models: the logical representations of the different branches of a discipline, with certain specified parameters that are used to understand their differences.

Disturbance: synonymous with perturbation, it identifies a process or event able to direct the structuring of an ecological system and to be incorporated by it.

Diversity: in general, the number of elements in a group. In ecology, the concept of diversity is strictly linked with information theory, as expressed by the Shannon equation. Presently, five types of diversity exist: α diversity (number of elements, usually species, per single site); γ diversity (number of elements usually species, per a group of sites); β diversity (ratio γ/α); η diversity (number of species per ecotope); τ diversity (the diversity of the standard classes of biological territorial capacity to which the landscape elements belong, in relation to their relative dominance in an ecotope or landscape unit). α, γ, β refers to the system of Whittaker, η to that of Brandmayr and τ to that of Ingegnoli.

Dominance: the degree to which one or a few elements predominate in a landscape in terms of number, biomass, dynamics or relative importance. It may be used as a measure of the information content of a system, in contrast to the equitability.

Ecobiota: generic term to indicate a group of biotic elements with an ecological sense; it is synonymous with ecological system and it is valid at all scales.

Ecocoenotope: a multifunctional entity corresponding to the biological organisation level existing between the population level and the landscape level. It represents the elementary unit of integration of the community, the ecosystem (functional characters) and the microchore (the spatial contiguity characters of the communities) in a definite geographic locality (site). In practice it is a part of the land that is uniform throughout its extent in landform, soil and vegetation. Its spatial dimensions may range from 10^2 to 10^8 m^2, the temporal ones from 10^0- 10^4 years. When referred to landscape ecology, it can be called a tessera.

Ecological density: or specific density, it express the ratio number of individuals/ own specific habitat surface. Thus it differs from geographical or planning density.

Ecological system: generic term to indicate each of the various levels of the biological spectrum. It must not be used as a synonym of ecosystem.

Ecomosaic: the most significant pattern of juxtaposed landscape elements (patches, corridors and matrix, natural or human, or tesserae and ecotopes) constituting a sort of ecological map of the territory.

Ecophysiognomy: the prediction of plant functioning (e.g. their water and energy budgets and their distribution in landscapes and regions) based on the plant form and climatic conditions.

Ecoregion: the ecological system composed of connected landscape systems with the same macroclimate and tied together by human activities. It is the ecological integration of biomes with their environment.

Ecosystem: in the classical definition, the ecological system formed by a group of biotic entities (usually populations) and by the reciprocal interactions among them and with the physical components of the environment. The definition congruent with this text is: the part of the ecocoenotope concerning the functional relations between its component communities and the abiotic factors of their environment.

Ecotissue: a multidimensional conceptual structure planned to well represent the hierarchical intertwining of the ecological lower (organisms, populations, communities/ecosystems) and upper (ecoregions) biological levels and of their relationships in the landscape.

Ecotissue model: the model proposed to understand not only the spatial but also the temporal and complex structure of a landscape as a specific level of the biological spectrum. Its elements are a basic mosaic (usually the vegetation one) and a set of other mosaics, and correlated arguments, of changeable number, partly simply thematic, partly deriving from the elaboration of spatio-temporal studies on landscapes (e.g. the mosaic of standard classes of BTC, the mosaic of human habitat and of natural habitat, the mosaic of landscape apparatuses…). Each single mosaic may comprise juxtaposed tesserae or overlapping ones or a mix. All these mosaics have to be correlated to, and/or intersected with, the main one in a hierarchical way, as the weft and the warp in weaving or the cells in an histologic tissue. The iterative integration of the results, guided by landscape anatomy and physiology, allow us to understand the ecological tissue of the studied landscape and can be rendered in an operative chart of integration, which is different from an ecomosaic or an ecological or a geographical map because of the ecotissue's intrinsic integration of the elements.

Ecotope: the smallest landscape unitary multidimensional element that has all the structural and functional characters of the concerned landscape. So it is the minimum system of interdependent ecocoenotopes (not less than two ecocoenotopes) or tesserae as determined by the topographical recurrence, the geomorphologic origin and the functional configuration and role in the landscape.

Engineer species: organisms that directly or indirectly control the availability of resources to other organisms by causing physical state changes in biotic and abiotic materials. Physical ecosystem engineering by organisms is the physical modification, maintenance, or creation of habitats (Jones et al. 1994)

Entropy: a measure of the absence of information about a situation, or the uncertainty associated with the nature of a situation; in a mathematical context, this concept is attached to dynamic systems, transformations between measured spaces, or systems of events with probabilities: it express the amount of disorder inherent or produced.

Equilibrium, dynamic: in physics, a condition in which several processes act at the same time to maintain a system in an overall state not changing with time.

Equilibrium, metastable: a state of pseudo-equilibrium having an higher free energy than the true equilibrium state (in which, by contrast, the free energy tends to zero). In physics it expresses a condition in which a system returns to equilibrium after small (but not large) displacements; it may be represented by a ball resting in a small depression on top of a hill.

Equitability: (H_{max}) the equipartition of the abundance of the considered element (species, tessera, ecotope or biomass) in a site; in other words, it indicates the same abundance for all the considered elements. So it represents both the maximum diversity and the minimum information.

Evenness: the ratio between the measured abundance or diversity of the considered element and the equitability (H_{max}) of the element itself.

Exportable characters: characters of one or more levels of the biological spectrum that are valid also for others levels of the spectrum itself. They are especially the chorological ones.

Fittest vegetation: the landscape ecology re-interpretation of the concept of potential natural vegetation. This new concept refutes the general notion of 'potentiality' as the possibility of the coming into existence, in the absence of man and for large territories, of a deterministic, a priori fixed vegetation type and interpreted as the best condition for a place, independent of all the other environmental and human factors and of time. The term *fittest* means to be of 'the right measure, shape, size for' or 'in the right, suitable time and place for' (Hornby, 1974). Thus, the concept of "fittest vegetation" indicates the most suitable or suited vegetation to: the specific climate and geomorphic conditions of a certain limited period of time in a certain defined place; the main range of incorporable disturbances (including man's); natural or not natural conditions. In addition its presence has to be intended as patchy.

Fluctuation: variation of the state of the system within a proper field (which corresponds to its metastable field) in response to the normal range of disturbances. When the disturbances pass the limits of the range, the system fits its organisation to the new conditions.

Focal species: a proper group of different species able to represent a sphere of spatial and functional demands that really include the needs of all the other species of the studied territory.

Fragmentation: the process of breaking up a landscape element or an aggregation of landscape elements into smaller parcels. It is one of the spatial processes in land transformation.

Functional configuration: a group of landscape elements, even those not connected, able to perform one or more particular functions.

Fuzzy logic: the logic of approximate reasoning, bearing the same relation to approximate reasoning that two-valued logic does to precise reasoning.

Fuzzy system: a process that is too complex to be modelled by using conventional mathematics methods and which gives rise to data that are, in general, soft, with no precise boundaries. A fuzzy set is an extension of the concept of a set in which the characteristic function, which determines membership of an object in the set, is not limited to the two values 1 (for membership in the set) and 0 (for non-membership), but can take on any value between 0 and 1 as well.

Generalist species: animal species showing multifunctional niche characters, and non specialised behaviour.

Heterogeneity: for a defined number of landscape element types, the variety of their patterns of arrangement. For each number of landscape element types (at least two), several patterns exist. In an ecotissue, heterogeneity can be structural, functional or ecosystemic. The condition of a sample of matter that is composed of particles or aggregates of different substances of dissimilar composition.

Holism: the view that the whole of a complex system is functionally greater than the sum of its parts.

Homeorhesis: the maintaining of a recurrent dynamic flux in a biological system.

Homeostasis: a process able to maintain a stationary state in a biological system.

Human habitat (HH): the areas where human populations live and work permanently, limiting the self-regulation capability of natural systems. In ecology, HH is not the entire territorial geographical surface, but it is limited to human and semi-human ecotopes and landscape units. Note that also the apparatuses pertaining to HH are intended in ecological sense, not in a geographical or urban planning sense, and that HH landscape elements contain a small percentage of natural habitat.

Incorporation of disturbances: the capacity of the organisation level of an ecological system to resist, or to condition itself to the consequences of, a disturbance(s) at a lower scale, using it/them to upgrade its structure.

Index: a numerical quantity, usually dimensionless, denoting the magnitude of some physical or ecological effect.

Information theory: the branch of probability theory concerned with the likelihood of the transmission of messages.

Intersection: the set of characteristics or properties that is in common to two or more landscape elements.

Iterative method: a process for reaching the desired results by successive approximations or repeated cycles of operations, which come closer and closer to the final outcome.

Key species: a data set of species that serves to uniquely identify a phytocoenosis, a plant association or a defined habitat. Ecologically, they are species occupying a critical position in the community as they are irreplaceable: so their removal, even if fortuitous, is fraught with consequences. Sometimes the key species are the ones giving their name to the coenosis, more often they are other species, even small or rare, but faithful to the coenosis itself.

Land: its etymology is to be sought in the Indo-European base *londh*. During its linguistic modifications, *londh* changed its form into *landam* from the Old Celtic *landa*, then into *lant*, *lann*, *llan* arriving at *land* in Old Saxon, but the meanings ranged from 'land, region, tract' (Old English, Old Norse, Gothic, Medieval Dutch, High German) to 'enclosure' (Gaelic, Welsh) to 'open space, region, plain' (English) and finally to 'heath' (Breton). In the neo-Latin group, Old Celtic *landa* led to the French *lande* and the Italian *landa*, as well as the Old Slavic *lędina* and the Russian *ljada*, but always indicated a heathland, a wasteland, or a fallow or moor land (De Vries 1962; Onions 1966; Partridge 1958). In German the word *land* is linked to the concept of land tilled or cultivated or to the countryside, in contrast with town, and to country in a geographical sense.

Landscape: a proper level of the hierarchy of life organisation, between the ecocoenotope level and the ecoregion level. The correct conceptual definition is: a system of ecocoenotopes in a recognisable configuration. But, for a more common understanding, following Forman and Godron (1986), in the text landscape is often defined as a 'system of ecosystems (i.e. biogeocoenosis)'. The range of the spatio-temporal scale is 10^6- 10^{10} m^2 and 10^2-10^5 years.

Landscape apparatus: a functional system of tesserae and/or ecotopes arranged in a specific configuration in the ecotissue of a landscape. Each different apparatus is characterised by a specific landscape function. It is not a district or sub-landscape, but a complex configuration of patches, even non-connected.

Landscape element: each elementary unit that is still a holistic unit at the proper range of scales of a landscape. In the ecotissue model landscape element is a general term which may indicate a tessera, an ecotope, a simple landscape unit, a complex landscape unit.

Landscape etymology: the relationship between the etymology of this term in the four main groups of Indo-European languages and its scientific concept have not been studied yet but, nonetheless, some very brief notes will be reported here. In modern Greek, landscapes are identified as τοπια, from the ancient τοποσ 'place, locality, region, space'. In the German languages group all the terms (e.g. German *Landschaft*, Dutch *Landschap*, English *Landscape*, Swedish *Landskap*) come from the Indo-European base *londh*. As to the second part of the term (ranging from *scípe* in Old English, to *skipi* in Old Saxon, *scaf* in Old High German, *skapr* in Old Norse, *scap* in Middle Dutch), its etymology is not clear: following Onions (1966) –*schip* is a suffix denoting the qualities or characters associated with the word itself. By contrast, an indication in Winters (1934) and in De Vries (1962) suggests to relate the German *schaft* to the ancient Greek σκῆπτρον, ου with the meaning of 'stick', 'staff': but the Greek word means also 'regal sceptre', 'domination', so implying for *landschaft* a government of the territory. In effect, in politics *landschaft* indicates also 'region' (Macchi 1978) while in geography it expresses the coexistence of physical and human characters in a specific territory (Migliorini 1942). What is clear (Onions 1966) is that the English term landscape is a loanword adopted from the Dutch *Lantscap* (the regular one in Middle Dutch) as a painter's term in the sixteenth to seventeenth century, with the meaning of 'picture representing natural inland scenery', following the fame of the Flemish and Dutch schools; the form *Landskip* represents the Dutch pronunciation. And 'natural inland scenery or its representation in painting' is the generalised sense of landscape (Bense 1939). In neo-Latin languages the used terms (like *paesaggio* in Italian, *paysage* in French, *paesajo* in Spanish) arise from the primeval Sanskrit root *pac* used to indicate 'to tie', 'to fasten'; this root developed into the Ancient Greek verb πήγνυμι with the meaning of 'to fix, to establish, to thrust, to plant' (Bolza 1852). The Latin *pagus* comes directly from that Greek verb in one of these two ways: as the translation of 'boundary stone fixed in the ground', to indicate a rural bounded territory (Olivieri 1961), that is a rural district in the Roman time (D'Anna, 1988); or through the Greek terms $\pi\eta\gamma\acute{\eta},\eta\sigma$ (or $\pi\alpha\gamma\alpha$) meaning 'source', 'spring', because the first primordial groups of huts were localised near springs (Nobile 1943). However, in Late Latin (Niermeyer 1976), *pagus* indicated also 'the countryside of a *civitas* as distinguished from the city' or 'a smaller district than the *civitas*' and gave *pagènsis* (belonging to the *pagus*) with many different shades of meaning, all implying 'the inhabitants of a country' or 'of a church district' or 'of a larger district': even in Germanic lands *pago* was used to indicate a district 'Gau' or one of the ethnic divisions. *Pagènsis* gave (around AD 1250-1350) *Paese, Pays, Païs, Paiz*. But what it is interesting for us is that the meaning of these latter words is 'a great extension of a cultivated and inhabited territory' (Cortellazzo and Zolli 1985) and 'a part of the inhabited lands … that we may consider different from the others for its climate, its language and for other local data' (Bolza 1852). So, when in the fifteenth to sixteenth century the terms *paysage* and *paesaggio* arose (Battisti and Alessio 1954; Cortellazzo and Zolli 1985), they were utilised to identify the object of the painting as a countryside, with all its different attributes, and the object of all the studies carried out on it. Among the Slavic languages, the landscape is related to the Sanskrit root *rāj* 'sovereign, governor' and *rājya (krāija)* 'ruled territory, kingdom' (Stchoupak et al. 1932), like in the Polish *krajobraz* from *kraj* 'country, land, home', but in this language for painting the word is *peizaż*. The Russians utilise *Пейзаж (peisak)* as locality or as picturesque landscape; *Ландшафт (landscaft)* as the morphology of a territory. In conclusion, it is evident that the present tendency to consider all the above-mentioned terms only as representative of the visual/scenic aspects of the territory is wrong, because the roots of almost all them imply an active relation with man as governor, and his consequent interest to study the territory itself to be able to manage it.

Landscape system: an arrangement of landscapes which have significant geographical and vegetational features at a sub-regional scale.

Landscape unit: or geo-bio-district is a part of a landscape, the peculiar structural or functional aspects of which characterise it as regards the entire landscape. It is not always a simple arrangements of ecotopes, even if it forms a connected patch of them, because it is intended in a non-deterministic sense: so its structure is not always immediately recognisable and the analysis of its range of functions is often consequently necessary. It may be simple, complex or merely operative.

Learning system: a system that is able to gather, process, store and recall information received through the senses and that is conscious of it.

Living system: approximately corresponding to ecological system. It includes both biological systems and environmental systems.

Matrix: the most extensive and most connected landscape element type present in a landscape; it plays the dominant role in landscape functioning and characterises the type of landscape.

Metastability: the state of a system oscillating around a central position, but susceptible to being diverted to another metastable equilibrium state.

Microchore: a spatial distribution of ecological elements in a defined site.

Model: a mathematical, geometrical or physical system obeying certain specified conditions the behaviour of which is used to understand a physical or biological system to which it is analogous in some ways.

Mosaic model: one of the models developed to understand the spatial structure of a landscape. It is more linked to a geographic and vegetational approach and includes two types: the 'patch-matrix' model (the elements of which are patches and corridors, both natural or human, in a sort of landscape matrix) and the 'ecomosaic' model (the elements of which are tesserae and ecotopes constituting a sort of ecological map of the territory). The elements of both the two types are juxtaposed: in a way it is a static model.

Movement of a system: the passage of a system from one state to another.

Natural habitat (NH): the natural ecotopes and landscape units dominated by natural components and biological processes, without direct human influence, capable of normal self-regulation. Note that a share of HH is present also in NH.

Negentropy: or information content, it's a numerical measure of the information generated in selecting a specified symbol or message, equal to the negative of the logaritm of the probability of the symbol (or message) selected.

Neophyte: plant that arrived in the studied locality after AD 1500.

Non-equilibrium thermodynamics: the domain of the multiplicity of solutions of non-linear equations, thus of the richness of behaviours in a coherent universe.

Operative chart of integration: the final chart of a landscape resulting from all the iterative hierarchic integrations of the ecotissue model, in which essential map information is combined with various other data critical to the intended use.

Operative landscape unit: merely work-based, it may be composed of the aggregation of simple landscape units or of parts of complex landscape units, even belonging to different landscape types, in accordance with the particular problem or function that must be investigated (e.g. the influence of a large airport or of an industrial plant).

Operator: a function between vector spaces or anything designating an action to be performed.

Patch: according to Forman and Godron (1981, 1986) and to Forman (1995) a patch is a scale independent concept indicating a non-linear portion of a territory, the aspect and/or the substance of which differs from the surrounding environment and which is relatively homogeneous (the internal microheterogeneity present is repeated in similar form throughout the area of the patch). In this volume a patch can be composed of a single tessera, by a group of tesserae, by a single ecotope or by a group of ecotopes.

Pathway: a real or virtual line along which something, somebody, or a process moves.

Pedon: the smallest unit or volume of soil that represents or exemplifies all the horizons of a soil profile.

Permeants: V.E. Shelford's term for the highly mobile animals, such as birds, mammals and flying insects. They move freely between strata and subsystems and between developmental and mature stages of vegetation that usually form a mosaic on most landscapes. They correspond to the nekton of aquatic ecosystems and often exploit the best of several worlds (Odum 1971).

Perturbation: any effect which produces a change in an ecological system.

Phytocoenosis: the entire plant population of a particular habitat.

Phytosociology: a discipline of Vegetation Science characterised by a qualitative and quantitative approach to the study of vegetation proposed by Braun-Blanquet.

Plant form: synonymous with vegetation form; the form characteristically taken by a plant at maturity.

Potential core area: in a remnant patch, it is the inside surface, after not less then three buffer belts, with an ecological potentiality for natural species of the original landscape type.

Potential natural vegetation: see Fittest vegetation.

Proper characters: characters belonging to a specific level of the biological spectrum that distinguish this level from the others.

Region: from the Indo-European root *rāj* (see also Landscape etymology in Slavic languages) meaning 'sovereign, governor' linked to *rāji* 'line, tract'. Two connected groups of terms, present only in the Celtic, Latin and Indian languages, derived from this root: *rex-regis* (Lat.), *re* (It.), *roi* (Fr.), *rey* (Sp.) *rajah* (Ind.), meaning king, and *regio-regionis* (Lat.), *regione* (It.), *region* (Fr.), *region* (Sp.) (Stchoupak et al. 1932). The first group is more linked to the interpretation as the unique religious-political authority given the power to trace out the linear boundaries of the town and to establish the rules of law; the second group is derived from the straight lines traced in the sky by the ancient augurs to subdivide it and assumed the meaning of limit of a portion of a territory. Dante Alighieri (AD 1321) defined region as a portion of the earth, sky or space with its own proper characteristics (Cortellazzo and Zolli 1985).

Remnant patch: rural and/or semi-natural residual areas of the original agricultural and/or open-forest matrix of the landscape, surviving after its excessive fragmentation.

Self-organising system: a system that is able to affect or determine its own internal structure.

Simple landscape unit: an interacting arrangement of ecotopes, the origin of which is ascribable to the same geomorphologic process, assuming a particular functional significance in its own landscape.

Spatial configuration of a landscape: the spatial arrangements of the pattern components of a landscape, that is of a complex landscape unit, simple landscape unit, ecotopes, tesserae.

Specialist species: animal species showing a specialised niche and a peculiar behaviour

Split method: analysing the behaviour of a complex system through several partial and separate models, each related to the combination of two or more aspects, then joining them together to reach the comprehension of the whole.

Standard habitat (SH) per capita: the reciprocal function of the specific density, it expresses the real relative proper habitat surface, even of different habitat types, at a specific biological entity's (even man's) disposal. The theoretic SH represents the minimum optimal habitat surface needed to sustain the individual.

Stationary state: an auto-state (energy state or *eigenstate*) of the energy operator so that the energy has a definite, not moving or changing (stationary), value.

Steady state: the state of a system in which the condition at each point does not change with time, that is, after initial transient or fluctuations have disappeared.

Succession: in general ecology, the most important process of transformation in which through serial stages an ecosystem, or a phytocoenosis, changes in a predictable way toward a final stage, called climax: after a perturbation succession returns the ecosystem to the climax through longer or shorter period of time. Non-equilibrium thermodynamics refutes this notion both in its deterministic, a priori fixed assumptions, in its independence from spatial and temporal scales and in its independence from all the other environmental and human factors. The process of succession may be valid if considered non-deterministic, non-linear and open to possible changes in direction, as a consequence of more ample fluctuations of the state of the system, and if considered limited to the single ecocoenotope scale, to the specific climate and geomorphic conditions of a certain period of time in a certain, defined, place and within the main range of incorporable disturbances (including man's). The final state generally is the fittest vegetation, but it can be reached through different pathways or not at all.

System: in general, a set of elements closely interacting, sometimes integrated to perform a/more specific function/s; a physical entity upon which some actions may be exerted through an input a and which, as a reaction, an output y derives from.

Territory: part of land, especially under one ruler or government. Also, in zoology it represents a defended area. It is a term of unclear etymology: it may derive from the Ancient Greek περιχωριον (*pericorion*) which indicated 'neighbouring region and/or people', coming from χωρα 'a tract of land, a district' (as a part of a τοποσ) and then 'region, country'; or it may be linked to the Latin *terra* (of restricted Indo-European origin), probably coming from a previous *ter-es* connected to the Gaulish *tir* (with the meaning of 'country') (Olivieri 1961).

Tessera: according to Forman and Godron (1981, 1986) it is the smallest homogeneous unit visible at the spatial scale of a landscape. In this text it is the landscape element corresponding to the ecocoenotope.

Time arrow: a simple representation of the irreversibility of time even at a cosmological scale, with a clear direction from past to future.

Umbrella species: the most ecologically exacting species and thus the first to rarefy or to disappear completely after the destruction of their habitat.

Variegation model: one of the models developed in order to understand the spatial structure of a landscape. It was proposed by zoologists and sees the landscape as a species-specific environment. The representation is an ecological mosaic variable in space and time, a sort of fuzzy-edged mosaic (the elements of which are tesserae with variable conformations) or, in other words, an overlapping series of different patch-matrix mosaics, one for each species or groups of species considered (steno- or euritopic). In a way it is a dynamic model.

Zero event: an out-of-scale disturbance so strong as to completely destroy an ecological system or a part of it and trigger the development of a new ecological system. Even the end of a continuous range of disturbances constitutes a zero event.

References

Allen TFH, Hoekstra TW (1992) Toward a unified ecology. Columbia University Press, New York

Anati E (1985) I Camuni. Jaca Book, Milano

ANPA (2000) Indice di Funzionalità Fluviale. Agenzia Nazionale per la Protezione dell'Ambiente, Roma

Bailey RG (1996) Ecosystem Geography. Springer Verlag, New York, Inc.

Bailey RG, Cushwa CT (1981) Ecoregions of North America. FWS/OBS-81/29, Washington DC

Bas Pedroli GM, Vos W, Dijkstra H, Rossi R (1988) Studio degli effetti ambientali della diga sul torrente Farma. Regione Toscana, Marsilio Editori Venezia

Begon M, Harper JL, Townsend CR (1990) Ecology, Individuals, Populations and Communities. Blackwell, London

Bennet G (ed) (1991) Towards an European ecological network. Ieep, Arnhem.

Bennet MD, Smith JB (1991) Nuclear DNA amounts in angiosperms. Phil. Trans. R. Soc. Lond. B. 334:309-345

Bense JF (1939) A Dictionary of the Low-Dutch Element in the English Vocabulary. The Hague Martinus Nijhoff

Berne RM, Levy MN (1990) Principles of Physiology, The CV Mosby Company

Bertrand G (1968) Paysage et géographie globale. Revue géographique des Pyrénées et du Sud-Ouest 39: 249-272

Bertrand G (1970) Ecologie de l'espace géographique. Recherches pour une science du paysage. Société de biogéographie, Compte rendus (19 décembre 1969) : 195-205

Biondi E, Allegrezza M, Ballelli S, Calandra R, Crescente MF, Frattaroli A, Gratani L, Rossi A, Taffetani F (1992) Indagini per una cartografia fitoecologica dell'altopiano di Campo Imperatore (Gran Sasso d'Italia). Boll Ass Ital Cartografia 86: 85-99

Blondel J (1986) Biogéographie évolutive. Masson, Paris

Bolza GB (1852) Vocabolario Genetico Etimologico della Lingua Italiana. Vienna

Bovet et al. (1969) Determinismo genetico dell'attitudine all'apprendimento. Science 163: 139-149

Box EO (1987) Plant Life Form and Mediterranean Environments. Annali di Botanica XLV:7-42

Bradshaw AD (1983) The reconstruction of ecosystems. J Appl Ecol 10: 1-17

Brandmayr P (1990) Le relazioni fra Zoologia ed Ecologia del Paesaggio terrestre. Congr. Naz. UZI Atti 53: 6-7

Brandt J, Tress B, Tress G (2000) Multifunctional landscapes: interdisciplinary approaches to landscape research and management. Centre for Landscape Research published, Roskilde

Braun-Blanquet J (1928) Pflanzensoziologie: Grundzüge der Vegetationskunde. Berlin

Brewer R (1988) The science of ecology. Saunders College Pub., Philadelphia

Buchwald K, Engelhart W (eds) (1968) Handbuch fur landschaftpflegeund naturschutz. Bd. 1 Grundlagen. BLV Verlaggesellschaft, Munich

Burel F, Baudry J (1999) Écologie du paysage: concepts, méthodes et applications. Technique & Documentation Editions, Paris

Canullo R, Spada F (1996) La memoria di un paesaggio scomparso nel comportamento di arbusti invasivi. In: Ingegnoli V, Pignatti S (eds) L'ecologia del paesaggio in Italia. CittàStudi Ediz, pp131-143

Caravello GU, Ingegnoli V (1991) Analisi ecologiche e comaparazione di due aree padane: i comuni di Veggiano e Torrile. SITE Atti 12: 493-496

Celesti Grapow L (1995) La flora. In: Cignini B, Massari G, Pignatti S (eds) L'ecosistema Roma: conoscenze attuali e prospettive per il Duemila. Fratelli Palombi ed., Roma, pp 47-53

Chen J, Franklin JF (1992) Vegetation responses to edge environments in old growth douglas-fir forests. Ecological applications 2(4): 387-396

CIP (1997) Statistisches Taschenbuch für Reinland-Pfalz.

Clark K (1976) Landscape into art. John Murray, London

Clements FE (1916) Plant Succession; analysis of the development of vegetation. Publ Carnegie Inst, Washington, 242: 1-512

Colantonio Venturelli R (2000) Die Nachhaltigkeit in der Landschaftsökologie. Atlas 19

Colinvaux P (1993) Ecology. Wiley & Sons, New York

Commission of the European Communities (1991) Corine biotopes manual. Habitats of the European Community. Office for Official Pubblications of the European Communities, Brussels and Luxembourg

Cortellazzo M, Zolli P (1985) Dizionario Etimologico della Lingua Italiana 4/O-R. Zanichelli

Dale VH, Haeuber RA (eds), Forman RTT (2001) Applying ecological principles to Land Management. Springer Verlag, New York

Daly HE (1999) Uneconomic Growth and the built Environment: In Theory and in Fact. In Reshaping the Built Environment, Kibert CJ (ed.) Island Press California pp 73-86

Darwin C (1859) On the Origin of Species by Means of Natural Selection. John Murray London

D'Anna G (1988) Dizionario Italiano Ragionato. Sintesi Firenze

Dawkins R (1976) The selfish gene. Oxford University Press

De Agostini G.I. (2001) Calendario Atlante De Agostini. Istit Geogr De Agostini, Novara

De Vries J (1962) Altnordisches Etymologisches Wörterbuch. E.J. Brill

Di Castri F, Hansen AJ (1991) Landscape Boundaries: Consequences for Biotic diversità and Ecological Flows. Springer-Verlag Berlin

Dodson SI, Allen TFH, Carpenter SR, Ives AR, Jeanne RL, Kitchell JF, Langston NE, Turner MG (1998) Ecology. Oxford University Press

Dokuchaev VV (1898) Writings (in Russian), Reprinted (1951) Vol. 6 Akad Nauk, Moscow

Duchaufour Ph (1984) Pédologie: sol, végétation, environment. Masson, Paris

Duvigneaud P (1977) Ecologia. In: Enciclopedia del Novecento, Enciclopedia Italiana Treccani, Roma

Ehrart H (1956) La genèse du sol en tant que phénomè géologique. Masson, Paris

Ekeland I (1995) Le Chaos. Flammarion Paris

Ellenberg H (1960) Grundlagen der Pflanzenverbreitung. E. Ulmer Verlag Stuttgart

Ellenberg H (1974) Zeigerwerte der Gefässpflanzen Mitteleuropas. E Goltze Verlag, Göttingen

Ellenberg H (1978)Vegetation Mitteleuropas mit den Alpien in oekologischer Sicht. Ulmer

Ellenberg H (1979) Zeigewerte der Gehfasspflanzen Mitteleuropas. Scripta Geobotanica Gottingen 9:1-97

Ermer K, Hoff R, Mohrmann R (1996) Landsschaftsplanung in der Stadt. E. Ulmer Verlag Stuttgart

Erz W (1980) Naturschutz Grundlagen, Probleme und Praxis. In Buchwald K, Engelhardt W (Eds) Handbuch für Planung, Gestaltung, und Schutz der Umwelt. Band 3 BLV Verlag München pp 560-637

Falinski JB (1986) Vegetation succession on abandoned farmland as a dynamic manifestation of ecosystem liberal, of long continuance anthropopression. Wiad Bot 30.1: 21-50 (part 1); 30.2: 115-126 (part 2)

Falinski JB (1994a) Concise geobotanical atlas of Białowieża Forest. Phytocoenosis vol. 6 (N.S.), Supplementum Cartographiae Geobotanicae 6, Warszawa- Białowieża

Falinski JB (ed) (1994b) Vegetation under the diverse anthropogenic impact as object of basic phytosociological map: result of the international cartographical experiment organized in Białowieża Forest. Phytocoenosis vol. 6 (N.S.), Supplementum Cartographiae Geobotanicae 4, Warszawa- Białowieża

Falinski JB (1998) Dioecious woody pioneer species in the secondary succession and regeneration. Phytocoenosis vol. 10 (N.S.), Supplementum Cartographiae Geobotanicae 8, Warszawa- Białowieża

Farina A (1993) L'ecologia dei sistemi ambientali. Cleup, Padova

Farina A (1998) Principles and methods in landscape ecology. Chapman & Hall, London

Farina A (2000) Landscape ecology in action. Kluwer Academic Publishers, The Netherlands

Finke L (1972) Die Bedeutung des Faktors Humusform für die landschaftökologische Kartierung. Biogeographica 1, S. 183-191

Finke L (1986) Landschaftsökologie. Verlags-GmbH Höller und Zwick, Braunschweig

Forman RTT (1995) Land Mosaics: the ecology of landscapes and regions. Cambridge University Press, Cambridge

Forman RTT, Collinge SK (1995) The 'spatial solution' to conserving biodiversity in landscapes and regions. In: DeGraaf RM, Miller RI (eds) Conservation of faunal diversity in forested landscapes. Chapman&Hall, London

Forman RTT, Godron M (1981) Patches and structural components for a landscape ecology. Bioscience 31: 733-740

Forman RTT, Godron M (1986) Landscape Ecology. John Wiley &Sons, New York

Forman RTT, Hersperger AM (1996) Road ecology and road density in different landscapes, with international planning and mitigation solutions. In: Evink GL, Garrett P, Zeigler D, Berry J (eds) Trends in addressing transportation related wildlife mortality. Florida Department of transportation, Tallahassee, pp 1-22

Forman RTT, Moore PN (1991) Theoretical foundations for understanding boundaries in landscape mosaics. In: Hansen AJ, Di Castri F (eds) Landscape boundaries : consequence for biotic diversity and ecological flows. Springer-Verlag, New York, pp 236-258

Forman RTT, Spierling D, Bissonette JA, Clevenger AP, Cutshall CD, Dale VH, Fahrig L, France R, Goldman CR, Heanne K, Jones JA, Swanson FJ, Turrentine T, Winter TC (2002) Road Ecology: Science and Solutions. Island Press, Washington DC

Frankel OH, Soulè ME (1981) Conservation and evolution. Cambridge University Press, Cambridge

Gandhi MK (1927) An autobiography or the story or my experiments with truth. Navajivan Publishing House, Ahmedabad

Gehu JM (1979) Pour une approche nouvelle des paysages végetaux : la symphytosociologie. Bull. Soc. Bot. Fr. 126, Lettres bot., pp 213-223

Gehu JM (1988) L'analyse symphytosociologique et geosymphytosociologique de l'éspace. Tthéorie et métodologie. Coll. Phytosoc. 17:11-46

Gell-Mann M (1994) The quark and the jaguar. Adventures in the simple and the complex. Freeman & C., New York

Giacomini V (1958) La Flora. Touring Club Italiano. Milano

Giacomini V (1965) Significato e funzione dei Parchi Nazionali. In: I Parchi Nazionali in Italia. Ist. Tec. E Prop. Agraria Roma pp 7-37

Giacomini V (1982) Risorse naturali, risorse biologiche. In: Enciclopedia del Novecento, vol IV, Istituto Enciclopedia Italiana, pp 171-187

Gleason HA (1922) On the relation between species and area. Ecology 3: 156-162

Godron M (1968) Quelques applications de la notion de fréquence en écologie végétale. Oekologia Plantarum 3 : 185-212

Godron M (1984) Ecologie de la végétation terrestre . Masson, Paris

Gödel K (1931) Über formal unentscheidbare Satze der Principia Mathematica und verwandter Systeme. In: Mh. Math. Phys. 38:173-198

Goldsmith J (1994) London Times, March 5

Goldsmith E, Allen R (1972) La morte ecologica. Laterza, Bari

Golley FB, Lieth H (1972) Summary. In Golloy PM, Golley FB (Eds) Tropical ecology with an emphasis on organic production. Univ. of Georgia, Athens

Golley FB, Vyas AB (1975) Ecological Principles. International Sc. Publ. Jaipur

Goodall J, Berman P (1999) Reason for hope. Soko Publications Ltd

Goode D (1998) Integration of Nature in Urban Development. In Breuste J, Feldmann H, Uhlmann O (Eds) Urban Ecology. Sprinter Verlag, Berlin pp 589-592

Gould SJ (1994) The evolution of life on Earth. Sc. American CCLXXI:84

Haber W (1979) Raumordnungs-Konzepte aus der Sicht der Ökosystemforschung. Forschungs und Sitzungsberichte Akademie für Raumforschung und Landesplanung 132:12-24

Haber W (1989) Using landscape ecology in planning and management. In: Zonneveld IS, Forman RTT (eds) Changing landscapes: an ecological perspective. Springer-Verlag, New York, Berlin, pp 217-232

Haber W (1990) Basic concept of landscape ecology and their application in land management. In: Ecology for tomorrow. Phys. Ecol. Japan. pp 131-146

Haber W (1998) Reflections on the ecological role of agriculture. In: Barron EM, Nielsen I (eds) Agriculture and Sustainable land use in Europe. Kluwer Law Intern, The Hague, pp 147-160

Haeckel E (1890-92) Storia della creazione naturale. Torino, UTET (original ed. Berlin 1869)

Hanski I (1983) Single-species metapopulation dynamics: concepts, models and observations. Biol. J. Linn. Soc. 42: 17-38

Hellrigl B (1990) Relaskop scala metrica CP, il relascopio a specchio. FOB, Salzburg

Hornby AS (1974) Oxford Advanced Learner's Dictionary of Current English. Oxford University Press, London

Hutchinson GE (1965) The ecological theatre and the evolutionary play. Yale University Press, New York

Ingegnoli V, Roncai L (1975) Ambiente e architettura rurale. In: Perogalli C, Alpago-Novello A, (eds.) Cascine del territorio di Milano. Milani Editrice, Milano, pp 23-42

Ingegnoli V (1980) Ecologia e progettazione. CUSL, Milano

Ingegnoli V (1981) Organizzazione agricola e casa rurale. In: Lombardia: il territorio, l'ambiente, il paesaggio. Electa ed, Milano, pp 27-64

Ingegnoli V (1986) Considerazioni sul rapporto fra transizione demografica e crisi ecologica. In: Ecologia e Longevità, Atti Conv. Naz. di Ecologia Umana, Firenze, Boll. SIEU: 53-63

Ingegnoli V (1987) Il territorio. I giardini. In: Ingegnoli V, Langè S, Suss F. Le ville storiche nel territorio di Monza. Pro Monza, Monza, pp 9-49, pp 177-213

Ingegnoli V (1989) Basi ecologiche del rapporto ville-ambiente. In: Brusa C (ed.) Ville e territorio. Lativa Edizioni Varese pp 17-24

Ingegnoli V (1991) Human influences in landscape change: thresholds of metastability. In: Ravera O (Ed) Terrestrial and aquatic ecosystems: perturbation and recovery. Ellis Horwood, Chichester, England, pp 303-309

Ingegnoli V (1993) Fondamenti di Ecologia del paesaggio. CittàStudi, Milano

Ingegnoli V (ed) (1997) Esercizi di Ecologia del Paesaggio. CittàStudi Edizioni, Milan

Ingegnoli V (1998) Landscape Ecological criteria as a Basis for the Planning of a Suburban Park in Milan. In Breuste J, Feldmann H, Uhlmann O Eds. (1998) Urban Ecology. Springer Verlag, Berlin, pp 657-662

Ingegnoli (1999a) Definition and evaluation of the BTC (biological territorial capacity) as an indicator for landscape ecological studies on vegetation. In: Windhorst W, Enckell PH (eds) Sustainable Landuse Management: The Challenge of Ecosystem Protection. EcoSys: Beitrage zur Oekosystemforschung, Suppl Bd 28:109-118

Ingegnoli V (1999b) Ecologia del Paesaggio. In: Baltimore D, Dulbecco R, Jacob F, Levi-Montalcini R (eds.) Frontiere della Vita. vol IV, Istituto per l'Enciclopedia Italiana G. Treccani, Roma. pp 469-485

Ingegnoli V (2001) Landscape Ecology. In: Baltimore D, Dulbecco R, Jacob F, Levi-Montalcini R (eds.) Frontiers of Life. vol IV, Academic Press, New York, pp 489-508.

Ingegnoli V, Gibelli MG (1992) Importanza dell'ecologia del paesaggio per la riqualificazione ambientale dei parchi storici: gli studi per il Parco di Monza. In: Mazzali B et al. (eds.) Atti del II° Conv.Naz. Parchi e Giardini Storici, Parchi Letterari: conoscenza, tutela, valorizzazione. Ministero Beni Culturali e Ambientali, pp 580-593

Ingegnoli V, Aquila C, Padoa-Schioppa, E (1995) Rapporto preliminare sullo studio dell'ecomosaico forestale del Gariglione. Atti 5° Workshop "Progetto Strategico Clima Ambiente Territorio nel Mezzogiorno" Amalfi 1993, Tomo I, Guerrini editore, C.N.R. Roma, pp 397-419

Ingegnoli V, Giglio E (1999) Proposal of a synthetic indicator to control ecological dynamics at an ecological mosaic scale. Annali di Botanica LVII: 181-190

Ingham DS, Samways MJ (1996) Application of fragmentation and variegation models to epigaeig invertebrates in South Africa. Conservation Biology10:1353-1358

Ipsen D (1998) Ecology as Urban Culture. In Breuste J, Feldmann H, Uhlmann O (Eds) Urban Ecology. Sprinter Verlag, Berlin

ISTAT (2000) Compendio statistico italiano. Istituto Nazionale di Statistica, Roma

Ives AR (1998) Population ecology. In: Stanley ID et al. Ecology. Oxford University Press, Inc.

Jones CG, Lawton JH, Shachak M (1994) Organisms as ecosystem engineers. Oikos 69: 373-386

Jongman RHG (1999) Landscape ecology and land use planning. In: Wiens JA, Moss MR (eds) Issues in Landscape ecology. Us-Iale, Colorado

Jongman RHG, Mander U (eds) (2000) Consequences of land use changes. Advances in ecological science 5, Computational Mechanics Public, Southampton, Boston, pp 11-38

Jongman RHG, Smith D (2000) The European experience: from site protection to ecological networks. In: Sanderson J, Harris LD (eds) Landscape Ecology: a top-down approach. Lewis Publishers, Boca Raton, Florida, pp 157-182

Kauffman SA (1993) The Origin of Order. Oxford University Press New York

Kauffman SA (1995) At home in the universe. Oxford University Press New York

Kimmins JP (1987) Forest ecology. Macmillan, New York

King AW (1997) Hierarchy Theory: A Guide to System Structure for Wildlife biologists. In Bissonette JA (ed) Wildlife and Landscape Ecology. Springer Verlag, New York, Berlin

Korten D (1996) The mythic victory of market capitalism. In: Mander J, Goldsmith (eds) The case against the global economy and for a return to the local. Sierra Books, San Francisco

Kosko B (1993) Fuzzy thinking: the new science of fuzzy logic. Hyperion

Kosko B (2000) Il fuzzy- pensiero: teoria ed applicazioni della logica fuzzy. Baldini & Castoldi, Milano

Kowarick I (1990) Some responses of flora and vegetation to urbanization in Central Europe. In: Sukopp H, Hejny (ed) Kowarick I (co-ed) Urban Ecology. SPB Academic Publishing bv, The Hague, pp 45-74

Lambeck RJ (1997) Focal Species: A Multi-Species Umbrella for Nature Conservation. Conservation Biol. 11:849-856

Landolt E (1977) Ökologische Zeigewerte zur Schweitzer Flora. Verhoff.Geobot.Inst. Rubel. Zurich 64

Leser H (1978) Landschaftsökologie. Uni-Taschenbucher 521, Stuttgart

Levin SA(1976) Population dynamic models in heterogeneous environments. Annual Review of Ecology and Systematics 7:287-310

Liebig J (1840) Chemistry in its application to agriculture and physiology. Taylor and Walton, London

Lieth H, Whittaker RH, Eds. (1975) The Primary Productivity of the Biosphere. Springer Verlag, New York, Berlin

Lorenz EN (1963) Deterministic nonperiodic flow. J Atmos Sci, vol 20: 130-141

Lorenz K (1969) L'aggression. Flammarion, Paris

Lorenz K (1973) Die Rückseite des Spiegels. Versuch einer Naturgeshichte menschlichen Erkennens. R Piper & Co. Verlag München

Lorenz K (1978) Vergleichende Verhaltensforschung: Grundlagen der Ethologie. Springer-Verlag, Wien

Lorenz K, Popper K (1985) Il futuro è aperto. Rusconi, Milano

Lovelock JE (1979) Gaia: A new Look at Life on Earth. Oxford Univ. Press, Oxford

Ludwig JA, Tongway DJ (1997) A Landscape Approach to Rangeland Ecology. In Ludwig J, Tongway D, Freudenberger D, Noble J, Hodgkinson K (eds) Landscape Ecology, Function and Management. CSIRO Australia pp1-12

Mac Arthur RH, Wilson EO (1967) The theory of island biogeography. Princeton University Press, Princeton, New York

Mac Arthur RH, Wilson EO (1972) Geographical Ecology: patterns in the distribution of species. Princeton University Press, Princeton, New York

Mac Harg I (1969) Design with nature. Natural History Press, New York

Mandelbrot BB (1975) Les objets fractals: forme, hasard et dimension. Flammarion, Paris

Margulis L, Lovelock JE (1974) Biological modulation of the Earth's atmosphere. Icarus 21: 471-489

Mariano C (1958) Nuovo Dizionario Italiano-latino. Soc. Editrice Dante Alighieri, Milano

Massa R (1999) I grandi progetti di conservazione. In: Massa R, Ingegnoli V (eds) Biodiversità, estinzione e conservazione: fondamenti di conservazione biologica. UTET Libreria, Torino, pp 332-347

Massa R, Fornasari L (1998) Birds as tools in Landscape Toxicology. Bird Numbers 1998, Cottbus

Massa R, Ingegnoli V (eds) (1999) Biodiversità, estinzione e conservazione: fondamenti di conservazione biologica. UTET Libreria, Torino

Meffe GK, Carroll CR (1999) Principles of conservation biology. Sinauer Associates, Inc Publ, Sunderland, Massachusetts

Merriam G (1984) Connectivity: a fundamental ecological characteristic of landscape pattern. In: Brandt J, Agger P (eds) Methodology in landscape research and planning. Proc 1st Int. Semin. Intern. Assoc. Landscape Ecology, pp 5-15

Migliorini B (1942) Appendice al Dizionario Moderno. In: Panzini A, Dizionario Moderno. Milano, pp 761-879

Moebius K (1877) Die Auster un die Austernwirtschaft. Berlino

Monod J (1970) Le hazard et la nécessité. Ed. du Seul Paris

Muir J (1901) Our national Parks. Houghton Mifflin, Boston

Myers N (1996) The biodiversity crisis and the future of evolution. The Eznvironmentalist 16: 37-47

Naveh Z (1987) Biocybernetic and thermodynamic perspectives of landscape functions and land use patterns. Landscape Ecology, vol 1, no 2: 75-83

Naveh Z (2000) Introduction to the theoretical foundations of multifunctional landscapes and their application in transdisciplinary landscape ecology. In: Brandt J, Tress B, Tress G Multifunctional landscapes: interdisciplinary approaches to landscape research and management. Centre for Landscape Research published, Roskilde

Naveh Z, Lieberman A (1984) Landscape Ecology: theory and application. Springer-Verlag New York, Inc.

Naveh Z, Lieberman A (1994) Landscape Ecology: theory and application. Springer-Verlag New York, Inc.

Neef E (1967) Die theoretischen grundlagen der Landschaftslehre, V.E.B. Haack Gotha

Niermeyer JF (1976) Mediae Latinitatis Lexicon Minus. Brill EJ, Leiden, pp 751-753

Nobile T (1943) La storia delle Parole. Macrì L., Bari

Noss RF (1992) The Wildlands Project: land conservation strategy. Wild earth (special issue)

Odum EP (1953) Fundamentals of ecology. WB Saunders, Philadelphia

Odum EP (1971) Fundamentals of ecology. 3rd ed,WB Saunders, Philadelphia

Odum EP (1983) Basic Ecology. CBS College Publ. USA

Odum EP (1989) Ecology and our endangered life-support systems. Sinauer ass. Publ. Sunderland, Massachussets

Oldeman RAA (1990) Forests: elements of sylvology. Springer-Verlag, New York, Berlin

Olmsted FL (1865) The great american Park of Yo-Semite. In: Yosemite Nature Notes (1954, reprinted).

Olmsted FL (1870) Public Parks and the enlargement of towns. Cambridge University Press

O'Neill RV, De Angelis DL, Waide JB, Allen TFH (1986) A hierarchical concept of ecosyestems. Princeton Univ. press, Princeton, NY

Onions CT (1966) The Oxford Dictionary of English Etimology. Clarendon Press, Oxford

Ott J (1995) Do dragonflies have a chance to survive in an industrialised country like Germany? In: Corbet PS, Dunkle SW, Ubukata H (eds) Proceedings of the International Symposium on the conservation of dragonflies and their habitats. Jap Soc Preserv of Birds, Kushiro: 28-44

Ott J (1997) Lo studio degli Odonati nella pianificazione del paesaggio. In: Ingegnoli V (ed) Esercizi di ecologia del paesaggio. CittàStudi Ed, Milano 105-129

Pabjanek P (1999) Land-use change on the Białowieża clearing (till 1989). Fotointerpretacja Geografii 30, Warszawa, 3-28

Panizza M (1988) Geomorfologia applicata: metodi di applicazione alla pianificazione territoriale e alla valutazione di impatto ambientale. NIS Nuova Italia Scientifica, Roma

Partridge E (1958) Origins: a short etymological dictionary of Modern English. Routledge & Kegan P., London

Passarge S (1912) Über die Herausgabe eines physiologisch-morphologischen Atlas. In: Verh 18 Dt Geogr Tages zu Innsbruck, Berlin, pp 236-247

Pedrotti F (1980) Il Parco nazionale di Bialowieza. Natura e montagna 3: 177-187

Pedrotti F, Venanzoni R (1994) Experiences of Italian authors (Map 1). In: Falinski JB (ed) Vegetation under the diverse anthropogenic impact as object of basic phytosociological map. Results of the international cartographical experiment organized in the Bialowieza Forest. Phytocoenosis 6, Suppl. Cartogr. Geobot. 4:31-37

Petersen RC (1991) The RCE: a riparian, channel and environmental inventory for small streams in agricultural landscape. Freshwater biology, March 1990

Pignatti S (1953) Introduzione allo studio fitosociologico della Pianura Veneta Orientale. Archivio Bot 28-29

Pignatti S (1980) Reflections on the phytosociological approach and the epistemological basis of vegetation science. Vegetatio 42: 181-185

Pignatti S (1988) Ecologia del Paesaggio. In Honsell E, Giacomini, V, Pignatti S: La vita delle piante. Utet, pp 472-483

Pignatti S (1994a) A complex approach to phytosociology. Ann. Bot. vol LII: 65-80

Pignatti S (1994b) Ecologia del Paesaggio. UTET, Torino

Pignatti S (1995) Ecologia vegetale. UTET, Torino

Pignatti S (1996a) Conquista della prospettiva e percezione del paesaggio. In: Ingegnoli V, Pignatti S (eds) L'ecologia del paesaggio in Italia. CittàStudi Edizioni, Milano, pp15-25

Pignatti S (1996b) Some Notes on Complexity in Vegetation. Journ, of Vegetation Sc. 7:7-12

Pignatti S, Dominici E, Pietrosanti S (1998) La biodiversità per la valutazione della qualità ambientale. Atti dei Convegni Lincei 145:63-80

Piussi P (1994) Selvicoltura generale. Utet, Torino

Poincaré JH (1906) La science et l'hypothèse. Flammarion, Paris

Poincaré JH (1908) Science et méthode. Flammarion, Paris

Popper KR (1962) The Logic of Scientific Discovery. Harper and Row, New York

Popper KR (1982) The open universe : an argument for indeterminism. WW Bartley III, London

Popper KR (1983) Realism and the aim of science. WW Bartley III, London

Popper KR (1990) A World of Propensities. Thoemmes, Bristol UK

Popper KR (1994) Alles Leben ist Problemlösen. Über Erkenntnis, Geschichte und Politik. R Piper & Co., München

Prigogine I (1988) La nascita del tempo. Edizioni Teoria, Roma

Prigogine I (1996) La fin dès certitudes: temps, chaos et les lois de la nature. Editions odile Jacob, Paris

Prigogine I, Nicolis G, Babloyatz A (1972) Thermodynamics of evolution. Physics Today vol 25, no11: 23-28; no12: 38-44

Prigogine I, Nicolis G (1977) Self-organization in non-equilibrium systems: from dissipative structures to order through fluctuations. Wiley, New York

Primack RB (1998) Essentials of Conservation Biology. Sinauer Associates Publishers, Sunderland, Massachusetts

Pulliam R (1989) Sources and sink complicate ecology. Science : 477-478

Rambler MB, Margulis L, Fester R (1989) Global ecology, towards a science of the Biosphere. Academic Press, Boston

Ramon y Cajal (1909) Histologie du système nerveux de l'homme e des vertébrées. Madrid

Renault Miskovsky J (1986) L'environnement au temps de la préhistoire. Masson, Paris

Repton H (1803) Observations on the Theory and Practice of Landscape Gardening. In: Loudon JC (ed) The landscape gardening and landscape architecture of the late H. Repton Longmans. Green and Co, London (1840)

Ricklefs RE (1973) Ecology. Freeman & Co., New York

Rivas-Martinez S (1987) Nociones sobre Fitosociología, biogeografía y bioclimatología. In : la vegetacion de España, Universidad de Alcalá de Henares, Madrid, pp 19-45

Rivas-Martinez S (1995) Clasificación bioclimática de la Tierra. Folia Bot Matritensis 16 : 1-27

Romani V (1988) Il paesaggio dell'Alto Garda bresciano: studio per un piano paesistico. Grafo ed., Brescia

Ruelle D (1991) Hasard et chaos. Éditions Odile Jacob, Paris

Ruelle D, Takens F (1971) On the nature of turbulence. Comm. Math. Phys. vol 20 pp 167-192

Sanderson J, Harris LD (eds) (2000) Landscape Ecology: a top-down approach. Lewis Publishers, Boca Raton, Florida

Shannon CE, Weaver W (1949) The Mathematical Theory of communications. University of Illinoiss Press Urbana

Shapiro AM (1979) Weather and the lability of breeding population of the Chekered White Butterfly Pieris Protodice. J of Research on the Lepidoptera 17: 1-23

Shelford VE (1913) Animal communities in temperate America. University of Chicago Press, Chicago

Sigman DM, Boyle EA (2000) Glacial/interglacial variations in atmospheric carbon dioxide. In: Nature, vol 407, 19 october: 859-869

Siligardi M (1997) Ecologia del paesaggio e sistemi fluviali. In: Ingegnoli V (ed) (1997) Esercizi di Ecologia del Paesaggio. CittàStudi Edizioni, Milano

Siligardi M, Maiolini B (1993) L'inventario delle caratteristiche ambientali dei corsi d'acqua alpini: guida all'uso della scheda Rce-2. Biologia Ambientale, Vol VII, 30: 18-24

Stchoupak N, Nitti L, Renou L, Dictionnaire Sanskrit-Francais. Librairie d'Amerique et D'Orient, Maisonneuve A, Paris

Storch I (1997) The importance of scale in habitat conservation for an endangered species: the capercaille in central Europe. In: Bissonette JA (Ed) Wildlife and landscape Ecology. Springer Verlag, New York, Berlin pp 310-330

Sukopp H (1984) Ökologische Charakteristik von Grosstädten: In Grundriss der Stadtplanung. Akademie für Raumforschung und Landesplanung Hannover pp 51-82

Sukopp H (1998) Urban ecology: scientific and practical aspects. In: Breuste J, Feldmann H, Uhlmann O, Urban ecology. Springer-Verlag, Berlin, Heidelberg, pp 3-16

Susmel L (1980) Normalizzazione delle foreste alpine. Liviana Editrice Padova

Tansley AG (1935) The use and abuse of vegetational concepts and terms. Ecol. 16:284-307

Teillard de Chardin P (1955) Le phenomòne humain. Edition de Seuils, Paris

Thoreau HD (1964) Walden. Rizzoli, Milano (original edition Boston, 1854)

Tjallingi SP, De Veer AA (eds) (1982) Perspectives in Landscape ecology. Centre for Agricultural Publishing and Documentation, Wageningen, Netherlands

Tricart J, Cailleux A (1972) Introduction to climatic geomorphology. St. Martin's Press, New York

Tricart J, Kilian J (1979) L'éco-géographie et l'aménagement du milieu naturel. Librairie François Maspero, Paris

Troll C (1939) Luftbildenplan und okologische bodenforshung. In: Zeitschrift d. Gesellschaft f. Erdkunde, Berlin

Troll C (1950) Die geographische Landshaft und ihre Erforshung. Studium generale, n.3 , Heidelberg, pp 163-181

Troll C (1963) Ueber Landschaft-Sukzession, Vorwort des Herausgebers. In: Bauer HJ, Landschaftökologische Untersuchungen im ausgekohten Rheinischen Braunkohlenrevier auf der Ville. Arbeiten zur Rheinischen landeskunde 19

Turner MG (1989) Landscape ecology: the effect of pattern on process. Annual Review of Ecology and Systematics 20: 171-197

Turner MG, Wu Y, Pearson SM, Romme WH, Wallace LL (1992) Landscape-level interactions among ungulates, vegetation and large-scale fires in Nothern Yellowstone National Park. In Plumb GE, Harlow HJ (eds.) University of Wyoming national Park Service Researche Center, 16th Annual Report, Laramie, pp 206-211

Turner MG, Gardner RH (eds) (1991) Quantitative methods in Landscape Ecology : the analysis and interpretation of landscape heterogeneity. Springer-Verlag New York Inc.

Tüxen R (1956) Die heutige potentielle natürliche Vegetation als Gegenstand der Vegetationkartierung. Angew. Pflanzensoziologie Stolzenau/Weser13:5-42

Tüxen R (1968) Pflanzensoziologie und Landschaftökologie. Intern Symp. Intern Ver. Für Vegetationskunde Stolzenau und Rinteln

Tüxen R (ed) (1978) Assoziationkomplexe. Ber. Intern. Symp. Veg., Rinteln. Cramer Verlag, Vaduz

Ulanowicz RE (1997) Ecology, the ascendent perspective. Columbia University Press, New York

Ulanowicz RE (2000) Quantifying constraints upon trophic and migratory transfers in landscapes. In: Sanderson J, Harris LD (eds) Landscape Ecology: a top-down approach. Lewis Publishers, Boca Raton, Florida

Ulgiati S (ed) 1998 Advances in energy studies: energy flows in ecology and economy. MUSIS, Roma

Urbanska KM, Webb NR, Edwards PJ (1997) Restoration ecology and sustainable development. Cambridge University Press, Cambridge

Von Bertalanffy L (1968) General System theory, Foundations, Development and Applications. George Braziller New York

Von Humboldt A (1846) Kosmos. Stuttgard

Vos W, Stortelder A (1992) Vanishing Tuscan Landscapes: landscape ecology of a Submediterranean-Montane area (Solano Basin, Tuscany, Italy). Pudoc Scientific Pub, Wageningen

Walter H (1973) Vegetation of the Earthin Relation to Climate and the Eco-Physiological Conditions. Springer-Verlag New York, Heidelberg, Berlin

Watt AS (1947) Pattern and process in plant community. J Ecol 35: 1-22

Weinberg GM, Weinberg D (1979) On the design of stable systems. J. Wiley & Sons, New York

Weiss PA (1969) The living systems: determinism stratified. In A Koestler and JR Smithies (Eds.) Beyond Reductionism: New Perspectives in Life Sciences. Hutchinson of London pp 3-55

Werner P, Sukopp H (1985) Development of flora and fauna in urban areas. Im Auftrag des Europarates Berlin

Westhoff V (1970) Vegetation study as a branch of biological science. Landbouwhogeschool Wageningen, Misc Pap 5: 11-30

Wiens JA, Stenseth NC, Van Horne B, Ims RA (1993) Ecological mechanisms and landscape ecology. Oikos 66: 369-80

Wiens JA (1995) Landscape mosaics and ecological theory. In: Hansson L, Fahrig L, Merriam G Mosaic landscapes and ecological processes. Chapman & Hall, London, pp 1-26

Wiens JA, Moss MR (eds) (1999) Issues in landscape ecology. US- IALE, Colorado

White F (1986) La végétation de l'Afrique; recherches sur les ressources naturelles. Orstom-Unesco, Paris

Whittaker RH (1975) Communities and ecosystems. 2nd ed., Macmillan Pub, New York

Wilson EO (1992) The diversity of life. Harvard University Press, Boston

Wood PA, Samways MJ (1991) Landscape element pattern and continuity of butterfly flight paths in an ecologically landscaped botanic garden, Natal, South Africa. Biological Conservation 58: 149-166

World Bank (1992) World development report, Development and the environment. Oxford University Press, New York.

Zadeh LA (1987) Fuzzy sets and applications. In: Yager, Ovchnikov, Tong, Nguyen (eds) Selected papers. Wiley, New York

Zonneveld IS, Forman RTT (1989) (eds) Changing landscapes: an ecological perspective. Springer-Verlag, New York, Berlin

Zonneveld IS (1989) Scope and concepts of landscape ecology as an emerging science. In: Zonneveld IS, Formann RTT (eds) Changing landscapes: an ecological perspective. Springer-Verlag, New York, Berlin, pp 3-20

Zonneveld IS (1995) Land ecology. SPB Academic Publishing, Amsterdam

Index

Druck: Strauss Offsetdruck, Mörlenbach
Verarbeitung: Schäffer, Grünstadt